P9-CCE-063

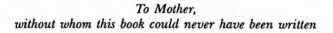

To Mother,
without whom this book could never have been written

Coal, steel, and the rebirth of Europe, 1945—1955

Coal, steel, and the rebirth of Europe, 1945–1955

The Germans and French from Ruhr conflict to economic community

JOHN GILLINGHAM

University of Missouri, St. Louis

CAMBRIDGE UNIVERSITY PRESS

PUBLISHED BY THE PRESS SYNDICATE OF THE UNIVERSITY OF CAMBRIDGE
The Pitt Building, Trumpington Street, Cambridge, United Kingdom

CAMBRIDGE UNIVERSITY PRESS
The Edinburgh Building, Cambridge CB2 2RU, UK
40 West 20th Street, New York NY 10011–4211, USA
477 Williamstown Road, Port Melbourne, VIC 3207, Australia
Ruiz de Alarcón 13, 28014 Madrid, Spain
Dock House, The Waterfront, Cape Town 8001, South Africa

http://www.cambridge.org

© Cambridge University Press 1991

This book is in copyright. Subject to statutory exception
and to the provisions of relevant collective licensing agreements,
no reproduction of any part may take place without
the written permission of Cambridge University Press.

First published 1991
Reprinted 1995
First paperback edition 2004

A catalogue record for this book is available from the British Library

ISBN 0 521 40059 7 hardback
ISBN 0 521 52430 X paperback

Transferred to digital printing 2004

Contents

Contents

Preface

On 9 May 1950 the foreign minister of France, Robert Schuman, proposed that "the entire Franco-German production of coal and steel be placed under a common High Authority in an organization open to the other countries of Europe." By pooling their heavy industry interests, he proceeded, the two powerful antagonists could end the competition that had caused armed conflict twice in the previous half-century and create the essential conditions for a new era of material growth and prosperity. On 18 April 1951, after nearly a year of negotiations, the representatives of France, West Germany, Italy, and the Benelux nations concluded the Treaty of Paris instituting the European Coal and Steel Community (ECSC), and on 10 August of the following year it began operations in Luxemburg.

Europe's "first great experiment in supranationalism" triggered an immense outpouring of scholarship. Inspired by the hope that a "new form of economic and political cooperation" had been invented, a number of leading social scientists set about meticulous and searching examination of the Schuman Plan and the ECSC, producing an impressive literature. Its overall conclusions were sadly disappointing, however. William Diebold reluctantly ended his masterful and immensely useful *Schuman Plan: A Study in Economic Cooperation* without having proven that the organization had had a measurable economic impact. Ernst Haas, whose *Uniting of Europe* presents the most thoroughly documented of the many cases argued for what might be called a dialectic of integration, felt obliged to introduce the 1968 reissue of his book with a thirty-seven-page apology for the failure of the "spillover effect" to work as predicted in the initial edition. Though unable to demonstrate that the ECSC had actually been a success, Haas, Diebold, and their contemporaries were loath to conclude that it had failed, instead maintaining a faith that – however mysterious its workings – a force called integration was transforming Europe from a continent of warring states into a new political entity.¹ Most others have continued to share this conviction: The ECSC has had remarkably few serious scholarly, or other, critics.² The organization itself soon lapsed into obscurity, playing only a minor role in the diplomacy leading to the Treaty of Rome concluded in 1957 and the founding a year later of the European Economic Community (EEC), into which it was incorporated. In 1968 the ECSC merged completed into the EEC and thus disappeared as a separate organization.

After nearly a generation of neglect, the coal–steel pool has emerged from the shadows. Books like Alan Milward's on reconstruction and Michael Hogan's on the Marshall Plan have generated a huge amount of recent interest in the reorganization of Europe after World War II,³ and as the target date of 1992 for economic unity approaches, historians and publics alike are paying special attention to the origins of European integration. Indeed, the once nearly forgotten heavy industry pool is now increasingly the subject of colloquiums, editorials, and pronouncements by television pundits. Yet knowledge about the European Coal and Steel Community has advanced little in the past twenty years. Though an important

¹ See William Diebold, *The Schuman Plan: A Study in Economic Cooperation, 1950–1959* (New York, 1959), and Ernst B. Haas, *The Uniting of Europe: Political, Social, and Economic Forces, 1950–1957*, 2nd ed. (Stanford CA, 1968). Among other important early works are Derek Curtis Bok, *The First Three Years of the Schuman Plan* (Princeton NJ, 1955); Henry W. Ehrmann, "The French Trade Associations and the Ratification of the Schuman Plan," *World Politics*. 6/ 1:1953, pp. 453–481; Louis Lister, *Europe's Coal and Steel Community: An Experiment in Economic Union* (New York, 1960); J. E. Meade, H. H. Liesner, and S. J. Wells, *Case Studies in European Economic Union: The Mechanics of Integration* (Oxford, 1962); Horst Mendershausen, "First Tests of the Schuman Plan." *Review of Economics and Statistics* 35/1:1953, pp. 269–288; Harald Jürgensen, *Die Westeuropäische Montanindustrie und ihr Gemeinsamer Markt* (Göttingen, 1955); William N. Parker, "The Schuman Plan: A Preliminary Prediction," *International Organization* 5/1:1952, pp. 381–395; Hans Potthoff, *Vom Besatzungsstaat zur europäischen Gemeinschaft: Ruhrbehörde, Montanunion, Europäische Wirtschaftsgemeinschaft, Euratom* (Hannover, 1964); F. Roy Willis, *France, Germany, and the New Europe, 1945–1967* (Stanford CA, 1968); Arnold Wolfers, "The Schuman Plan and Ruhr Coal," *Political Science Quarterly*, 57/4:1952, pp. 503–520; K. W. Zawadzki, "The Economics of the Schuman Plan," *Oxford Economic Papers*, new ser. 5:1953, pp. 157–189.
² Among the rare critics and skeptics are Bernard Lavergne, *Le Plan Schuman: Exposé et critique de sa portée économique et politique*, 2nd rev. ed. (Paris, 1952), and Altiero Spinelli, *The Eurocrats: Conflict and Crisis in the European Community* (Baltimore, 1966).
³ Michael J. Hogan, *The Marshall Plan: America, Britain, and the Reconstruction of Western Europe, 1947–1952* (Cambridge, 1987); Alan S. Milward, *The Reconstruction of Western Europe, 1945–1951* (London, 1984).

collection of conference papers concerning the Schuman Plan negotiations appeared in 1988,[4] there has up to now been no full-scale
study of either the diplomacy leading to the creation of the ECSC or
the organization itself. Though lip service is often paid to the
founding of the coal–steel pool as the birth act of the new Europe,
historical understanding of its origins, operations, and consequences
is lacking, and basic questions remain unanswered. If, as the evidence of the social scientists suggests, the ECSC did none of the
things it was designed to do, how could the pool have contributed to
a Franco-German reconciliation or to European unification? And
how, if at all, might these developments have otherwise occurred?
Was the heavy industry community, as purported, something truly
novel, or did it have important precedents? And who, or what, advanced the integration process? Was it essentially the work of great
men or powerful forces, and when and where did it begin: before,
during, or after World War II, in France, Germany, or possibly
someplace outside of Western Europe? And how, finally, is one to
assess the ECSC's overall historical impact?

This book will demonstrate that the coal–steel pool was created to
solve a particular historical problem, did so by building constructively on antecedent tradition, and left a legacy that included a number of minor failures but a single huge accomplishment of
overriding importance. The *problem* was of course the Ruhr, command of whose resources conveyed economic supremacy in Western
Europe during the first half of the century. Conflict over this industrial heartland, often cited as the cause of the two world wars, is
more accurately described as a chief source of peacetime instability,
standing in the way of both Franco-German reconciliation and the
restoration of a sound system of international trade and payments
after each of the two world wars. The *solution* was different from
what anyone expected or could have predicted. Jean Monnet, with
strong American backing, devised the ECSC in an attempt to reform, and thus tame, Ruhr industry. The diplomacy that led to the
formation of Europe's "first supranational institution," as well as its
actual operations, resulted instead in its restoration to power. Yet
the Ruhr, and the Germans, did not become a threat to the peaceful
and progressive Europe beginning to emerge but became instead its
mainstay and guardian. The integration that grew out of the founding of the European Coal and Steel Community should therefore be
understood not as a process set in motion by some new mechanics of

4 Klaus Schwabe (ed.), *Die Anfänge des Schuman-Plans, 1950/51* [*The Beginnings of
the Schuman Plan*] (Baden-Baden, 1988).

economics or politics but as the successful result of a kind of cooperative diplomacy made possible by fundamental changes of attitude
occurring over a generation. It continues to be successful because
the most powerful West European nation is committed to making
integration work, and the former great power west of the Rhine
accepts the leadership of the neighbor to the east. This result is
ECSC's greatest *accomplishment.* Time has proven that the new departure promised in the epochal announcement of 9 May 1950 was
neither the invention of publicists nor the fantasy of scholars but the
very bedrock of the new Europe.

The organization of this book is guided by the ideas outlined
above. The first chapter recounts the failure of successive interwar
attempts to solve the Ruhr Problem but also depicts the origins of
the antecedent tradition, the institutionalized cooperation upon
which the coal–steel community would eventually be built. Chapter
2 describes how the economics of warfare made the United States a
power on the European scene (while also measuring its newly acquired strength against that of the main European nations), and the
following chapter reveals how America, in eventual partnership with
Europe, developed the constructive policies that led to the formation of the ECSC. The fourth chapter discusses the hardships of
Ruhr industry in the postwar years, revealing linkages to the past
but also a determination to make new beginnings. Chapter 5, in
tracing the diplomacy that led to the coal–steel pool, shows how it
became part of a larger settlement with Germany – a substitute
peace treaty – but also demonstrates that its results were far different from those Monnet had sought. The final chapter describes the
Ruhr restoration, considers the impact of external events on the integration process, and assesses the contribution of the coal–steel
pool to its future development.

Grants awarded me in 1984 by the Friedrich Ebert Foundation and
the Hoover Institution made it possible to begin this project. The
Deutsche Forschungsgemeinschaft funded my research the following year as part of the larger Westintegration program supported
by the Volkswagen Foundation, which Ludolf Herbst then directed
at the Institut für Zeitgeschichte in Munich. A semester as Visiting
Professor at the European University Institute enabled me to begin
publishing from my material. A grant from the Weldon Springs
Fund of the University of Missouri provided the necessary means of
completing archival investigation and beginning the book itself. The

National Endowment for Humanities, through its Fellowships for College Teachers Program, secured the release time from lecturing needed to complete the manuscript. Even with such generous and unstinting support, the book would not yet be finished were it not for the flexible scheduling and course reductions arranged at my parent institution by our Chairs Neal Primm and William Maltby in cooperation with Edwin Fedder, Director of the Center for International Studies.

I cannot cite by name everyone whose assistance was required to research this book or make personal mention of all those whose acts of kindness made doing the job a pleasure. A mere enumeration of archives visited is a paltry acknowledgment of appreciation to the many sympathetic and highly professional colleagues who went to unusual lengths to see this project to completion; I can only hint at the full measure of services graciously rendered. One must, however, make beginnings by thanking the entire staff of the archives of the Fondation Jean Monnet pour l'Europe, who tirelessly integrated me into the life of the Ferme de Dorigny, aroused what threatens to become a lifelong fascination with Jean Monnet, and Xeroxed tirelessly on my behalf, saving me months of research. Mme. Bonazzi and Irogouin at the Section Contemporaine of the Archives Nationales de France helped me locate two of the most valuable sources used in this study, the Bidault Papers and the French Plan's files concerning the ECSC. Without the competent assistance of Frau Kossol, chief archivist at Kloeckner A.G., a large part of this book would be missing; the Guenter Henle Papers are surely the richest single available nonpublic source of material concerning Westintegration.

To thank any single individual on the staff of the Bundesarchiv for the assistance offered over what has now been nearly two decades would be absurd; credit can only be granted collectively to what is the best-financed, most competently organized and soundly managed national documents repository in Europe. The frequency of citations in this book from the U.S. National Archives as well as the Government Records Center in Suitland, Maryland, is disproportionate to the limited amount of time I was able to spend in these places; this is a testimony not only to the high quality of the material and to the efficiency of the classification system but to the great physical strength of John Butler, who provided me with a constant flow of documents cartons during my relatively short stay. For patience, devotion to duty, and great personal concern I would like to thank above all the staff of the Institut für Zeitgeschichte. It was reassuring to know that the paper-snagging copier would always soon be restored to working order, expense reimbursements promptly forthcoming, and 1 DM worth of food stamps from the Bavarian Ministry of Agriculture issued without fail at the end of every month for each day worked.

The staff of the Truman Library was unfailingly helpful, as ever, and the same was true of the archivists and technicians at the Institut für Weltwirtschaft in Kiel who guided me through the relevant files of their extensive collection of newspaper clippings. I would also like to mention the courteous and efficient assistance provided by several institutions where I worked only briefly, the Public Record Office in Kew, the library and archives at the League of Nations in Geneva, the staffs of the Deutsches Museum and the Bayerisches Bibliothek in Munich as well as of the library at the European University Institute in Fiesole.

Several persons deserve special thanks. Werner Bührer, a colleague in the "Herbst Project," unselfishly permitted me to photocopy extensively from the Salewski Papers, which he discovered while researching *Ruhrstahl und Europa*. Catherine André, a doctoral student at the European University Institute, offered access to the material from the archives of the European Community that she had gathered for a forthcoming dissertation on the canalization of the Moselle River. Dr. Hans von der Groeben invited me to his home for an interview and loaned the portion of his papers bearing on the ECSC. Dr. Axel Plagemann, chief of public relations at the Wirtschaftsvereinigung der Eisen- und Stahlindustrien enabled me to become the first scholar to examine the as then unclassified records of the association's board meetings. Jean Martin left the papers of her late husband, James Martin, in my custody for a period of several years. Hans-Günther Sohl furnished his privately published autobiography, *Notizen*. Reinhard Kuls and Boris Ruge provided copies of their masters' dissertations, as did Isabel Warner the galley of an unpublished article concerning decartelization and Regine Perron a chapter from her forthcoming dissertation concerning the European coal problem after World War II. Others who have helped me prefer to remain anonymous.

A book of this scope cannot be based entirely on archival research but must rest in part on the work of others. In order to prevent the manuscript from becoming too long, I forswore annotating footnotes and can only hope that the citations indicate some measure of my immense scholarly gratitude. I would like particularly to thank

Alan Milward, who inspired this book, and Ludolf Herbst, who encouraged me to complete it. Donna Palmer and Mary Hines helped type the manuscript, partly from nearly illegible handwritten notes, and for discharging this laborious task have earned a greater reward than any I can bestow. My tireless editors at Cambridge, Barbara Palmer and Janis Bolster, may also merit canonization. Anne Blanchardon and France Pelletier eliminated numerous mistakes in French from the manuscript, as did Peter Helmberger in German. Ambassador William C. Vanden Heuvel found time in his crowded schedule to read the book in draft, and he too has helped spare me the embarrassment of errors and omissions. All remaining shortcomings of fact and interpretation are my own fault: The buck, as we say in Missouri, stops here. The acknowledgments cannot, however: During the several years needed to research and write this book, the author has been half-husband to a loving wife, half-father to three dear children. May they be as happy as he is that the job has now been completed.

Clayton, MO
October 1990

1

Mending a broken world: coal and steel diplomacy between the wars

How could one have put back together that which World War I had broken up, the "delicate complicated mechanism . . . through which alone the European peoples can employ themselves and live"?[1] Although John Maynard Keynes, who thus described the European economy, quite properly blamed the Allied heads of state for having done too little at the June 1919 Versailles peace conference to restore it to normal operation, neither they nor those who succeeded them on the diplomatic stage could have done this by themselves. The thing was broken, and no amount of patchwork could repair it. New mechanisms were needed as well as time to test them. Only then could the European economy work properly again.

Two serious attempts *were* made between the wars to mend the broken world, neither successful in the short run, both future contributors to the Schuman Plan settlement. The first, an important subcurrent in French diplomacy, aimed at fitting Germany into a framework of international economic agreements which, though initially tilted radically in France's favor, could eventually be reworked into a mechanism for the equitable adjustment of industrial and financial relationships between the two countries. Yet the French were then too weak, and German industry too bent on revenge, to have made a policy of even limited accommodation politically feasible. Abandoned in 1923 after the decision to occupy the Ruhr, the theme recurs in French policy after World War II.

Business and finance made the second attempt. Wall Street's "private Marshall Plan" was part of it. The huge influx of American capital into Germany that began with the adoption of the Dawes Plan in 1924 stimulated a boom and gave rise to an enthusiasm for placing affairs of state in the hands of the directors of banking and industry. Though it vanished after the 1929 crash on Wall Street as rapidly as it had once appeared, the vogue for business involvement

[1] John Maynard Keynes, *The Economic Consequences of the Peace* (London, 1919), p. 2.

in public problems did have one important consequence, the creation in 1926 of the International Steel Cartel (ISC). In some respects similar to counterparts in other branches of production, the ISC soon came to be regarded outside as well as within the industry as a possible forum for the settlement of political differences between France and Germany.

Though the depression soon wrecked the International Steel Cartel, it revived in the 1930s and developed into a permanent fixture, thanks to both the adoption of Ruhr-style organized capitalism in Western Europe and an increasing moderation of German producers. As the 1930s reached their tragic end, the ISC became the most important remaining link between the Third Reich and Europe, or so the governments of both France and Great Britain at least thought. In a desperate effort to appease Hitler, they tried to broaden the close relationships among international coal and steel producers, form transnational industry pacts, and launch joint ventures culminating in a community of interest strong enough to have eliminated any reason for his wanting to start a war. Though this "business diplomacy" failed, the ISC lived on, albeit under a different guise, as the organizational scaffolding for raw materials rationing and order placement through which the economy of the wartime New Order would operate.

1.1   A broken world

World War I resolved nothing. It was a defeat for Germany only in a military sense; in every other sense save possibly the moral one, the Germans were stronger in 1918 relative to the French and British than in 1914. This was above all the case with regard to the economy. More than ever before, the welfare of the Continent depended upon restoring Germany to economic health. The victory parades of 1918 concealed this fact only from the publics of the militarily triumphant nations. French policymakers knew even then that their country needed reparations not merely for recovery but to remain a great power.[2] Only three means were available for securing them after the United States, as soon happened, withdrew from Europe. The French could have either pleaded for the support of their inconstant allies, the British, tried to coerce the Germans into making

---

[2] Marc Trachtenberg, *Reparation in World Politics: France and European Economic Diplomacy, 1916–1923* (New York, 1980), p. 1.

present concessions as a first step to future reconciliation, or, as a last resort, risked a nihilistic act in order to compel both Britain and Germany to do their bidding. The Germans understood all of this quite well. Even during the months of desperation following defeat, they were ready to raise the stakes in the contest with France, confident that time worked in their favor. And they eventually won: French weakness, not French strength, led to the Ruhr occupation of January 1923. Its failure ended France's bid to reorganize Europe until, after 1945, it would be resumed, this time with American support.

France's German policy during and after World War I is in the throes of revision, and though the final word has yet to be written historians will never again be able to describe it as merely punitive in character.³ It amounted to far more than exploiting the harsh terms of the Treaty of Versailles in order to "squeeze the German lemon until the pips squeak." The treaty itself did not govern French policy; it was never more than a vehicle in the service of the national interest. This called not for an economically crippled Reich but for one which, though industrially and financially sound, would be politically shackled. The French had two kinds of impediments in mind. The first of them, detachment of the Rhineland, was of obvious military inspiration and should be understood more as a policy of harassment, vicious and destructive though it was, than as a major attempt to dismember the Reich. It coexisted in an uneasy relationship with a more serious and sustained effort to coerce Germany into accepting agreements that would serve the twofold purpose of supplying France's immediate needs and creating the basis for a more equitable long-term relationship; by demonstrating their goodwill, the Germans could be raised from economic subjection to industrial partnership. This notion would provide the gist of French *Ruhrpolitik* after World War II as well.⁴

³ Jon Jacobson, "Is There a New International History of the 1920's?" *American Historical Review* 88:1983, pp. 617–645; Marc Trachtenberg, "Reparation at the Paris Peace Conference," *Journal of Modern History* 51:1983, pp. 24–55; Walter McDougall, "Political Economy versus National Sovereignty: French Structures for German Economic Integration after Versailles," *Journal of Modern History* 51:1979, pp. 4–24; Charles Maier, *Recasting Bourgeois Europe: Stabilization in France, Germany, and Italy in the Decade after World War I* (Princeton NJ, 1975), passim.

⁴ Trachtenberg, *Reparation*, p. 2; Walter A. McDougall, *France's Rhineland Diplomacy, 1914–1924: The Last Bid for a Balance of Power in Europe* (Princeton NJ, 1973), pp. 10–11.

4   Coal, steel, and the rebirth of Europe

The author of this policy was Minister of Commerce Etienne Clémentel, the man in charge of France's foreign trade and supply during World War I.[5] His ideas would have a profound influence on the architect of the future European Coal and Steel Community, Jean Monnet. In the second week of September 1914, with the battle of the Marne still raging, Monnet, then a twenty-five-year-old foreign sales representative of his family's cognac firm, managed to arrange an audience with Prime Minister René Viviani in the hope of persuading him to recognize that French military strategy was based on false premises: The war would be long, not short, and something completely overlooked in prewar planning – foreign supplies – would make the difference between victory and defeat. Viviani was duly impressed: Monnet soon found himself serving in London as chief of trade liaison with the British and reporting to Clémentel in Paris.[6]

Along with Walther Rathenau and V. I. Lenin, Clémentel was among the first to recognize that the organizations set up to administer industry, agriculture, and trade during the war could serve as the basis of a peacetime economy whose priorities would be determined by the state rather than the marketplace. Underlying Clémentel's predilection for government intervention was less dislike of liberal economics than fear that France's survival as a great power made it necessary to appropriate the resources of the world in general, and Germany in particular, until recovery had occurred. The French minister of commerce first presented his proposals at the Paris peace conference of June 1916, which had been convened for postwar planning. At this gathering he secured British agreement to maintain the inter-Allied commissions set up in 1914 to purchase and distribute world raw materials supplies. Complementing these proposals in the international sphere were others for domestic reform. According to Marc Trachtenberg, "The organization of French industry completed the Ministry of Commerce plan. Under the guidance of the state, firms within an industry would cooperate with one another, sharing technical knowledge, dividing markets, each perhaps specializing in the manufacture of particular products."[7] A national producer association, animated by the same spirit, was to underpin this structure.

Coal and steel were central concerns of Clémentel and French

[5] Etienne Clémental, *La France et la politique économique interalliée* (Paris, 1931).
[6] Jean Monnet, *Memoirs* (London, 1978), pp. 48–49.
[7] Trachtenberg, *Reparation*, p. 4.

policy makers generally. They had a number of worries. Most obvious was physical damage to plant: overworked or flooded mines and missing, wrecked, and obsolete mill machinery. Even more important were the problems arising from the return of Alsace-Lorraine. This doubled France's theoretical steel capacity, equaling Germany's in 1913, thus perilously increasing its already heavy dependence on Ruhr coke and coking coal. Secretary-General Robert Pinot, of the steel producers' association, the Comité des Forges, estimated in October 1915 testimony before the Senate Committee on Economic Expansion that after the war France would require an additional seven million tons of the former and thirty million tons of the latter.[8] The annexation of the Saar would make no difference in this respect, he added, because the coal of that region was unsuitable for coking: The mills of German Lorraine, which France would acquire, drew their supplies of combustible from the Ruhr. Neither Pinot nor anyone else from the Comité appears to have had a clear-cut plan for dealing with this problem. The closest thing to one was a vague steelmakers' resolution of July 1916 to the effect that "any extension of French control beyond Alsace-Lorraine and beyond the Saar could only simplify the problems that recovery of Lorraine would create for France."[9] Far from being a demand for the seizure and annexation of the Ruhr mines, this amounted to a plea that the government devise ways to pressure their operators into supplying the raw material needed to keep French mills in operation.

Clémentel's plans had been frustrated even before the peace conference convened in Versailles. The British would not go along with them because they profoundly resented the advantages these conferred on France and recognized as well that any appropriation of German coal output would reduce the sales of their own mines. The United States, which by early 1919 was hastily dismantling its temporary national commissions for supply and transport, objected in principle to maintaining wartime controls.[10] Partly because of business objections to further "government interference," Clémentel's power even within France was limited; authority passed instead into the hands of the minister of armaments, Louis Loucheur, a visionary industrialist. In April 1919 Clémentel and Monnet both resigned and Monnet joined the secretariat of the League of Nations, where he hoped to keep alive Clémentel's plans for international control of raw material supplies.[11] Although this aim was also

[8] McDougall, *Rhineland Diplomacy*, p. 19.   [9] Ibid.
[10] Trachtenberg, *Reparation*, p. 38.   [11] Monnet, *Memoirs*, p. 79.

largely frustrated, the most essential component of Clémentel's approach, his Ruhrpolitik, remained a key element of French policy. In the future, however, France would be obliged to accept as interlocutors not the official representatives of the Weimar government but the real powers behind it, the magnates of heavy industry.

Their influence was due partly to the consistent French preference for reparations in kind, whose delivery only producers could assure, over financial transfers. Clémentel feared that the latter would be inflationary and dreaded that they would turn France into a nation of consumers rather than producers, another sixteenth-century Spain. In his view, as in that of successive leading economic advisors, financial reparations were a lever of policy, not an end in themselves. As Trachtenberg puts it, the "nominal demand for vast payments was the instrument by which vague 'concessions' could be extracted from Germany, and by which recalcitrant allies might be induced to favor the idea of a 'world fund' to finance the rebuilding of the devastated areas."[12] Clémentel's successor, Loucheur, who had been Prime Minister Clémenceau's chief adviser on financial and industrial matters at Versailles, secured the really important concessions from the Germans. He managed to have incorporated into the treaty stipulations requiring German delivery of twenty-seven million annual tons of coal (or its equivalent), as well as two clauses. One assigned France proprietorship of the Saar mines for fifteen years; the other removed German tariffs on goods produced in Alsace or Lorraine for a period of five years.[13] Like Clémentel, Loucheur sought broader solutions. Even before the treaty had taken effect, he launched an initiative aimed at a more permanent settlement of Western European coal–steel problems. On 1 August 1919 he proposed to a German delegation that an international steel cartel be set up, which was to include the producers of Belgium and Luxemburg along with the producers of France and the Reich and, in addition, was to be coupled with an agreement for the exchange of French iron ore and Ruhr coal. Representatives of one of the biggest Lorraine producers, Schneider-Creusot, conducted negotiations with the Germans on this basis during the final months of 1919.[14] In undertaking such initiatives the French may well have had in mind something that "went far beyond a mere business arrangement with Germany . . . and was in fact [aimed] at some kind

---

[12] Trachtenberg, *Reparation*, p. 18.   [13] McDougall, *Rhineland Diplomacy*, p. 104.
[14] Trachtenberg, *Reparation*, p. 112.

of political arrangement."[15] They may also have merely been dangling the possibility of such a settlement before German eyes in order to obtain fuel.

The coal crisis of 1919–1920 was little short of catastrophic for France. It stemmed directly from the dislocations of war and peace, the destruction of the mines, the exhaustion of the rails, and labor shortages and unrest. In 1919 France had to import half the coal it consumed, no less than 70 percent of which was from Britain. The high figure was due in part to the fact that the delivery provisions of the Treaty first took effect in April 1920. After the war U.K. coal arrived at French ports at what seemed extortionate prices, seven times more than in 1913. In April 1920 the British cut France's quota of their total coal exports from 60 to 40 percent, and the French government foolishly retaliated by reducing the official price of import coal; supplies soon virtually disappeared.[16] Not surprisingly, "Unemployment spread for lack of fuel . . . Paris utilities rationed heat and light [and] the newly lighted foundries were largely extinguished."[17] Disappointment mingling with disgust, the French government turned from Britain to the Ruhr.

The big steelmakers of the eastern provinces supported this policy shift. Unlike the main Ruhr producers, few of them controlled their own mines; most purchased fuel on the market and derived no cost, or any other, benefits from reparations coal. Since April 1916 the Law of Perequation had regulated prices of all solid fuels sold in France, using an equalization formula that favored small consumers. For domestic political reasons the rules could not be changed.[18] It therefore made good economic sense for Lorraine steelmakers to establish an independent supply relationship with the Ruhr mines. In April 1920 the secretary-general of the Comité des Forges, Robert Pinot, again approached the Germans with proposals for entente and requests that discussion begin concerning French participation in the Ruhr mining industry.[19]

If in 1919 and 1920 the German concerns of the Comité des Forges extended little beyond the immediate problem of the coal supply, the same cannot be said about those of the undersecretary for commercial relations at the Ministry of Foreign Affairs, Jacques Seydoux. This physically feeble font of intellectual energy was the

[15] Ibid., p. 86. [16] McDougall, *Rhineland Diplomacy*, pp. 106–107.
[17] Ibid., p. 109. [18] Ibid., pp. 104–105. [19] Trachtenberg, *Reparation*, p. 112.

tireless author of numerous elaborate schemes for working French reparations policy into eventual détente with Germany. The so-called Seydoux Plan, presented to the general public in a 12 December 1920 article entitled "How to Make Germany Pay," featured a revolving fund to commercialize the reparations debt in order to give German manufacturers an economic stake in French reconstruction. Seydoux also envisaged so-called joint boards (*bureaux mixtes*) to administer the transfer of product between the two countries. Such an arrangement would have required German acceptance of foreign economic controls, at least in the near term. This he expected to gain by creating the hope that in time the machinery for collecting reparations could eventually be regeared to promote economic collaboration with France. Although prominent on the agenda of the Brussels conference held from December 1920 to January 1921, Seydoux's ingenious scheme had no immediate practical importance in French policy.[20] Nor, for that matter, did the bold initiative of Loucheur that resulted in the Wiesbaden Accords concluded in October 1921. This was another clever plan for the large-scale transfer of German product to France. The French were to place orders through a centralized, technically private body representing the interests of industrial groups. It would then transfer the orders to a counterpart German organization, which would distribute them to producers. Prices were to be set jointly. Although Weimar Foreign Minister Walther Rathenau signed the agreement, it never took effect.[21]

There are many reasons why the Seydoux Plan and the Wiesbaden Accords failed. Not only were they overly complex, but a great deal of ground had to be crossed before Ruhr industry would be ready to accept the verdict of battle. If there had been an equivalent after World War I to the post-1945 trials of major industrialists held at Nürnberg, the representatives of some of the Ruhr's most prominent firms would surely have found themsleves in the dock. Since the publication of Fritz Fischer's compendious study of German war aims during World War I, *Griff nach der Weltmacht,* no serious doubt remains that as a group the leaders of the Ruhr Konzerne did indeed wage aggressive warfare, despoil the economy of the occupied territories, and promote the exploitation of slave labor. German big businessmen did far more than merely support the annexationist policies of their government; they devised them, exercised political

[20] Ibid., pp. 156–157.   [21] McDougall, *Rhineland Diplomacy,* pp. 167–168.

pressure to secure their adoption, and propagandized for them tirelessly until the end of the war.[22]

Demands that the war yield a rich material harvest first cropped
up spontaneously as German troops advanced across the plains of
northern France in late summer 1914. Heavy industry's interest
centered on the rich ore beds of Longwy-Briey, which German capital had begun to penetrate shortly before the war. The Ruhr was
unrelenting in the pursuit of its objectives. In the final days of August and early September 1914 the Thyssen firm, the Röchling
brothers, Stinnes, Krupp von Bohlen, and Emil Kirdorf (of the coal
producers' syndicate) each made representations to Chancellor
Theobald von Bethmann-Hollweg concerning Longwy-Briey. In a
fall 1914 effort led by big business, Alfred Hugenberg of Krupp set
about organizing a massive national lobby for a peace of conquest,
one "worthy of the immense sacrifices" being made for the war.
Early the following year, Hugenberg put the lobbying effort on a
permanent footing by organizing a national league of producers
and managers to exert constant pressure for expansionist policies.[23]

In March 1915, and later in May, it presented the famous Petition
of the Six Economic Organizations to the chancellor. The document
called for the creation of "a colonial empire adequate to satisfy Germany's manifold commercial interests" and the annexation of large
portions of the Baltic and White Russia. As befit the organization's
origins in heavy industry, the petition also contained a very specific
list of demands in Western Europe. These would have brought no
less than fifty thousand square miles and eleven million new inhabitants within the Reich. Belgium, it stated, was to become a vassal
state. As for France, "the Six Associations demanded the coastal districts, including the hinterland, as far as the mouth of the Somme
[and] the district of Briey, the coal country of the Departments of
Nord and Pas de Calais, and the fortresses of Verdun, Longwy and
Belfort."[24]

This was a program motivated by more than mere concern for the
Reich's future military security: German heavy industry wanted to
eliminate peacetime competition. The events of the occupation, especially in Belgium, provide proof of this fact. The French were
a bit luckier than the Belgians. Longwy-Briey came under civil administration in October 1914, and a committee of prominent industrialists with interests in Lorraine adminstered the French ore

[22] Fritz Fischer, *Griff nach der Weltmacht* (Düsseldorf, 1961).
[23] Hans W. Gatzke, *Germany's Drive to the West* (Baltimore, 1950), pp. 37–38.
[24] Ibid., p. 45.

properties located there. For the rest, the industry of occupied France stood idle, left to rust. Though this inactivity was less evidence of a German plot than of a shortage of raw materials, it was far from unwelcome to Ruhr producers. A comprehensive and detailed survey of the region's production facilities conducted under the auspices of the army supreme command in 1916 concluded gloatingly that the war had set back France's industry by many years – nothing was to be feared from its future steel competition. The damage done to the mines also meant, it added, that the French would be far more dependent than previously on supplies from the Ruhr. Best of all, postwar reconstruction of the manufacturing industry was expected to provide ample contracts for German firms, which in 1914 had already controlled almost 20 percent of the French market for production machinery. The report observed in closing that even without the annexation of new territory, something its author considered desirable, opportunities for German producers would indeed beckon in postwar France.[25]

In Belgium, Ruhr industry took a direct hand in the exploitation of the economy. There were two schools of thought as to how to deal with this historically anomalous nation. Military Governor Baron F. W. von Bissing wanted to incorporate it into the Reich as a dependent territory but not to strip it bare and reduce it to penury. German big business belonged to a more hard-line party that lobbied for truly drastic policies in the interests of victory and expansion. An ugly truce prevailed in the beleaguered country during the first half of the war. The leaders of Belgian business refused to work for the Reich; German industry, suffering from shortages of raw material, had no intention of utilizing capacities in occupied territory when this would have required leaving their own factories idle. The adoption in August 1916 of the Hindenburg Plan to increase war production introduced the phase of *exploitation à outrance*. One consequence of it was the organization of Industriegesellschaft 1916 m.b.H., which had analogues in the fields of transportation and agriculture. Among shareholders in the new company were Alfred Hugenberg of Krupp, Hugo Stinnes, Emil Kirdorf of the coal syndicate, Paul Reusch of Gutehoffnungshütte, and Wilhelm Beuckenberg of Phoenix. Along with the other government-sponsored, producer-owner consortiums, the new enterprise was supposed to help the Reich take over controlling shareholdings in the main pro-

---

[25] Ibid., pp. 152–153.

duction units. Industriegesellschaft soon liquidated Belgium's gas, water, and electrical companies at knock-down prices. Actual payment was never made.[26] Exploitation à outrance also extended to labor. At a 16 September 1916 meeting attended by leading producers (among them Klöckner, von Siemens, Springorum, Vögler, and Röchling), the chemical industrialists Carl Duisberg and the armaments czar Walther Rathenau jointly proposed deporting labor from Belgium in order to relieve shortages in German factories. The intervention of Hugo Stinnes was necessary to overcome the vehement opposition of Military Governor von Bissing to this policy.[27]

Industry extremism lasted until the end of the war, its mood glaringly evident in the petition submitted by the president of the steel producers' association, Albert Vögler, to the chancellor. Germany must have Briey, Vögler claimed, as otherwise domestic ore supplies would hold out for only fifty years. Paper guarantees were worthless! If the opportunity to annex the area were passed up, he warned, "The German people will be doomed to extinction in a future war."[28] This was sheer hysteria. The Ruhr would have no future need for French minette because during the war it had shifted to the consumption of richer Swedish ores. Even in the commodities-glutted 1920s, few minerals would be as superabundant as the low-grade French product. Not even in World War II, when steel outputs reached historic highs, did the Reich suffer from lack of access to iron ore.

In 1911 Hugo Stinnes boasted at a reunion of his former classmates that if given three or four years of peace he could secure German steel predominance in Europe.[29] World War I did little to sober these ambitions. The producers did not accept the loss of the Lorraine mills; aided by generous compensation from the government, they began at once to erect modern new installations in the Ruhr and with grim determination set about recapturing former customers. In this respect the inflation was, as put in a famous *Simplicissimus* cartoon, "catastrophically favorable." It washed away debt, whittled down wage gains, and reduced real prices, enabling German producers to increase their shares on the shrunken international steel markets of the postwar years (the earnings from which rapidly piled up in Swiss banks).[30]

---

[26] Ibid., pp. 85–86, 154–155.　　[27] Ibid., p. 156.　　[28] Cited in ibid., p. 246.
[29] Ibid., p. 34.
[30] Gerald Feldman, *Iron and Steel in the German Inflation* (Princeton NJ, 1977), pp. 280–281.

Stinnes was the largest beneficiary of the perverse economics of the inflation and the uncrowned king of the Ruhr during the four years after the war. Between 1919 and 1924 he pieced together the largest combine ever amalgamated in Germany, the Siemens-Rhein-Elbe-Schuckert Union. Though troubled by the social and political consequences of inflation, the rest of heavy industry supported Stinnes's attempt to wreck the one initiative that might have stabilized the Reichsmark. This was the November 1921 credit action of the Reich Association of German Industry. It would have pledged Germany's foreign assets in order to secure a hard-currency recovery loan.[31]

They also backed Stinnes's sabotage of French efforts to reach a reparations settlement. "It must come down to this," he said, "above all, in the next few years, we pay nothing. A settlement of our liabilities must be put off as long as possible, because the development of political and economic relations in the world is moving in a direction favorable to us."[32] Like his colleagues, he expected the steel industry of French Lorraine to choke on excess ore and blast furnace capacity. Other considerations increased the Ruhr's reluctance to treat with the French. One was the dissolution of cartels and producer associations, an inevitability during the readjustment period. A second was the steel industry's construction program, which would not be completed until the end of 1924, and a third the tariff provisions of the treaty that allowed Lorraine manufactures to pass onto German markets duty free until 1925. Thus Stinnes walked out of the July 1920 Spa conference, flatly refusing all future cooperation, and Ruhr industry declined to be represented at the Brussels conference held some five months later, whose agenda included the Seydoux Plan. Private Franco-German talks *were* held in the aftermath of the Wiesbaden Accords, but the Germans manifested no intention to enter serious negotiations.[33]

The French initiatives to German business were bound to fail for other reasons as well. The British naturally looked askance at any industrial entente from which they were excluded. Not the Germans but Prime Minister David Lloyd George caused the breakdown of the Brussels conference by demanding the immediate establishment of a lump-sum figure for the reparations debt. The British also objected to the Wiesbaden Accords on the ground that they gave France an excessive share of the reparations yield.[34] Even the

[31] Ibid., p. 285.   [32] Cited in Trachtenberg, *Reparation*, p. 185.
[33] McDougall, *Rhineland Diplomacy*, p. 211.   [34] Trachtenberg, *Reparation*, p. 190.

French Chamber opposed the Wiesbaden Accords. The Bloc National, which held power until April 1924, was unwilling to countenance any compromise based on the principle that the Reich might not be able to make good the material losses of the war. The final obstacle to a settlement was Weimar. Was the Reich's shaky democracy really to be saddled wtih responsibilities that might cripple it, plunge Germany into chaos, and possibly end in dictatorship?

This was a risk the French were reluctant to take; yet run it they did. In January 1923 Prime Minister Raymond Poincaré occupied the Ruhr. He had decided to embark upon this course in July 1922, after six months of futile overtures to the German heavy industrial leadership, desperate to achieve by force what persuasion had repeatedly failed to secure. Poincaré wanted a heavy industry delivery agreement that could eventually be worked into economic and politic rapprochement.[35] Poincaré occupied the Ruhr in spite of the skepticism of other leading policy makers and in the absence of any agreed-upon plan of economic exploitation, yet in full awareness of the plan's seriousness. Predicting that French occupation would plunge Germany into chaos, Seydoux recommended taking the bold stroke in a memorandum of 21 November 1922 because "there is no reason to fear that such a state of things should have repercussions in France: the French government is solid enough to resist, and it will . . . remain for us to use Germany's political situation to prevent her from harming us. In the event that Germany is left to herself and the inevitable catastrophe follows, the populations of the left bank of the Rhine will accept, with satisfaction . . . , the assistance . . . we provide them, and which would not go beyond the occupied territories and ourselves."[36]

The French owed their failure as much to themselves as to the Germans: They were simply unprepared to pursue the logic of their policy to the bitter end. Though ready to embargo Ruhr sales to the rest of the Reich and ship off booty coal by the trainload, France's leaders shied away from the really ruthless measures needed to break the passive resistance of the subject population. General Degoutte, the commander in charge of the operation, refused to execute the orders of Minister of War Maginot to cut off industry from its raw materials, allow the mines to flood, and take over the power stations. Starving the population was never considered.[37]

---

[35] McDougall, *Rhineland Diplomacy,* p. 214.
[36] Cited in Trachtenberg, *Reparation,* p. 267.   [37] Ibid., p. 296.

French producers were also too ill organized to make a collective
settlement with their German counterparts. By midsummer at least
one faction of Ruhr industrialists was prepared to take the first steps
toward cooperating with the so-called MICUM commissions installed to direct coal deliveries back to France. Though the existence
of MICUM provided a starting point, any permanent arrangement
would have had to involve French industry as well as German. Here
severe problems arose, even though in August 1923 a changing of
the guard at the Comité des Forges brought a de Wendel-led activist
faction into power. In October, the Comité presented its first comprehensive program for dealing with the Ruhr. It demanded the
right to seize as well as to purchase assets. The de Wendel faction
also recommended cutting the Ruhr off from the rest of Germany
by a customs barrier and confiscating all Reich-owned mines.[38] The
French coal industry refused to countenance such new competition,
and the general secretary of the Comité des Houillères, Henri de
Peyerhimof, would not support the steelmakers' demand for a takeover of the mines. The delay proved to be fatal. As 1923 drew to a
close, the value of the franc plunged. In November Seydoux wrote
Monnet, "There is no use hiding the fact that we have entered on
the path of the financial reconstruction of Europe. We will not deal
with Germany as conqueror to vanished; rather Germans and
Frenchmen will sit on the same bench before the United States."[39]

1.2   Attempts to mend it

It took nearly five years for the French to fail in their bid to mend
the broken world, and it would take another five for big business
and high finance to fail in theirs. The new phase began with the
Dawes Plan, a U.S.-led bankers' scheme to finance the payment of
German reparations. It would advance by means of new cooperative
economic arrangements between France and the Reich, among
which the International Steel Cartel formed in 1926 was preeminent, and reach a kind of culmination in the World Economic Conference of 1927 sponsored by the League of Nations. The collapse
of 1929 ended the vogue for private interventions into public problems but left a substantial legacy. Heavy industry cooperation in
Western Europe was not blown away as easily as the precarious
prosperity of the late 1920s; in the decade leading from depression

---

[38] McDougall, *Rhineland Diplomacy*, pp. 339–340.   [39] Cited in ibid., p. 345.

to war, cooperation between coal and steel producers was one of the few remaining links between the nations of an economically and politically shattered continent.

Only the United States possessed the financial resources required to settle the economic problems arising from the war. The American commitment was made grudgingly. The United States did not bury its head completely in the sand after the Senate's failure to ratify the Versailles Treaty, but from the standpoint of Europe it might just as well have done so. Whether adherents of the laissez-faire ideas inherited from the previous century or adepts of the newfangled doctrines of business self-government propagated by Secretary of Commerce Herbert Hoover, Americans recoiled with genuine aversion from governmental intervention into the spheres of industry and finance. Few could discern merit in such potentially beneficial and profitable reconstruction proposals as the May 1919 plan drafted jointly by the Morgan partner Thomas Lamont and the banker Norman Davis, which envisaged "long-term financing organized by private experts on a cooperative, multinational basis."[40] The Wall Street plan called specifically for the organization of a European bankers' committee to administer a general scheme of credits in cooperation with an American financial consortium acting under the supervision of the Department of the Treasury. But there was the rub: Neither Treusury nor Congress was ready to make long-term commitments of any kind to European reconstruction. According to Michael Hogan, both "wanted to avoid new international programs that might promote state management or impair their domestic stabilization plans. [They] intended to cease government lending and return international finance to private channels . . . . Treasury officials [took] the position that Allied countries possessed sufficient assets to attract long-term credits without special inter-Allied economic arrangements based on loans or debt cancellations."[41]

The United States consequently refused to participate in the reparations conferences of the postwar years and made only perfunctory efforts to encourage private financial solutions. The Department of the Treasury even denied moral support to the conference of international bankers held in Amsterdam in 1919 to consider reconstruction problems. It also did nothing to further attempts made by the International Chamber of Commerce during

[40] Michael J. Hogan, *Informal Entente: The Private Structure of Cooperation in Anglo-American Economic Diplomacy, 1918–1928* (Columbia MO, 1977), p. 29.
[41] Ibid., p. 24.

its biannual conventions of 1921 and 1923 to find financial solutions to the reparations problem. Not until January 1923, after the French had actually entered the Ruhr, does one encounter the first American step toward an eventual solution of the financial issues arising from the reparations problem, the settlement of Britain's U.S. war debt.[42]

After a couple of months of total impasse in the Ruhr, the long period of strict American nonintervention into the reparations problem finally came to an end. In October 1923, President Calvin Coolidge agreed, at the request of the Reparations Commission, to dispatch experts to advise the two new committees it had formed to find ways of balancing the budget and stabilizing the Reichsmark. The American president did not realize what would come of his decision. Interpreting their mandate liberally, the two committees produced a general reparations settlement bearing the name of the chief American delegate, Charles G. Dawes.[43] It opened the floodgates to a torrential flow of dollars, buoyed up the economy of Europe for four generally prosperous years, and for the first time ever during peace put the future of the continent in American hands.

Events would fulfill Seydoux's pessimistic prediction of late 1923: At the meetings of the Dawes committee between January and April 1924, Germans and Frenchmen did indeed sit "on the same bench before the United States." Though not heedless of French security requirements, the primary concern of the financial experts was to restore the Continent to prosperity. In their view a reparations settlement should aim neither to right a historic wrong nor to pay off a debt but to create the business confidence needed to stimulate American investment in both Germany and Europe. Refusing to determine the defeated nation's overall payments obligation, the U.S. experts tried to arrive at a compromise based on the ability to pay that would provide receipts large enough to satisfy creditor demands without at the same time either destroying the equilibrium of the currency or impoverishing the public.

The main feature of the Dawes Plan was a five-year schedule of payments, the amounts to increase annually but fluctuating within an established range in order to adjust to changes in both levels of national income and the gold value of the Reichsmark. An international loan was to be raised to prevent overburdening the budget during the first year after the currency stabilization. The federal

[42] Ibid., pp. 21–22.
[43] Ibid., pp. 68–69; Joseph S. Davis, *The World between the Wars, 1919–1939: An Economist's View* (Baltimore, 1975), pp. 56–57.

railways were to be mortgaged to increase liquidity. Finally, "a fairly complicated system of controls, in which foreigners would have a large role, was designed to protect Germany as well as her creditors."[44]

The Dawes Plan loan was soon oversubscribed, and between 1925 and 1929 some $2.5 billion of private capital, more than the reparations payments of those years, poured into Germany. This money financed the rationalization of Germany industry and deserves much of the credit for the high growth rates enjoyed by the Weimar Republic in the late 1920s.[45] Politically as well as economically, the Dawes Plan was a success. In France the elections of 11 May 1924 ended the stranglehold of the Bloc National on policy toward Germany, and a succession of cabinets, beginning with that of Edouard Herriot formed in June 1924, pledged themselves to normalizing relations with Germany. By then many Frenchmen were having trouble remembering why France had persisted for so long in its hard-line Ruhr policy. An investigation conducted by a special committee of the Chamber, while concluding somewhat limply that public opinion had been misled by propaganda, ridiculed France's "directing circles" for culpable ignorance of German realities. A sequel to it laid blame at the feet of the French government of occupation, the High Commissariat, which it depicted as being oversized, incompetent, and staffed with interpreters ignorant of German as well as political agents afraid to report the truth. The policies that led to the Ruhr occupation had few public defenders in France.[46]

The pro-settlement parties east of the Rhine had a comparatively difficult time. They actually lost the spring 1924 Reichstag elections. Credit for holding together the unstable coalition endorsing rapprochement with the French belongs uniquely to one man, Gustav Stresemann. Although he was a warhawk and annexationist during World War I and a vacillator between hard and soft lines toward France in the turbulent years following it, the Ruhr occupation persuaded him that any attempt to overthrow the Versailles settlement by force would be futile and any effort to undermine it by a spurious policy of "fulfillment" certain to fail in the long run. Once committed to détente, Stresemann was unswervingly loyal to it. As chancellor from 13 August to 1 December 1923, he negotiated France's painful withdrawal from the Ruhr with consummate fi-

[44] Davis, *World between Wars*, p. 65.
[45] See William L. McNeil, *American Money and the Weimar Republic: Economics and Politics on the Eve of the Great Depression* (New York, 1986).
[46] McDougall, *Rhineland Diplomacy*, p. 367.

nesse, at the same time managing with tact and firmness the immense and highly risky operation of stabilizing the currency. As foreign minister from then until his untimely death in 1929 Stresemann's purpose was to solidify the pro–rapprochement Reichstag coalition by demonstrating that his policies brought tangible benefits.

The Locarno Agreement is the best-known monument to the Stresemann–Briand era. The pact consisted of five treaties, of which the Rhineland Pact declaring Germany's western boundaries "inviolable" was the most important; it conceded the permanence of the loss of Alsace-Lorraine. The other treaties were arbitration conventions of a secondary nature. For its part, France agreed to evacuate the northern half of the Rhineland at once and withdraw from the remaining areas of the occupied territory by 1933. Locarno had both a predecessor in coal and a sequel in steel. At the London conference of December 1924 the Germans agreed to supply combustibles as part of their Dawes Plan annuities in amounts substantially greater than previously required. Actual imports of coke and coal increased steadily until 1930.[47] Where the coal agreement removed an impediment to cooperation, the founding of the International Steel Cartel in 1926 would, it was hoped, pave the way to an overall economic settlement in Western Europe. Foreign Minister Stresemann hailed the steel agreement as a "landmark of international policy the importance of which cannot be overestimated," adding movingly, "It has been the object of my life to realize in the political field what has been accomplished by economics in this pact. Groups of industries which a short time ago regarded their interests as irreconcilably opposed, met to bridge their differences and to regulate their production in conformity with their mutual requirements and the demands of the steel market."[48]

Would not the new ISC, Stresemann wondered, provide something more than a solution to past problems? Might not an agreement in steel, *the* measure of national power, lead to similar ones in other branches of industry and with other countries, he asked, once it had been demonstrated that tension and conflict arising from the struggle over markets could be replaced by gentlemanly agreement to share them? Stresemann was by no means alone in cherishing such hopes. Not only the French and German publics greeted the new understanding; the socialist government of Belgium welcomed

---

[47] René Girault and Robert Frank, *Turbulente Europe et nouveaux mondes, 1914–1941*, Histoire des relations internationales contemporaines, 2 (Paris, 1983).

[48] Cited in Ervin Hexner, *The International Steel Cartel* (Chapel Hill NC, 1943), p. 221.

it, as did both management and labor in Luxemburg.[49] The representative of the U.S. Commerce Department in Berlin praised "the conclusion of the European steel agreement as a first step towards the formation of an 'Economic United States of Europe.' "[50]

The belief that market-sharing agreements could be worked into a new system of trade and diplomacy owes a good deal to the mystique of the Dawes committee. Had not practical, hardheaded men of affairs succeeded where dreamers and schemers had failed? Should governments not therefore have the good sense to keep their hands out of things they could only muddle up and let business "get on with the job"? The late 1920s were indeed years of benign neglect in which the governments of Europe and the United States in some cases waved benedictions and in others tactfully averted their eyes as the controllers of factory, field, and finance designed and assembled private structures of cooperation to regulate economic activity. Another name for them is international cartels, and steel was but one of them.

Though the purpose behind the formation of cartels, control of supply, was always the same, they varied substantially in form, effectiveness, and rapaciousness. Though most were purely private arrangements, governments participated in a few of them as either sponsor or member. Their mechanisms ran the gamut from tacit cooperation to verbal pacts ("telephone cartels"), to written agreements, and on through more formal arrangements culminating in veritable private governments administered by large management staffs. A few cartels behaved as if aware of public responsibilities but many were simply greed-driven. And though some of them worked, at least for a while, most others did not.

Cartels in Action by George W. Stocking and Myron W. Watkins, along with its two sequel volumes, is the one great case-by-case analysis of interwar cartels.[51] Their studies make it possible to compare the Ruhr and its Western allies with the men and interests guiding the fortunes of other industries. One must conclude in a rough manner of speaking that as both businessmen and statesmen Ruhr coal and steel industrialists were no worse than most and better than

[49] Ibid., p. 220.
[50] Cited in John Gillingham, "Coal and Steel Diplomacy in Interwar Europe," in Clemens A. Wurm (ed.), *Internationale Kartelle und Aussenpolitik* [*International Cartels and Foreign Policy*] (Stuttgart, 1989), p. 88.
[51] Geroge W. Stocking and Myron W. Watkins, *Cartels in Action: Case Studies in International Business Diplomacy* (New York, 1947); idem, *Cartels or Competition? The Economics of International Controls by Business and Government* (New York, 1948); idem, *Monopoly and Free Enterprise* (New York, 1951).

many of their peers from other industries. Stocking and Watkins show, first of all, that like other producers of basic commodities they faced problems far too great to overcome by merely putting up institutional barricades. Gluts of both nitrates and sugar, for instance, overwhelmed all cartelization attempts in the 1920s. Results were only slightly better in rubber. Set up with the encouragement of the British government in an attempt to restrict plantings in Indonesia, the so-called Stevenson Plan saddled Commonwealth producers with export quotas but raised prices only marginally. The United Kingdom's scheme proved to be self-defeating. In Malaya restriction of supply created ant armies of native tappers who brought in latex from trees growing wild in the jungle; in Indonesia it stimulated a big increase in cultivation; and in the United States as well as in Europe it diverted capital toward research into the development of an artificial product.[52]

The most effective international cartel arrangements were for sharing patents and processes. Though in many high-technology, capital-intensive fields such agreements merely formalized the existence of the industrial status quo, the lightbulb cartel created a world monopoly. The December 1924 Convention for the Development and Progress of the Incandescent Electric Lamp Industry joined together the producers of the United States, Germany, the Netherlands, Hungary, Britain, and France, as well as the American-controlled Overseas Group – the dominant internationl interests – in a pact whose two purposes were to enforce noncompetition in established markets and to regulate conditions of sale in the rest of the world. Underpinned by well-established traditions of patent exchange, the cartel operated through a jointly owned Swiss corporation named Phoebus SA Compagnie pour le Développement de l'Eclairage.[53]

Phoebus was one of the largest employers of international technocratic talent in the Helvetic Confederation. The Administrative Board, which commanded separate divisions for accounting, sales, advertising, and technology, served as the cartel's executive organ. A Board of Arbitration consisting of a Swiss professor of law, a Swiss federal judge, and a technical expert on international cartels acted as advisor to the Administrative Board. Thanks to these strong internal controls as well as a policy of waging war ruthlessly against competitors as varied as the high-technology factories of Swedish

[52] Stocking and Watkins, *Cartels in Action*, pp. 171–172, 118–119, 14–15, 56–57.
[53] Ibid., pp. 304–305.

cooperatives and the low-cost workshops of Japanese artisans, the lightbulb convention succeeded in maintaining excessive prices on national markets, regulating sales on the others, and frustrating the development of a cheap, efficient, and long-burning electric lamp.[54]

The Aluminum Alliance formed in 1926 and renewed thereafter was the strongest of international cartels between the wars, thanks to the dominant position of the Mellon interests in the industry. Joined by the big companies of Britain, Germany, France, and Switzerland and led by the American Alcoa and its Canadian affiliate Alted, this financially powerful group created chronic shortages by massive purchase of stocks overhanging the market. At times it even managed to jam up prices during periods of fallimg demand. The cartel helped make aluminum the most profitable industry in the world during the 1920s and 1930s.[55]

Finally, there was the International Steel Cartel (ISC), which though far from being the most predatory producer entente of the 1920s was similar to the others in purpose, its immediate objective having been "to eliminate competition . . . and thereby insure a better price in world markets."[56] The ISC also served as a crucible for forging analogous organizational structures, common interests, and even similar outlooks in the national steel industries of Western Europe. The Ruhr provided the model. Although the "Germanization" of the producers of France and the Low Countries would last for over a generation, the late 1920s was the period of critical beginnings. The ISC itself soon proved to be a perilously leaky vessel; it never floated properly and frequently threatened to capsize. The steel producers of Western Europe nonetheless thought well enough of the 1926 arrangement to redesign and rebuilt it after the depression had dashed it onto the rocks.[57]

The history of the ISC began in the Ruhr, with the death in April 1923 of Hugo Stinnes and the collapse of his vast empire, the Siemens-Rhein-Elbe-Schuckert Union. The steel properties of this ramshackle conglomerate were taken over by a new combine whose organization was completed in 1926, Vereinigte Stahlwerke (VS). VS manuactured between 40 and 50 percent of the main German steel products. Run by no single person, the huge complex was directed by boards representing management along with the main compo-

---

[54] Ibid., pp. 333–334.   [55] Ibid., pp. 216–217.   [56] Ibid., p. 161.
[57] Hexner, *International Steel Cartel*, pp. 82–83.

nent firms, which in addition to the former Stinnes interests included those of the Thyssen group, Phönix AG, and Rheinische Stahlwerke AG (itself controlled by the IG Farben chemical combine).[58]

Vereinigte Stahlwerke was less a dynamic motor of progress than a convenient point of crystallization for the rest of the steel industry. There were five large, mainly family-controlled combines – Friedrich Krupp AG, the Gutehoffnungshütte Oberhausen AG, the Hoesch Eisen- und Stahlwerke AG, the Klöckner Werke AG, and the Mannesmann Röhrenwerke. Though challenged by Hitler in the late 1930s, by the Anglo-American occupants after 1945, and by Monnet during the negotiations surrounding the Schuman Plan, this basic structure of interests would endure until eroded by the tidal economic changes set in motion during the 1950s.

The organization of Vereinigte Stahlwerke in spring 1924 was only the first of the steel industry's preparations for an anticipated era of stabilization in which excess capacity was expected to be the main problem facing producers. Their concerns are readily understandable. In 1925, when prewar steel outputs of twenty-six million tons were again reached, some ten million tons of Western European capacity went unused. To cope with the problem, Ruhr producers naturally sought high tariffs and low freight rates and borrowed dollars heavily to modernize plant and equipment, improve layout, reduce fuel consumption and handling costs, and replace technically obsolete or poorly located plant. This so-called rationalization program increased the indebtedness of German iron and steel producers by some 600 million Reichsmarks between 1925 and 1928.[59]

Cartelization went hand in hand with the new investment strategy. Except for the pig iron syndicate (Roheisenverband GmbH), none of the prewar cartels had survived the postwar instability and inflation. The "cartelless" period ended on 1 November 1924 with the formation of the crude steel syndicate (Rohstahlgemeinschaft). The new organization, according to Gerald Feldman, "served as a parent cartel for a host of tightly organized product syndicates that were to be established during the ensuing months to encompass 90 percent of the industry's production as against the 43 percent that had been so organized in the prewar period."[60] The crude steel syndicate reg-

[58] Robert A. Brady, *The Rationalization Movement in German Industry: A Study in the Evolution of Economic Planning* (Berkeley CA, 1933), p. 108.
[59] Gillingham, "Coal and Steel Diplomacy," p. 85.
[60] Feldman, *Iron and Steel*, pp. 451–452.

ulated overall outputs by assigning production quotas factory by factory. In December 1925 a new institutional wing was attached to this edifice. This was the so-called AVI Agreement (*Arbeitsgemeinschaft der Eisenverarbeitenden Industrie*), a shrewd arrangement enabling steel producers to profit from tariff protection without at the same time pricing German manufacturers out of growing export markets. AVI provided for payment of rebates by steel producers to exporting manufacturers, the amounts based on the price differential between foreign and domestic markets. Once these arrangements were in place German heavy industry was ready, though still not eager, to enter negotiations with the French for an international steel settlement.[61]

The fall of the French franc, which set in during the Ruhr occupation and accelerated in 1925 and early 1926, forced Reich producers back to the negotiating table. Private Franco-German heavy industry discussions had begun in December 1924 in Paris but broke off inconclusively in the new year. They resumed in March 1925, with the French intent upon protecting as much as possible of their inflated sales on the German market. Due to the fall of the franc these increased even after 10 January 1925 when the provision for duty-free entry of Lorraine products lapsed and was replaced with high new tariffs. For their part, Ruhr spokesmen hoped that by offering quota guarantees the French could be coaxed into a broader agreement to share international markets. In discussions of May 1925 chaired by Emile Mayrisch of the dominant Luxemburg steel producer ARBED, Fritz Thyssen and Ernst Poensgen, both future directors of Vereinigte Stahlwerke, first formally proposed the creation of an international cartel designed along the lines of the German Crude Steel Producers' Association.[62]

The success of this initiative also depended upon satisfying other claimants to the German market, above all the Luxemburgers. Steel was the only important industry in the Grand Duchy, over 90 percent of whose output was normally sold elsewhere. A member of the Zollverein prior to 1918 and barred from the French tariff area by its steel industry thereafter, the Grand Duchy had no choice but to join a customs union with Belgium in 1919, even though that nation too was a major steel exporter. To aggravate the firm's difficulties, ARBED had begun to diversify before the war into both coal (owning Eschweiler Bergwerke in the Aachen basin) and manufacturing (controlling the large electrical equipment producer in Köln, Felten-

[61] Hexner, *International Steel Cartel*, p. 68.
[62] Gillingham, "Coal and Steel Diplomacy," p. 86.

Guilleaume). The restoration of German tariff sovereignty threatened to cut the steel production units off from these holdings. ARBED, whose rolling mill operations actually stretched across the border into France, was also more exposed to French dumping than the Ruhr.[63]

In June 1925 Managing Director Mayrisch worked out the terms of a quota deal with the Germans in which foreign producers were to be allowed to sell 1.75 million tons on their market. No less than a tenth of this amount was to be reserved to ARBED as a *pourboire* to Mayrisch, the rest being split between the French and the Saar, which on 10 January 1925 had reentered the French customs zone. Although agreement on this deal was reached on 17 June 1925, negotiations were stalled over the next six months while delegates from the Auswärtiges Amt and the Quai d'Orsay haggled in Paris over the terms of both the Franco-German trade treaty and the Saar customs settlement. It was not until Christmas 1925 that Thyssen again repeated his proposal of a year earlier, not until June of the following year that negotiations were concluded, and not until 1 September 1926 that the agreement actually took effect.[64]

The negotiations turned less on the merits of Thyssen's proposal than on the ability of two national groups, the French and Belgian, to put it into effect, Luxemburg having had too few producers to present organizational difficulties. Belgium was the more serious of the two problem children. If laissez-faire can be said to have had a home on the Continent it was there, the strength of the tradition due doubtless to the nation's long-standing dependence on exports. The Belgians normally sold about one-half of their three million tons of annual steel production abroad and thus were a major force on international markets.[65] During the 1920s, the giant holding company Société Générale and the Banque de Bruxelles–Brufina–Cofinindus complex patterned on it increased their control over both Belgian heavy industry and the economy generally. In addition to possessing large shareholdings in Luxemburg's mills, the two holding companies together controlled 80 percent of Belgian crude steel output. The remaining 20 percent was in the hands of

[63] Ibid.
[64] Jacques Bariéty, "Das Zustandekommen der Internationalen Rohstahlgemeinschaft (1926) als Alternative zum misslungenen 'Schwerindustriellen Projekt' des Versailler Vertrages," in Hans Mommsen, D. Petzina, and B. Weisbrod (eds.), *Industrielles System und politische Entwicklung in der Weimarer Republik* (Düsseldorf, 1974), pp. 554–555.
[65] John Gillingham, *Belgian Business in the Nazi New Order* (Ghent, 1977), p. 16.

feisty independents, and though Belgians held leading roles in a few export cartels such as those for rolled wire products, the domestic market was mostly unregulated. The so-called rerollers, who bought cheap crude on the open market and did low-cost finishing, were a particular problem. They refused to be represented in the steel negotiations; the Société Générale rather than the weak producer association actually conducted these on behalf of the Belgian national group. The giant holding company habitually played a double game, threatening domestic opponents with the power of the international cartel on the one hand, while using their existence to pry concessions out of it on the other. "[The] tenacious methods of bargaining and the obstructive tactics of Belgian producers in establishing and modifying international agreements," according to Ervin Hexner, "became proverbial."[66]

French industrial organization was also relatively underdeveloped. In the 1920s the Comité des Forges, in which only representatives of steel participated, served as an informal political directorate for industry as a whole. Another producer association known as the Comptoir Sidérurgique exercised actual cartel responsibilities for the mills. Suspended in 1921, the Comptoir resumed operations after 1926 so that France could participate in the International Steel Cartel. The new French organization was far more impressive than its predecessor and closely resembled the Stahlwerksverband: "The national *comptoir* consisted of a general policy-determining division and of particular sub-*comptoirs* for rails, tubes, wire rods, structural shapes, and semi-finished steel. Each of the particular *comptoirs* was divided in a section for the domestic market and another for the export market."[67] On French markets too, successful cartelization would depend in part on the strength of ties to German producers.

The structure of the 1926 International Steel Cartel (Internationale Rohstahlgemeinschaft) represented a compromise between the Ruhr's ambitions and economic reality: After long and wearying discussion its delegates had to give up the effort to control pricing and concede that the cartel's responsibilities should be limited to regulating market share. Even this presented serious difficulties, since some of the participating nations did not have domestic cartels in all product lines. Nor were there international agreements for each of them. Eventually, the only possible solution was adopted, with

---

[66] Ibid., p. 18.   [67] Hexner, *International Steel Cartel*, p. 67.
[68] Ibid., p. 123; see also Henry W. Ehrmann, *Organized Business in France* (Princeton NJ, 1957), p. 18.

each founder-member receiving a crude steel quota, divisible by product on a prorated basis, that covered sales both at home and abroad. To enforce respect for the quotas, a scheme of penalties and compensations for excess, or under-, production was to apply. Germany received a quota of 40.45 percent, France, 31.89 percent, Belgium, 12.57 percent, Luxemburg, 8.55 percent, and the Saar, 6.54 percent. Penalties were set at four dollars per ton, compensations at two dollars per ton. It was hoped that as export demand increased supply to home markets would be reduced and domestic price levels would also rise.[69]

The International Steel Cartel was set up for motives similar to those underlying the formation of other producer ententes and the directors of Western European coal and steel undeniably played no more than a modest role in the events leading to Locarno. Yet the ISC would be the first industrial agreement used as a bridge between the nations of Western Europe. In December 1927 Emile Mayrisch, who was then chairman of the ISC, spelled out his thoughts on this subject in an address entitled "Les Ententes économiques internationales et la paix." Asserting that the governments of the era had not only failed to master the economic problems arising from Versailles but would continue to do so, Mayrisch urged the leaders of Western European industry to seize the initiative in restoring foreign trade and, with this in view, to set about organizing international pools and holding companies.[70] Unlike many of his compeers, the Luxemburg steel magnate knew that "business diplomacy" could not work without public support. Doubting that politicians would want or be able to generate it, he recommended that producers circumvent them and appeal directly to the material interests of the electorate. Mayrisch organized and financed the Comité Franco-Allemand d'Information et de Documentation, composed of prominent academic and business figures, to publicize his approach but perished tragically in an automobile accident before the group had a chance to make much progress.[71]

Ideas similar to Mayrisch's had increasingly wide currency toward the end of the decade. The cartel question was a topic of serious concern at the League of Nations World Economic Conference of

[69] Gillingham, "Coal and Steel Diplomacy," p. 87; see also Günther Kiersch, *Internationale Eisen- und Stahlkartelle* (Essen, 1954), pp. 161–162.
[70] Emile Mayrisch, "Les Ententes économiques internationales et la paix," in Centre de Recherches européenes, *Emile Mayrisch: Précurseur de la construction de l'Europe* (Lausanne, 1967).
[71] Gillingham, "Coal and Steel Diplomacy," p. 89.

1927. The chief *rapporteur* was Reichstag delegate Dr. Clemens Lammers, whose analysis was accepted as a basis for discussion by all parties including the United States. According to Lammers, disinvestment and misinvestment during the war, economically ill-advised provisions of the peace treaty, and dislocations resulting from the formation of new states had together frustrated repeated attempts to restore foreign trade by means of commercial treaties, "which on account of the uncertainty [have been] invariably of short duration."[72] Thus, he proceeded, leading business figures had begun to deal with Europe's economic problems on their own, initially for informational purposes but thereafter with a view to devising new methods of problem solving. It soon became apparent to them, Lammers added, that strong cartels had to be formed domestically "to interlock the various interests as much as possible from within," and further that these groups should rationalize on an international scale in order to bring about "a thoughtful rearrangement of European production along lines of economic requirements."[73] Some form of public supervision, the rapporteur emphasized, should be introduced to curb monopolistic abuses by producer ententes. He advised that they could be effective only against a background of a "sensible trade policy." In conclusion Lammers warned that no high court or other supreme authority could make international cooperation a reality, as this required full recognition of a community of economic interests on the part of participating countries.[74]

The ISC disappointed the hopes of businessmen like Mayrisch as well as politicians like Lammers: It began to fall apart almost at once. Economic conditions were largely to blame. Disappointing growth in the export trade caused considerable capacity to remain idle, over 30 percent on the average during the second half of the decade. With the exception of brief periods in 1928 and 1929 the Western Europe steel industry failed to turn a profit. Stimulated by rapid growth of the domestic market, German overproduction broke the ISC. Reich steelmakers naturally resented paying the heavy fines imposed for this offense and, in effect, subsidizing competitors; at same time, their partners in the cartel were displeased

[72] Clemens Lammers, "The Cartel Question at the World Economic Conference," *American Academy of Political and Social Sciences* 129–134:1927, p. 147; see also Paul de Rousiers, *Cartels, Trusts, and Their Development* (Geneva, 1927); D. H. MacGregor, *International Cartels* (Geneva, 1927); World Economic Conference, *Final Report* (Geneva, 1927); Eugene Grossmann, *Methods of Economic Rapprochement* (Geneva, 1927).
[73] Lammers, "Cartel Question," p. 148.   [74] Ibid., p. 149.

by selling that deprived them of sorely needed exports. An attempt
to compromise differences by splitting the Reich's share into foreign
and domestic components failed in April 1927. In May 1929 German producers formally renounced the ISC agreement, a superfluous gesture as it happened: In the following months the downturn
on steel and other commodities markets began that would signal the
coming depression. Even before the great Wall Street crash in October 1929, it appeared to contemporaries as if "the individualist
laissez-faire system [of steel marketing] would be restored."[75] Subsequent attempts to hold bits and pieces of the cartel together were
futile. Liquidation proceedings for the ISC began in March 1931,
and by the middle of the year it had ceased to exist.

1.3   Business diplomacy in the depression decade

Business diplomacy did not die with the collapse of the ISC but took
on a new life in the 1930s, at first as a means of combating the
depression and later as a way to forestall a plunge into war. These
uses are difficult to applaud: Cartelization almost invariably involves
a shift of costs to others, and industry's attempt to restrain Nazi expansionism failed utterly. At the same time, businessmen were not
mainly responsible for the persistence of the world trade slump during the depression: Real blame rested with their governments, especially the United States.[76] In the case of coal and steel, moreover, the
international syndicates mitigated the worst of the depression's effects. The economic appeasement of Hitler at the end of the decade
was no less effective than any other method of staving off war and in
any case was used only as a last resort. One final thing can be said in
behalf of business diplomacy: It reflected a perception, however
dim, that functional integration was at work in the heavy industry of
Western Europe. Despite the setbacks of the depression, a tradition
of economic cooperation took hold in the heavy industry of Western
Europe during the 1930s.

In 1933 the International Steel Cartel was revived and strengthened. This time it would not fall apart. In addition to the four
founder-members – Germany, France, Belgium, and Luxembourg –
the new organization would include Britain (which joined on 30
April 1935 as an associate), Czechoslovakia, Hungary, and Poland
(which entered as members the following year), and the United

---

[75] J. W. F. Rowe, *Markets and Men* (Cambridge, 1936), p. 18.
[76] Charles Kindleberger, *The World in Depression* (Berkeley CA, 1973), p. 134.

States (which affiliated with it in November 1937). Among major
steel producers, only Japan and Sweden remained outside the ISC.
The new cartel controlled over 85 percent of the steel export trade
and succeeded in maintaining price levels throughout the decade
while protecting the market share of its members. It deserves credit
for having contributed to the overall profitability of the Western European steel industry at a time of no real growth in international
trade.[77]

In the 1930s a thickening network of coal agreements comple-
mented those in steel. These were not easy to arrive at. Whereas the
big steelmakers were all from the same general region and had
roughly similar cost structures, coal arrangements involved different
sets of principals and problems. The main exporters were Germany,
Great Britain, Poland, and the Netherlands, each of whose competitive position was quite different. Belgium had an approximate
balance between exports and imports. The Netherlands, whose few
mines were among the most productive in Europe, depended
heavily on Ruhr imports and was an important transshipper of both
British and German coal. Starved for hard currency, Poland was a
major exporter whose prices depended not only on demand for coal
but on the current balance of payments. The British were the largest international exporter of solid fuel under normal circumstances,
but the rule admitted of numerous exceptions. The Ruhr was the
main supplier of western Germany (the Silesian mines being generally dominant in East Elbia), the Bel-Lux union, and the heavy industry of eastern France. The French were the world's largest
importer of coal, though also important producers of it. In November 1936 an international coke cartel was concluded, and two years
later a similar arrangement in coal. The critical issue of regulating
German fuel supplies to France required the use of a different kind
of mechanism, intergovernmental agreements. These were difficult
to arrive at because of the collapse in trade and payments that occurred as a result of the depression. However, the Franco-German
coal convention in effect during the 1930s did prevent disruptive
interventions of the type that in other sectors contributed to the
breakdown in relations between the two nations.[78]

The success of these arrangements can be traced back to the discredit cast on liberal economics by the depression. The era's buzz-

---

[77] Gillingham, "Coal and Steel Diplomacy," p. 90; see also Kiersch, *Eisen- und
Stahlkartelle*, p. 123.
[78] John Gillingham, *Industry and Politics in the Third Reich: Ruhr Coal, Hitler, and
Europe* (New York, 1985), pp. 92, 93–94, 102–103.

word was "industrial self-government," by which was meant that, instead of competing, producers should do business amicably through special associations set up to regulate markets and simultaneously promote modernization. Germany, home of "organized capitalism," had traditionally provided the model of industrial self-government. The dramatic economic recovery and the triumphs of Hitler's foreign policy made this even more true in the 1930s than before.[79]

German producers did not have to pay a high price of admission into the Third Reich. Hitler had no intention of stripping business of its powers and prerogatives but was prepared to leave well enough alone so long as it performed up to his expectations. For business, the process of "administrative coordination" (Gleichschaltung) by which Germany's main social, political, and economic institutions were linked to the Nazi leadership involved primarily nominal changes. Existing industrial associations were allowed to remain unaltered, with new ones being formed where none existed before. Renamed "business groups" (Wirtschaftsgruppen), these later assumed foreign trade and rearmament responsibilities on behalf of the state. Capping the entire edifice was Reichsgruppe Industrie, the legal successor to the Reichsverband der deutschen Industrie of the Weimar years.[80]

As for the cartels, after an initial hostility the regime vigorously encouraged their formation throughout industry. The German economy soon became more completely regulated by private agreement than either before or since. The regime had no discernible policy toward *international* cartels other than an apparent willingness to allow producers to handle the matter. The steel industry welcomed this loose hand for, as Poensgen reminded skeptics during discussions about ratifying the ISC in 1933, the only alternative to renewed international agreement would be high tariffs, which would put producers at the mercy of the regime.[81] Poensgen's

---

[79] Clemens A. Wurm, "Politik und Wirtschaft in den internationalen Beziehungen," in Wurm, *Internationale Kartelle*, pp. 17–18.

[80] Gillingham, *Industry and Politics*, pp. 32–33; see also K. D. Bracher, W. Sauer, and G. Schulz, *Die nationalsozialistische Machtergreifung*, 2nd ed. (Cologne, 1962), pp. 186–187; Henry Laufenberger and Pierre Pflimlin, *La nouvelle structure économique du Reich: Groupes, cartels, et politique de prix* (Paris, 1938), p. 126; Arthur Schweitzer, *Big Business in the Third Reich* (Bloomington IN, 1964), pp. 420–421; Robert A. Brady, *The Spirit and Structure of German Fascism* (London, 1937), passim.

[81] Gilingham, "Coal and Steel Diplomacy," p. 90; see also J. W. Reichert, *Nationale und Internationale Kartelle* (Berlin, 1936); Ernest Poensgen, "Hitler und die Ruhrindustriellen" (unpublished ms., 1946).

guarded attitude proved wise. In late 1936 Ruhr heavy industry found itself in conflict with the regime over the issue of domestic low-grade ore consumption. The refusal of producers to contemplate the use of this high-cost material caused Hitler to create the huge Hermann Göring Works, a new state-subsidized competitor and feared harbinger of the so-called German socialism featured in regime propaganda. Faced with such a threat, the Ruhr welcomed foreign allies.

To outsiders unaware of the seriousness of the steel industry's conflicts with the regime it seemed as if producers in the Reich had never had it so good. Nowhere was the German example more important than in Britain, because nowhere was the rejection of laissez-faire as explicit or consequential as in the island nation.[82] Though British advocates of industrial policy at the Treasury, Board of Trade, and the Bank of England had found it difficult to chip away at economic orthodoxy in the late 1920s, the depression tilted the balance in their favor for a generation. The Conservatives won the 27 October 1931 election on the issue of trade protection, which soon became the policy of the National government. Berndt-Jürgen Wendt has described what followed as being no less momentous than the adoption of free trade in 1846. It amounted to a government-encouraged shift from liberal to organized capitalism. The Customs Duty Act of 17 November 1931 was the main lever of the new policy. Industries desiring protection (duties of up to 50 percent could be imposed) were required to devise rationalization schemes, which were then vetted by an informal but powerful directorate, the Import Duties Advisory Committee (IDAC).[83]

The 1930s witnessed significant improvement in British industrial organization. In 1926 Britain had been excluded from negotiations for the ISC on grounds of associational underdevelopment (*Kartellunfähigkeit*). A first effort to set up a steel export syndicate in 1929 failed. Between 1927 and 1936, however, some forty-two million pounds' worth of excess pig iron capacity was written off, IDAC stepped in to reorganize the industry, and by 1932 vertical combines controlled 47 percent of what remained of it as well as 60 percent of

[82] See Berndt-Jürgen Wendt, *Economic Appeasement: Handel und Finanz in der britischen Deutschland-Politik, 1933–1939* (Düsseldorf, 1971), passim; Clemens A. Wurm, "Handelsdiplomatie in der Weltwirtschaftskrise: Internationale Kartelle, Stahl, und Baumwolltextilien in der Aussenpolitik Grossbritanniens, 1924–1939," in Wurm, *Internationale Kartelle*, pp. 107–108.

[83] Wendt, *Economic Appeasement*, p. 33.

crude steel capacity. In 1934 the government encouraged producers to form the British Iron and Steel Federation, which "took over the price-fixing functions of earlier sectional associations."[84]

In France depression conditions advanced the organization of industry more rapidly in fact than in theory. The one attempt made by a French government to enact a law for compulsory cartelization, the Marchendeau project of 1935, failed in the Senate. Yet in spite of employer and politician lip service to laissez-faire, voluntary agreements, which were called either *comptoirs* or *ententes*, had regulated conditions of sale in many branches of large-scale production, especially coal and steel, since World War I.[85] The number and importance of such cartels increased still further after 1929, partly through international arrangements. As a result, according to Henry Ehrmann, "French *comptoirs* and *ententes* became frequently akin in functions and powers to German cartels: their activities now centered around price-fixing and the determination of production quotas for member firms."[86] Noting the difficulties involved in determining how, and how well, individual agreements operated, Ehrmann adds that between one and three thousand of them were in effect during the 1930s.[87]

Individualism and chronic suspicion of the state caused French business adoption of industrial self-government to begin quite late. It came in response to the path-breaking Matignon Agreement of June 1936 by which factory owners, unable to end the rash of sit-down strikes paralyzing production, recognized unions as bargaining agents. This humiliation gave birth to the modern employers' movement in France, an event commemorated by rebaptizing the Conféderation Générale de la Production Francaise as the Conseil National du Patronat Francais (CNPF). The new CNPF was comprehensive (included small and medium-sized employers as well as large ones), centralized, and administered by a large permanent professional staff. It was also committed by its new leader, Claude Gignoux, to providing national leadership in economic policy.[88]

Although the reorganization of CNPF betokened a new employer awareness that "an unavoidable internal evolution had killed liberalism," no single new corporatist doctrine emerged but rather several intellectual variants of one. The Nouveaux Cahiers circle composed

[84] A. J. Youngson, *The British Economy, 1920–1957* (London, 1960), p. 106.
[85] Anita Hirsch, "Cartels et ententes," in Albert Sauvy (ed.), *Histoire économique de la France entre les deux guerres*, vol. 4 (Paris, 1975), pp. 49–78.
[86] Ehrmann, *Organized Business*, p. 31.   [87] Ibid.   [88] Ibid., p. 3–4.

of progressive businessmen and technocrats advocated Swedish approaches to settling labor disputes, a thoroughgoing organiza-ion of industry, and the resolution of international conflict, espe-cially between France and Germany, by means of international producer agreement. Another organization, the Comité Central de l' Organisation Professionelle (CCOP), which represented the views of certain medium-sized firms, urged the state to turn over policy-making responsibilities to "vocational organizations," to be built upon future state-reinforced cartels. Finally, a Comité de Prévoyance et d'Action Sociale, headed by the former finance minister M. Germain-Martin, worked parallel to the CNPF. Vociferously antiunion, the committee was outspoken in its admiration for foreign authoritarian and corporatist regimes, praising "Nazi and fascist youth organiza-tions . . . , as setting a desirable example for France."[89] Regardless of the differences between them, each of these schools of thought held that economic agreements between the organizations of French and German business could bridge the political gap dividing the two na-tions. Their model was the international steel export cartel.

Advances in industrial self-government were indispensable pre-requisites to the new ISC agreement of 25 February 1933. As Hex-ner puts its, "The establishment of strong national groups was a condition upon which the new cartel was based."[90] Unlike the ear-lier agreement, this one made a sharp distinction between domestic and foreign markets, a fact reflected in a modification of the cartel's name to the Internationale Rohstahlexportgemeinschaft (IREG). The new organization's competence specifically excluded home sales; they were to be handled by national associations, some of which were new. One had been set up in Belgium, where after 1929 underbidding by independents had contributed to the collapse of the 1926 cartel. This activity was temporarily tolerated by the big financial holding companies in the belief that it would lead to bank-ruptcies. This indeed happened, the Société Générale bought out former competitors at fire sale prices, and in Belgium too the bal-ance tilted definitely to the side of "organized capitalism." Accord-ing to one commentator, "The year 1933, if we except the period of the war, has been the most eventful in the history of the Belgian steel industry [and] has witnessed a revolutionary change in the principles upon which [it] has been conducted since its foundation, the relentless suppression of individualism, and the imposition of a

---

[89] Ibid., p. 45.   [90] Hexner, *International Steel Cartel*, p. 83.

Table 1.1. *Total annual crude steel shares of exported commodities by percentages at designated production levels*

	6.8 million metric tons or less	11.5 million metric tons or more
Germany-Saar	29.2	33.7
Belgium	29.0	26.0
France	20.6	23.5
Luxemburg	21.2	16.8
	100	100

Source: Guenther Kiersch, *Internationale Eisen- und
Stahlkartelle* (Essen, 1954), p. 29.

system of international cooperation, almost socialistic in theory,
upon the steel works of the continent, of which the Belgian works
had been the most independent section."[91] Although the big banks
continued to represent Belgium in the ISC, a "cooperative commercial company," or cartel (Comptoir de Vente de la Sidérurgie Blege,
or COSIBEL), was formed on 31 May 1933 for the domestic market.
The way was now clear for the setup of a pure export cartel.

The trend toward organized capitalism enabled new features to be
built into the 1933 ISC agreement that are at least partly responsible for its success. Control of domestic output first of all led to the
adoption of a sliding quota for foreign deliveries, which as a corrective to practices that had wrecked the earlier agreement favored
export-dependent Luxemburg and Belgium when sales abroad were
low but enabled producers from France and Germany, nations with
large domestic economies, to profit as they rose (see Table 1.1). The
revival of strengthened sectional cartels, along with the formation of
new ones, also permitted a much more thoroughgoing market control than previously. These existed for structural shapes, merchant
bars, thick plates, medium plates, wire rods, hot-rolled bands and
strips, cold-rolled bands and strips, piled sheets, black sheets, galva-
nized sheets, rails, flexible wire, tubes, and tin plate. Although expressed in tons of crude steel, quotas were subdivided by specific
product, each assigned a crude steel equivalency, the remainder
then normally being absorbed in crude steel sales. ISC-controlled

[91] Cited in ibid., p. 87.

comptoirs, another new feature, monopolized wholesaling in various product lines. Thanks to these innovations, distribution on most export markets occurred through tied-in dealers on exclusive contract.[92]

Business diplomacy also played a role in the success of the ISC. Though certain key French officials deserve credit for having been the first to recognize the need for some form of Western European heavy industry settlement, the Germans, who as the largest producers also inevitably bore the greatest responsibility in the matter, proved to be good learners. In official histories of the Ruhr, the 1926 cession of the south German quota to its cartel partners usually features as the first great act of national statesmanship on behalf of the West European industrial community and may well have constituted a turning point; the negative German producer attitudes of the early 1920s never returned. The 1930s would witness acts of genuine Ruhr statesmanship, beginning in 1931 with a decision to protect Belgian steel. The British decisions of September 1931 to devalue the pound by 30 percent and of February 1932 to impose a 33 1/3 percent ad valorem tariff on steel threatened this industry with catastrophe, eliminating about one-third of its traditional export market. To ease the pain of the inevitable adjustments, the Ruhr browbeat independents on behalf of the big bank holding companies and refrained from underbidding Belgian producers while tolerating their frequent violations.[93]

Hitler's seizure of power did not change the behavior of the German national group toward its Western European cartel partners. It continued to act with moderation, following a policy of enlightened self-interest. In 1935, when the Saar returned to the German customs area, the Ruhr absorbed the output previously sold on the French market.[94] When in the same year the British attempted a forced entry into the ISC by threatening a further increase of 50 percent in the steel tariff, a measure that would have eliminated remaining exports from France and Belgium, the Ruhr forced the

[92] Kiersch, *Eisen- und Stahlkartelle*, p. 25.
[93] Auswärtiges Amt (AA), Sonderreferat Wirtschaft, Kohle/Bd. 8, "Deutschbelgisches Kohlenabkommen 9.9.33," and "Kohlenfrage, 2.10.33"; AA T120/1988/E655505, "Aufzeichnung Wingen, 20.3.37"; Fernand Baudhuin, *Histoire économique de la Belgique, 1914–1939*, vol. 2 (Brussels, 1946), p. 30.
[94] AA Botschaft Paris 8906/Bd. 13, Dohle to Ambassador, 29 January 1935, and "La réglementation des échanges franco-sarrois," *Journée Industrielle*, 15 March 1935; AA T120/1638/E021741, "Die deutsch-französischen Wirtschaftsverhandlungen mit Frankreich" (n.d.).

United Kingdom to accept import quotas of 525,000 to 670,000 tons per year.[95] Between 1937 and 1939, years of capacity operations in Germany, the Ruhr steel industry ceded portions of its unfillable export quotas to the mills of Belgium and Luxemburg without asking for compensation. Finally, after the occupation of the Sudetenland, Reich producers granted portions of the former Czech export quotas to their West European partners. Far from being divisive within the ISC, German expansionism paid handsome dividends to the French, Belgian and Luxemburg national groups.[96]

While after 1933 steady progress was made in forming a Western European steel bloc, one encounters in coal a trend toward cooperation which though often wavering *did* eventually culminate in the economic appeasement attempts that occurred in the months after Munich. On the producer side the initiative came from the United Kingdom, along with the Germans the main European exporter. The United Kingdom had a special incentive to seek cooperation with the Reich: As a result of the commercial convention in effect between the two nations, their collieries enjoyed substantial quotas on the markets of Hamburg, Bremen, and their hinterlands that could be expanded in line with the volume of bilateral trade. In contrast to the overall decline in British coal exports in the 1930s, sales on the German market rose after 1933. This set the stage for broader U.K. initiatives in both coal and steel. These began in 1935 but ran into difficulties until March of the following year when the Central Collieries Commercial Council set up district selling syndicates. In December 1936 the protocols for a new agreement limited to coke were signed. Germany was to receive an export quota of 5.6 million tons, or 48 percent of the authorized total, and Britain some 2.4 million tons, or 20.88 percent, the remainder to be divided later among Belgium, the Netherlands, and Poland.[97] The agreement also empowered a business committee to administer prices and arbitrate disputes. His Majesty's government hoped that the coke pact

[95] Sidney Pollard, *The Development of the British Economy, 1914–1967* (London, 1969), pp. 166–167; Youngson, *British Economy*, pp. 99–100; Bundesarchiv (BA), R13/1272, "Der europäische Eisenpakt," *Die deutsche Volkswirtschaft*, 2 December 1935; Gutehoffnungshütte (GHH), 40000090/11, "Kurzbericht über die Sitzung der Verbändekommission," 10 March 1936.

[96] BA R13/1273, "Sitzung des Joint Coordinating Committees," 15 February 1939.

[97] Wendt, *Economic Appeasement*, p. 301; Gillingham, *Industry and Politics*, pp. 97–98.

would serve as a stepping-stone to a broader settlement with Hitler, but because of differences over the allocation of quotas, more than two years of negotiation were needed to reach an understanding even over coal. On 28 January 1939 agreement was reached on a British-German coal export ratio of 50:30, with 1937 serving as base year.[98]

A different type of coal arrangement gave rise to economic appeasement in France, though progress would be even slower there than in Britain. In addition to public hostility toward things *Boche*, a French coal deal with Germany had to surmount a complex of commercial problems arising from the depression. In order to earn a payments surplus from Germany for interest payments to holders of Dawes and Young bonds, France found itself in the unenviable position of having either to encourage additional imports from the Reich at a time of high unemployment or to restrain exports to the Reich, the latter possibility being politically less difficult. Either way, the result was "L'équilibre par le bas," the incessant ratcheting downward of bilateral trade that reduced French exports to Germany by four-fifths between 1929 and 1936 and German exports to France by six-sevenths over the same years. By 1939 coal was virtually the only commodity being shipped between the two countries.[99]

The French decision of 1936 to reequip the armed forces created a high demand for imported coking coal. Curiously loyal to the coke negotiations then underway with the Ruhr, the British did not compete for this business, forcing France to turn to Germany. The French then found themselves caught in a thicket, having little to sell and little to buy and with political problems cropping up left and right. The Ruhr no longer normally needed large quantities of minette, yet the public objected strongly to exporting this supposedly "strategic" good to Germany, especially after the reoccupation of the Rhineland. At the same time, the government faced union pressures to increase ore sales, to which it eventually succumbed.

[98] GHH 400101320/88, "Gremium, 10.8.38"; Documents on German Foreign Policy (DGFP), Ser. D 3263, German Economic Mission to Great Britain to Foreign Ministry, 7 November 1938; DGFP, Ser. D 03, Ambassador in Great Britain to Foreign Ministry, 28 January 1939.

[99] AA Botschaft Paris 8906/Bd. 13, "Les relations commerciales franco-allemandes," *Petit Matin*, 22 February 1935; Hirsch, "Cartels et ententes," passim; Allen T. Bonnell, *German Control over International Relations* (Urbana IL, 1940); AA Botschaft 892a/Bd. 21, "Aktenvermerk Ritter," 5 June 1937.

In fall 1936 the mayor of Nancy met in Berlin with Minister of Economics Hjalmar Schacht and Four-Year Plan plenipotentiary Hermann Göring, who "gave the impression that Germany is predisposed to collaborate with France in the closest possible way."[100] On 1 November 1936 Premier Blum declared a readiness to export ore if Germany renounced "aggressive intentions towards other nations" and agreed to deliver coke in exchange, and on 10 July 1937 Schacht signed a new Franco-German commercial convention at the opening of the Paris Exposition.[101] Its nucleus was a coal–ore understanding. Germany was to increase monthly minette imports from 490,000 to 600,000, France those of German coke from 116,000 to 275,000, it being agreed that "the French will fill all of their coke import requirements from the Reich."[102]

The deal had a temporarily electrifying effect on French opinion, being heralded by many in government, industry, and the general public as the dawn of a new era in relations with Germany. The head of the negotiating team, Minister of Commerce Chapsal, declared that "it went far beyond usual trade agreements and would result in the settlement of many outstanding questions ... [it] provides an opportunity to inaugurate a period of trust and closer relations ... and to create a smoother, more amicable and calm atmosphere."[103] Foreign Minister Delbos regarded the arrangement as a point of departure for "joint cooperation and the conclusion of further agreements, above all for a *collaboration plus large* on the international level ... a pact of mutual understanding."[104] Deputy Paul Elbel, taking a still broader view, declaimed in the Chamber that "Germany ... once satisfied ... will cease to appear 'a nation of dispossessed' and, with prosperity, shall return to a love of calm and a will to collaborate." Thus, he concluded, "the question of raw

[100] AA Botschaft Paris 892a/Bd. 20, "Une délégation d'industriels est l'objet d'attentions officielles multiples," *Paris-Midi*, 19 December 1936.
[101] AA Botschaft Paris 892a/Bd. 20, L. Allera, "La paix par le minerai? La France va désormais livrer à l'Allemagne 600,000 tonnes de minerai de fer par mois," 8 August 1937.
[102] BA R7/621 RWM NrIII 20574/37III, "Kohlenverständigung," 15 April 1935; see also United States National Archives (USNA), T120/2 6101E 4110115, "Aktenvermerk Ritter," 2 July 1935; AA Botschaft Paris 892a/Bd. 21, "Lothringische Minette und Ruhrkoks," 12 July 1937; USNA RG1575 Sr. 6, "New Franco-German Trade Agreement," 16 July 1937.
[103] AA Botschaft Paris 892a/Bd. 21, "Brieftelegramm W912," 9 July 1937.
[104] AA Botschaft Paris 892/Bd. 3, "Aufzeichnung Hemmen," 13 July 1937.

material, its distribution and transformation, moves out of the limited technical realm to determine, in truth, the question of war or peace."[105]

The hopes raised by the 10 July 1937 treaty were cruelly deceived. Total French exports to the Reich, which had risen to 154 million Reichsmarks in 1937 not only failed to increase but actually fell to 141 million Reichsmarks the following year. Worse yet, monthly German coke deliveries, targeted at 275,000 tons slipped to 113,000 tons in the first four months of 1938 and dwindled to 90,000 tons over the next year. The decline stemmed from an economic slowdown in France, which reduced demand for coke. In the Reich, by contrast, demand for minette remained surprisingly high. The French met it out of fear of increasing unemployment, even though doing so wiped out the trade surplus the Germans needed in order to service their franc creditors. To avoid political embarrassment the French compensation office made these payments secretly.[106] In July 1938 France reluctantly agreed to renew the agreement of the previous year, the Quai being resigned to the fact that "in economic matters even the most mediocre arrangement is better than quarreling."[107] If nothing else, the 10 July 1937 trade treaty seemed to have prevented German use of the coal weapon as a measure of economic warfare, preserving at least a ray of hope that economic agreements, if properly arrived at, could bridge Franco-German political differences.

Neither British nor French but German action thrust economic appeasement into the forefront of the European diplomacy, both London and Paris having long, and fruitlessly, sought to work the solution of trade and financial problems into the framework of general détente. As a first step towards impressing their skeptical publics, the Foreign Office and the Quai both had to secure official German commitments, however vague, to maintaining peace. Britain got one on 30 September 1938, shortly after Munich, in the form of an Anglo-German declaration that underlined the importance of upholding good relations between the two countries and provided for joint consultation with regard to all problems affecting either of them. The French needed another two months of intense negotiations, which were complicated by the murder in Paris of the Ger-

[105] AA Botschaft Paris 892/BD. 3, Paul Elbel, "France-Allemagne," *L'Oeuvre*, 10 November 1937.
[106] Gillingham, *Industry and Politics*, p. 105.
[107] "L'accord commercial Franco-Allemand," *L'Usine*, 11 August 1938.

man diplomat vom Rath and the ensuing anti-Semitic disorders known as the Reichskristallnacht, to conclude an understanding. On 6 December 1938 Ribbentrop signed a declaration of goodwill in Paris, and the French soon found themselves competing with their British ally for the hand of German big business.[108]

On 7 November 1938 Sir Frederick Leith-Ross, chief economic advisor to the British government, proposed to a German trade mission returning from Dublin that the coal talks, which had been proceeding in desultory fashion through most of the year, be resumed. This was more than a casual suggestion; a coal agreement was intended to be a springboard to bigger deals with Reich industry, or so the chief British negotiator, president of the Board of Trade Oliver Stanley, described it on 28 January 1939 at the gala dinner held to commemorate the new pact: "The coal trade talks," he pontificated, "have been valuable precursers to the wider talks now scheduled to start. From them we can draw many lessons and much encouragement. . . . It might be possible to look back upon their conclusion as a turning point, not only in the methods of Anglo-German industrial relations, but also in the history and hopes of the world."[109]

It would soon become clear what the Foreign Office had in mind. At a mammoth 16 March 1939 meeting the peak organizations for business, the Federation of British Industry and Reichsgruppe Industrie, concluded the so-called Düsseldorf Agreement. An accompanying public statement noted concurrence on twelve points, some of them merely banal: Existing cooperative agreements should be strengthened, exports be promoted to raise living standards, and destructive competition banished from the earth. Yet the understanding contained additional provisions that, if enforced, would have created a globe-girdling partnership. It was agreed, notably, that British producers should organize themselves as desired by the national government and also, as with coke and coal, that outsiders should be invited to enter Anglo-German conventions only after the two parties had worked matters out between themselves. Individual branches of industry in the two countries were further encouraged to begin negotiations at once to form bilateral cartels. The two peak organizations also pledged themselves to invoke the powers of their

[108] Anthony Adamthwaite, *France and the Coming of the Second World War* (London, 1977), p. 284.
[109] "Anglo-German Trade," *Times*, 29 January 1939.

respective governments when necessary in order to force third countries into compliance with the terms of the Anglo-German understanding.[110]

French economic appeasement was cranked up more slowly after Munich than the British but soon rose to the same dizzy heights. Acting at the direction of Foreign Minister Bonnet, Count Renom de la Baume (head of the Commercial Relations Section at the Quai d'Orsay) entered negotiations with a German trade delegation on 7 December 1938. De la Baume's office would work overtime during the next three months. The numerous important proposals produced by it fell under three headings. They called first for a trade increase. In exchange for a German readiness to accept more agricultural exports, France promised to import fifty million francs' worth of synthetic nitrogen as well as provide German firms with public contracts (machine tools, diverse machinery, scientific instruments, etc.) in the value of ninety-five million francs.[111]

The proposals also involved the formation of joint ventures. De la Baume suggested "in a general way" harbor improvements in South America, bridge and road building in the Balkans, railway construction in Africa, and the setup of a Gallo-Teuton consortium to handle repair and recovery work in Franco's Spain. With regard to the French empire, a note of 11 March 1939 proposed opening the Conakry mine for mutual exploitation, establishing a joint paper manufacturing project, and expanding the Moroccan manganese mines to meet rising Reich demand. Other collaborative efforts were to include a deal between Société française de Châtillon-Commentry and Vereinigte Stahlwerke to barter ten million francs of machine tools for the ore mine at Halouze (Orne) against delivery over a two-year period of half the planned production of 300,000 tons per year. A third set of proposals involved the immediate "adaption of existing industrial agreements to present conditions and the creation of ententes for new categories production."[112]

Planning for discussions between the Confédération Générale du Patronat Francaise and the Chambre de Commerce de Paris on the one hand and Reichsgruppe Industrie on the other was to begin at once for a general interindustry pact. "These diverse propositions," the paper stated, "are but a first step toward more active inter-

[110] Wendt, *Economic Appeasement*, p. 574.
[111] DGFP, Ser. D. v.IV/71, "The Franco-German Economic Discussions in Paris on 7 December."
[112] AA Botschaft Paris, "Note 11 mars 1939 (Abschrift WII 11594)."

change between our two countries," adding that the "initial program will serve as an example in the future for the conclusion of new deals. . . . The French Government hopes that the negotiations now about to open between the industries and Governments of the two countries will [create] the bases for large-scale collaboration favorable to the economies of [both France and Germany]."[113] On 22 February 1937 a "Centre Economique Franco-Allemand" opened. Its task was "to promote Franco-German economic relations with all available means [as well as] centralize and broaden them. The 'Center' is being founded by well-known parliamentarians and the presidents of the largest French Chambers-of-Commerce."[114]

On 15 March 1939 the Wehrmacht marched on Prague, the Düsseldorf Agreement became a dead letter even before it was initialed, and the far-reaching plans of de la Baume's office were buried in the files. Public opposition in both France and Britain ruled out future discussion of further large-scale deals with the Reich. Economic appeasement would nonetheless have failed even without a renewed act of Hitlerian aggression because German business was disinterested in playing the role assigned it by London and Paris. The breakdown in trade during the depression was so complete that no strong lobby existed in the Reich for promoting exports to either Britain or France. On 2 March 1939 the business manager of Reichsgruppe Industrie reported that "after circularizing all branches of industry . . . we received no serious expressions of interests in entering negotiations with the French. Where on the German side market agreements are felt to be necessary they are working well. As for the rest, we have little to contribute to the discussions with French industry."[115] The close relationships between the steel producers of the two countries in the 1930s proved to be the exception rather than the rule. The same was true in the case of Germany and Britain, even if one includes coal as well. Political obstacles in the way of business diplomacy were even greater. Moderates in the economic administration of the Third Reich such as Minister of Economics Schacht and his successor Funk may have favored collaboration like that proposed in Paris or London in the belief that a German-led economic recovery could eventually lead to domination of the Continent. Yet nothing whatsoever indicates that Hitler was prepared to follow such a strategy or capable even of un-

[113] Ibid.
[114] AA Botschaft Paris 892c/Bd. 1, "Bevorstehende Gründung eines 'Centre Economique Franco-Allemand,'" 22 February 1939.
[115] AA Botschaft Paris 892c/Bd. 1, Koppen to Campe, 2 March 1939.

derstanding how it might have worked. After 1936 his regime was officially committed to a policy of autarchy (*Autarkiepolitik*), which, though its operative importance can easily be exaggerated, nonetheless constituted a crippling blow to international economic problem solving. The political context could hardly have been less favorable to "European" settlements.

The failure of economic appeasement should not be allowed to blind one to certain positive developments during the 1930s. The Franco-German post–World War I coal and steel rivalries were all but absent in the depression decade. In Western Europe commercial warfare did not accompany Hitler's aggressive diplomacy. The coal weapon was never unsheathed in an effort to sabotage French rearmament. Germany's trade relations with Great Britain actually improved as war approached. More important yet, a Western European bloc developed within the International Steel Cartel that enabled the four founder-members to operate successfully over a period of years as a single business unit vis-à-vis third parties.

One can easily sneer at economic appeasement, whose shortcomings are obvious and whose failure was immediate, glaring, and total. Yet the policy recognized at least one truth often overlooked by historians: In the 1930s commerce and diplomacy did not normally go hand in hand. The British enjoyed better commercial relations with the Germans than with their French ally: Trade conflict did not accompany Hitler's aggressive foreign policy. Not Hitler but senior officials and leading businessmen were in charge of German commercial policy. Their concern was to balance payments, secure the necessary raw materials for an expanding economy, and hold on to export markets where possible, through surrogates if necessary. Strategic considerations counted for little, especially in Western Europe.

Germany's neighbors had no real alternative to its economic leadership. The British turned their back on Europe in fall 1931 without fully realizing what they had done. An accidental consequence of this move, minor from London's perspective, had critical implications for heavy industry: It put the Belgian steel industry at the mercy of the Ruhr, forcing it to break with its free trading traditions. The combination of imperial preference and devaluation had deep reverberations elsewhere as well. The American abdication from responsibility had even graver consequences. The story has become depressingly familiar. It begins with the Smoot-Hawley Tariff Act of spring 1930 when at the first hint of falling agricultural prices Congress raised duties to historic highs. It follows with President

Hoover's rejection of Secretary of State Stimson's urgent plea of July 1931 to rescue the imperiled Reichsmark, leaving Chancellor Brüning with a Hobson's choice of risking a recurrence of inflation by devaluing the currency or erecting barriers against imports. He opted for the politically safer and economically less sound deflationary alternative. The United States initially stood aside from the last major international attempt to organize recovery from the depression, the World Economic Conference of 1933, then eventually wrecked it by unilaterally devaluing the dollar. With some 40 percent of international markets thus cut off from the exports of the rest of the world, global trade could not recover during the 1930s. The United States also nearly disappeared as foreign lender during the decade, American international recovery policy consisting of little more than admonitory speeches by Secretary of State Hull to multilateralize trade relations.

The Reich lacked both the strength and the desire to fill the vacuum in international commerce (except possibly in southeastern Europe) and could not have served as the motor of economic revival. The development of a growing sense of community within the heavy industry of Western Europe was one of the 1930s' few accomplishments, but it was by no means an unmixed blessing. Though it is difficult to find fault with the depoliticization and normalization or the German-French coal traffic, the international cartelization of steel is a more ambivalent matter. Though the ISC was not an instrument of Nazi aggression, or even a vehicle for securing Reich steel hegemony, it *was* a cartel and as such of questionable merit as an instrument of policy. The steel entente raised prices (although not unreasonably so), beat back new entrants ruthlessly, made no provision for representing labor, included no mechanism for adapting to technological change, and could not be made accountable to the public. The ISC also had little difficulty in adapting to the coloration of an increasingly authoritarian, not to say fascist, political environment. The experience of Germany and Western Europe during World War II would reveal the ill uses to which the newly developed functional integration could be put. Yet the ISC would provide the framework around which, after a generation of conflict, failure, and little overall progress toward solving the Ruhr Problem, a satisfactory Franco-German settlement would be built. This would require an American international commitment to Europe as well as a French and German readiness to break with the past. These things would occur as a result of World War II.

2

The greater and lesser wars

There was a lesser and a greater Second World War. Hitler started the first of the two in September 1939 in an attempt to conquer Europe by defeating opposing armies piecemeal in short Blitzkrieg campaigns in which the tank, supported by mechanized infantry, was the main element of attack. To wage warfare on this scale was not economically painful for the Reich. Though stocks of raw materials had to be built up and national self-sufficiency increased in fuel and other war-critical supplies, it was only necessary to mobilize industry and agriculture partially; guns and butter could be kept in reasonable equilibrium and battle done with the technologies of the past. Hitler did not have to reorganize Germany in order to sustain a new type of combat; however radical his aims, his approach to waging war was socially conservative. The United States launched the greater war in summer 1940, actually a year and a half before either of the two simultaneous strategic turning points, the Japanese attack on Pearl Harbor and the Wehrmacht's defeat at Rostov-on-Don, had been reached. The American war against fascism was total in character: global, fought in the air and at sea as much as on land, technology-driven, and potentially revolutionary in its political, social, and economic implications. Success in this greater war depended less on diplomatic or military strategies than on vast industrial and agricultural power that made it possible to conduct hostilities on a scale far more immense than anything Hitler could have imagined in 1939. Economically, the greater and lesser wars might have been fought in different epochs. The lesser war was one of coal and steel; they were the very bases of Hitler's enterprise. Coal provided over 80 percent of the Reich's peacetime energy requirements, even more in war, and was the source of raw material for both artificial petroleum and most electrical current. Without coal the trains upon which the movement of goods and men in Germany depended would not have run, nor could steel have been produced; acute shortages of the commodity would have brought the Nazi war economy to a shuddering halt. Without steel, the tank and the battleship were unthinkable. As the "lead metal" in weapons production, its allocation steered the armaments economy. German strategists thus used the output of ferrous metal as the prime index of national war-making potential[1]

This was an error. In the greater war waged by the United States neither coal nor steel was of overriding importance. American solid fuel reserves are nearly inexhaustible. The United States never suffered coal shortages during the war because petroleum, of which the United States was the world's largest producer and exporter, met the bulk of the nation's fuel consumption requirements. The American war machine, whose thirst for liquid fuels was unslakable, consumed over half of the world total between 1939 and 1945.[2] The war's only noteworthy fuel shortage, one due to a short-lived German threat to coastal shipping in 1942, was soon overcome by the construction of a new pipeline connecting the eastern seaboard to the Texas fields. Nationwide gasoline rationing was introduced only temporarily during the war, and its purpose was less to save fuel than to economize on the use of rubber by reducing civilian automobile travel.[3] American ingot capacity was rated at 81.6 million tons, greater than that of Europe and Great Britain combined. This was not quite enough to meet the requirements of the ambitious factory and ship construction programs planned for 1942; an additional 14 million tons of capacity had to be erected. Except for shortages caused by administrative bungling, the steel supply was more than adequate during the rest of the war.[4] The central importance of air power in American strategy in fact made nonferrous metals, particularly aluminum, far more war-critical than iron and steel.[5] The existence of the atomic bomb rendered all assessments of national power in terms of heavy industry outputs obsolete.

The greater and lesser wars were not mutually exclusive but overlapped at various times, in different ways, and with dissimilar consequences. The United States, which launched the greater war, would be thrust into a world prominence it had not sought, adaptation

[1] Johann Sebastian Geer, *Der Markt der geschlossenen Nachfrage: Eine morphologische Studie über die Eisenkontingentierung in Deutschland, 1937–1945*, Nürnberger Abhandlungen zu den Wirtschafts- und Sozialwissenschaften, 14 (Berlin, 1961).
[2] Alan S. Milward, *War, Economy, and Society. 1939–1945* (Berkeley CA, 1977), passim.
[3] Donald M. Nelson, *Arsenal of Democracy: The Story of American War Production* (New York, 1946), pp. 303–304.
[4] Harold G. Vatter, *The US Economy in World War II* (New York, 1985), p. 28.
[5] Nelson, *Arsenal*, p. 173.

to which would require wrenching domestic political changes. Britin had the opportunity to participate more fully in the greater war than any Continental power but complacently relied on the Americans to win the battle of production, thus weakening the economy, undermining the reform plans of the postwar Labour government as well as Britain's status as an international power, and increasing the dependence of Western Europe on Germany. The Reich, for its part, escalated from lesser to greater war with remarkable speed and, though the Ruhr mines were in bad shape, would emerge from the war with both its industrial and its human capital enlarged and modernized. Placed under a harsh and economically profitless occupation in 1940, France would suffer through five barren years. Yet, of desperation was born renewal.

When Hitler had the greater war inflicted on him at the end of 1941, no choice remained but to switch from partial to total mobilization and from a strategy of "armament in width," in which the manufacture of weaponry could be limited to the requirements of a particular campaign, to one of "armament in depth," involving the conversion of the entire economy to war production. To maximize outputs control had to be exercised at every stage of the manufacturing process, which also had to be simplified by reducing fabrication types and standardizing components. An organizational revolution was called for.[6]

Albert Speer directed it on behalf of the führer. His instruments were Germany's businessmen; they would run the new committees for finished categories of weapons as well as the so-called rings for components and subassemblies through which the armaments economy was administered. Speer and his associates succeeded in increasing total munitions output nearly two-and-a-half-fold within two-and-a-half years. They also made the first, belated wartime attempt to incorporate the industry of the occupied nations into an overall production plan, to create a European armaments economy.[7] The Speer initiatives could prolong the death agonies of national socialism but not alter its fate. By 1942 American economic superiority was overwhelming and still growing. By mid-1943 the Third Reich's domestic economy began to break down irreversibly, a pro-

---

[6] See Alan S. Milward, *The German Economy at War* (London, 1965), passim; Burton H. Klein, *Germany's Economic Preparations for War* (Cambridge MA, 1959), pp. 220–221; Hans Kehrl, *Kriegsmanager im Dritten Reich* (Düsseldorf, 1973), pp. 185–186, 244–245, 310–311.

[7] Albert Speer, *Inside the Third Reich* (New York, 1970), passim.

cess most advanced at the mines. Yet in some respects the war was not an economic loss for Germany. In spite of the bombing, the national stock of capital goods increased between 1939 and 1945, and the same was true of human capital. Thanks to extensive vocational instruction, labor skills improved substantially. Postwar Germany could draw on reserves of trained manpower absent elsewhere in Europe. The war also increased management experience and strengthened the hand of big business. In prosecuting the war on behalf of Hitler the leaders of industry and finance took on new responsibilities at home and abroad and both extended their reach and strengthened their grip on German society.[8]

In Great Britain too, the greater and lesser wars overlapped. Economically speaking, the war began for the United Kingdom not on 3 September 1939 but after Dunkirk, when no choice remained but to switch from the "peacetime scale of rearmament" conducted since the post-Munich failure of appeasement to full mobilization. As in Germany, this required a shift from armament in width to armament in depth. In the United Kingdom, however, the changeover did not precipitate a far-reaching reorganization of production but created a new dependence on the United States. The Americans would supply all the means of payment, much of the food, raw materials, and components, and many of the machines as well as the weapons required to sustain warfare against Germany.[9] U.S. aid was not an unmixed blessing. Britain lost an opportunity to refurbish its shopworn economy, whose true deterioration during the war was all too easy to overlook, thanks to apparent additions to national income stemming from increased armaments output. The new production often involved only low-skilled finishing operations, however, far too much of which was done at excessive cost.[10] Had Britain been Germany, inefficiency would have ended the Nazi threat. Nowhere does this appear more glaringly than in the case of the coal-mining industry, a very sick sector which became critically ill during the war.[11] The greatest of world exporting industries before World War I, unreliable and increasingly uncompetitive in the 1920s and 1930s, the British mines dropped out of foreign markets altogether after 1945, leaving France and the Low Countries at the mercy of the Ruhr.

The French and the other nations of occupied Western Europe

[8] Gillingham, *Industry and Politics*, pp. 141–142.
[9] M. M. Postan, *British War Production* (London, 1952), p. 121.
[10] Ibid., pp. 228, 232, 366, 391.
[11] See William H. B. Court, *Coal* (London, 1945).

participated only in the lesser part of the lesser war and thereafter
went into a kind of industrial hibernation. Cut off from markets and
sources of imported raw materials; truncated territorially; saddled
by onerous levies of goods, services, and financial assets; administered by occupation authorities lacking independent governing
authority; and ruled ultimately by a distant madman driven by
obscure criminal fantasies, their lot could only have been misery.
Yet misery can vary. Western Europe never became Poland. Though
for Frenchmen, Dutchmen, or Belgians, political accommodation
to a Hitlerian New Order was ultimately unthinkable, economic collaboration seemed feasible within limits. The Reich needed arms,
consumer goods of all kinds, food, and raw materials. Were not the
Germans the obvious replacements for buyers in lost markets?
Might they not keep industry at work and men employed?

There were at least some signs that Germans might assert real
economic leadership. After the onset of occupation, representatives
of the Wehrmacht set up agencies like those in the Reich for the
distribution of raw materials, which served as pressure groups vis-
à-vis the central authorities, provided liaison with German buyers,
and took over many tasks formerly discharged by cartels. The raw
materials agencies became analogues, in other words, to organizations across the Rhine.[12] This remodeling was not enough to bring
prosperity without massive German order placement. In the absence
of any overall Hitlerian plan to utilize the economic potential of
Western Europe, and in the presence of administrative and political
obstacles of all kinds, such a thing was not easy to arrange. There
was no lack of effort by the businessmen of Festung Europa to make
deals; and some attempts were of breathtaking grandiosity. Though
most German administrators in fact had previous associations
(sometimes through international cartels) with their charges in the
occupied territories, few new business arrangements of great importance actually materialized. The prevalent shortage of fuel also contributed to this result. The Reich refused to export coal during the
war, and outputs in occupied Western Europe plunged. During the
years from 1940 to 1944 coke was too scarce for steel outputs to rise
above much more than one-half of prewar levels, and manufacturing activity thus languished as well.[13]

[12] See Alan S. Milward, *The New Order and the French Economy* (Oxford, 1970), pp.
    45–46; Gillingham, *Belgian Business*, pp. 39–40; Gerhard Hirschfeld, *Fremd-
    herrschaft und Kollaboration: Die Niederlande unter deutscher Besatzung, 1940–1945*
    (Stuttgart, 1984), pp. 117–118.
[13] Gillingham, *Industry and Politics*, p. 141.

50   Coal, steel, and the rebirth of Europe

However dark their situation between 1940 and 1944, the French saw glimmers of the greater war. In political economy as in other areas of French life, defeat resulted in reappraisal. It went on in several places simultaneously: at Vichy, within the resistance movement, and in exile circles. Reform proposals evincing a positive attitude toward economic growth and rejecting the compound of frustration and complacency known as Malthusianism proliferated.

Jean Monnet found a cure for France's malady, and he administered it in two doses. The first was the Plan de Modernisation et d'Equipement of 1946, or Monnet Plan, a program for renovating and rationalizing industry; the second the European Coal and Steel Community (ECSC). These institutions were by-products of Monnet's wartime experiences as chief of the French purchasing mission in London, delegate-at-large on the British Supply Council in Washington, and influential adviser to American policy makers. Monnet had realized at the very outbreak of World War I that its outcome would depend on the economic strength of France and Great Britain, and he understood long before Hitler's invasion of Poland in September 1939 that the fate of Europe would be determined by the industrial and agricultural power of the United States. The war drove home the lesson that the future would belong to those who could feed, clothe, fuel, and arm the big battalions. The fall of France proved to Monnet that not only industrial modernization but political federation were indispensable prerequisites for the maintenance of national sovereignty, and he recognized that this held true for Britain as well as for his native land.[14] The short, unprepossessing Frenchman learned still more in Washington, where he served as administrator of the Lend-Lease program: It was on the Potomac that he discovered how to operate as a politician, there too that he conceived his postwar program and found the necessary backing for it.

In the 1930s Monnet got enmeshed in a couple of unsuccessful business ventures (consumer banking in the United States and economic development in China) far removed from the scene of diplomatic conflict in Europe. He returned to public affairs immediately after Munich, convinced that France and Britain needed vast numbers of U.S.-made fighter aircraft in order to survive. Though the British were not yet ready to go along with Monnet's order-placement plan, using the Banque de France's gold hoard as a pledge he closed the first big U.S. airplane deal in December 1938.[15] After the outbreak of war Monnet again found himself serving as France's

[14] Monnet, *Memoirs*, pp. 21–22.   [15] Ibid., pp. 107–108, 116–117.

representative – this time as director rather than deputy – on the Anglo-French committee responsible for official purchasing in the neutral nations. He soon began lobbing intensely for far-reaching co-ordination of economic strategy.[16] Monnet also devised and pressed for the adoption of a balance sheet of armaments programs listing estimates weapon by weapon of required manpower, raw materials, and machinery in one column and, for purposes of comparison, available inventories for these items on the other. Production requirements could then be set, shortfalls determined, and provision made for overcoming them.[17] On the very eve of the fall of France Monnet proposed a formal political union of Britain and his native land. Though to Churchill this amounted to little more than a propaganda appeal intended to stiffen French morale, Monnet's warning that only through such a partnership could the island nation also preserve its stature as a great power should have been heeded.[18] The British survived the fall of France militarily but not economically.

Their ability to remain in the war owed much to Monnet. As delegate-at-large on the British Supply Council, to which Churchill had appointed him, he helped secure the American aid necessary for the United Kingdom's survival. President Roosevelt may have made the decision to adopt Lend-Lease on his own, as well as designed the political package necessary to sell the idea to the American public, but Monnet deserves much of the credit for making it work. His instrument was once again the balance sheet, now officially known as the Anglo-American Combined Statement.[19] It also provided the conceptual underpinnings of the Victory Program, the mobilization plan for the American economy. Monnet's scheme was more than a planning device; it served several political purposes as well. One was to confront American policy makers with the full magnitude of the economic challenge lying ahead and to win their commitment to an all-out program for meeting it. Another purpose was to maneuver British military and civilian officials, who even after Dunkirk insisted upon preserving "freedom of action," into a relationship with the United States that provided desperately needed aid without causing loss of face.[20] The Combined Statement provided a kind of template for the Combined Boards through which Lend-Lease

---

[16] W. K. Hancock and M. M. Gowing, *British War Economy* (London, 1949), p. 185.
[17] Ibid.; Monnet, *Memoirs*, p. 126.   [18] Monnet, *Memoirs*, p. 21.
[19] H. Duncan Hall, *North American Supply* (London, 1955), pp. 161, 252, 293, 322, 334.
[20] Nelson, *Arsenal*, pp. 130–131.

goods were given to Britain.[21] Though the effectiveness of this machinery was hampered by American administrative confusion and British economic inefficiency, through it streamed increasing proportions of the United Kingdom's total requirements for food, raw material, machinery, components, and weaponry.

Monnet never held public office in the United States. His influence in Washington was exercised behind the scenes through power brokerage.[22] He was on a basis of "total trust" with columnist Walter Lippmann, a formidable intellectual influence on opinion making for nearly thirty years, as well as united in "deep friendship" with Supreme Court Justice Felix Frankfurter, the most revered man on the American bench, the leading educator of a generation of American jurists, and a powerful informal adviser to Roosevelt. Monnet also developed, or maintained, close personal contacts with many of the men who would shape future American foreign policy, among them Averell Harriman, Dean Acheson, John Foster Dulles, and John J. McCloy. It is difficult to determine precisely how much influence Monnet had on the thinking of such future policy makers because, according to all accounts, there was a true meeting of minds at the frequent get-togethers where the Frenchman discussed his ideas with his American friends.[23] One can easily see how the atmosphere of wartime Washington influenced Monnet, however. He observed a U.S. president preparing his country for war while denying that this was happening, was witness to an economic mobilization conducted on a staggering scale that was successful beyond expectations, and participated first-hand in the rise of American global power. In the United States Monnet learned how to organize Europe.

2.1   The Nazi economy at war

The Third Reich began its economic preparations for war on 18 October 1936 with the publication of a law creating the misleadingly named Four-Year Plan, which was not a comprehensive scheme for economic development along Soviet lines but a bureaucratic apparatus vested with responsibility for reducing Germany's dependence on imported sources of raw material that had been aggravated by

[21]   Hall, *North American Supply*, pp. 339–340.
[22]   Monnet, *Memoirs*, pp. 153–154.
[23]   Dean G. Acheson, *Present at the Creation* (New York, 1969), p. 117.

severe shortages of foreign exchange.[24] The Four-Year Plan had limited powers and in no sense provided overall direction of rearmament. Nor for that matter did anyone else: The armed services took care of their own respective needs, and Minister of Economics Hjalmar Schacht held a vague mandate to coordinate the civilian with the armaments economy but never exercised direct control over production. The Four-Year Plan *did* acquire certain regulaory functions, setting priorities for trade and payments policy, and its competence also soon extended into rationing and wage and price matters. The Four-Year Plan never became a solid, monolithic authority governing the economy but served instead as a kind of draping for prominent factions and interests. Though it housed the board that through steel allocation directed the rearmament process, control of this agency was in the hands of the armed forces, industry, the party, and other groups. Independent ceners of influence also formed within the Four-Year Plan, especially around synthetics industries like chemicals and armaments-related industries like light metals, both of which benefited from lavish investment subsidies. Yet not all producers were either happy about the Third Reich's official autarchy policy or well served by it.[25]

The leaders of Ruhr industry were distinctly of two minds after the 1937 conflict with Hitler over the smelting of previously unused German low-grade ores. The consumption of such minerals would have required massive new investment in blast furnaces and played havoc with costs. The producers' refusal to contemplate its use precipitated Hitler's decision to break the power of "monopoly capitalism" by creating a fearsome new competitor in steel, the Reichswerke Hermann Göring AG. The new enterprise, the corpulent Reichsmarschall announced, would serve as an instrument of national policy, need not be bound by the need to make a profit, and could expect to receive both generous subsidies from Berlin and tribute from industry in the furtherance of its mission. The Reichswerke was capitalized on the basis of a forcible subscription imposed upon Ruhr steel producers who, though initially the largest shareholders in the new company, had no influence over its management.[26] By 1940 the Göring works had become the largest single producing complex in the Reich. The construction of the "monster

[24] See Dietmar Petzina, *Autarkiepolitik im Dritten Reich* (Stuttgart, 1968).
[25] Gillingham, *Industry and Politics*, pp. 50–51.
[26] See Matthias Riedel, *Eisen und Kohle für das Dritte Reich: Paul Pleigers Stellung in der NS Wirtschaft* (Göttingen, 1973).

of Salzgitter" consumed huge amounts of steel and drained reserves
of skilled labor when because of the rearmament boom both were in
short supply. Its operation saddled the overburdened wartime rail
system with massive additional deliveries of coal. The Ruhr steel
industry quite properly feared that the Göring works were a harbinger of Hitler's long-promised "German socialism." Thanks to
forcible acquisitions in the annexed and occupied lands, the works
bearing the Reichsmarschall's name had become the largest conglomerate in the world by the eve of the German invasion of Western Europe.[27]

The mining industry's relations with the regime were no better
than those of steel. The coal operators were not eager to expand
operations, nor did they solicit investment funds from the Four-
Year Plan, instead keeping the organization at arm's length. The
mines were largely bypassed by the successive expansion program of
the prewar years. This was fatal for the German war economy.[28]
The growth in Reich fuel consumption requirements soon outpaced
the ability to raise coal outputs even after big reductions in exports.
By the end of 1937 the industry was operating at full tilt, and further gains could be made only by overworking men and machinery.
Productivity soon began to decline. In June 1939 the workday was
lengthened by forty-five minutes in a futile attempt to increase production. Though aggressive recruitment of young miners raised
overall employment by twenty thousand in the two years before the
war, the increase in quantity masked a decrease in quality. Giant
projects such as the Göring works lured skilled labor away from the
pits with offers of fat wages. Miner recruits were thus predominantly
"young persons whose inferior capabilities make them otherwise
unemployable."[29] Mine wages were insufficient to offset the disagreeable character of the work. The military draft also siphoned
off productive workers. By June 1939 the age structure had become
badly skewed. The 14–21-year-old group was underrepresented by
53 percent, the 22–25 group by 29 percent, and the 26–30 group by
27 percent; at the same time, the older categories were overrepre-
sented, 31–35 by 41 percent, 36–40 by 58 percent, and 41–45 by 14
percent. Miner health deteriorated alarmingly during the twenty-
four months before the war because of a decrease in the quality of

[27] Ibid.; see also Richard J. Overy, "Heavy Industry and the State in Nazi Germany: The Reichswerke Crisis," *European History Quarterly* 15:1985, pp. 313–
337.
[28] Gillingham, *Industry and Politics*, p. 51.   [29] Cited in ibid., p. 54.

food consumed. Bad morale, as reflected in a rise of absenteeism, became a pressing concern of both operators and political officials over the same period.[30]

Shortages of raw material such as wood, steel, and equipment for conveyancing compounded the difficulties of the mines. Coal shortages estimated at between seven and eleven million annual tons became the rule after 1938. They could have been overcome only by reducing either allocations or the quality of coal delivered, both of which would have created serious political difficulties for the industry. In summer 1939 an official of the German Labor Front named Paul Walter was appointed "productivity plenipotentiary" for coal. Though he was demonstrably incompetent, his authority increased steadily. In April 1940 he became Reich coal commissar and thus, at least in theory, "czar" of the industry. As a result of Walter's rise the mines were the first sector of production to face the threat of de facto expropriation.[31]

The disharmony in relations between heavy industry and the regime had a number of important wartime consequences. Production problems could not be overcome, coal policies were regressive, and forced mine labor was worked to death en masse in a crude but successful attempt to maintain Ruhr outputs at prewar levels, which, however, were still too low to meet the rising demand. To prevent a fall in output nonetheless required the adoption of *Raubbauwirtschaft*, an economy of attrition. Its human costs were even greater than its material ones. Though still mostly operable, the mines were a wreck at the end of the war, and there was no one left to work in them. Tens of thousands of slave miners had perished. Nearly all of those who survived had fled. The exhausted German miners and supervisors who remained on the job feared retribution from those they had abused. For them the pits had also become a hell, and they too soon left in droves. The labor force decreased by 125,000 between January and May 1945.[32]

The lack of common purpose between heavy industry and regime so evident on the domestic scene turned the occupied countries into a place of frequent power struggles. They began with the carving up of spheres of influence in summer 1940 and continued intermittently thereafter, with industry gradually gaining the upper hand but never quite winning. To the men of the Ruhr, the coal and steel

[30] Ibid., pp. 55–56.   [31] Ibid., pp. 62–63.
[32] Werner Abelshauser, *Der Ruhrkohlenbergbau seit 1945: Wiederaufbau, Krise, Anpassung* (Munich, 1984), p. 45.

producers of occupied Western Europe were welcome allies in the struggle against Göring and others like him. Industry in France and the Low Countries in turn welcomed the protection of their German counterparts. During the dismal years between 1940 and 1944 the members of the former International Steel Export Cartel, and coal producers as well, formed a kind emergency alliance (Notgemeinschaft) against both the willful interventions of the regime and the vicissitudes of war. It afforded a measure of security to the unwilling industrial inhabitants of Hitler's economic New Order as well as a form of reinsurance to Ruhr producers in the event of German defeat.[33]

If by the time the war broke out it was too late for the massive investment needed to produce enough coal for the growing requirements of the war economy, Berlin did not lack imaginative expedients for dealing with the problem of fuel scarcity. In 1940 coal "czar" Walter was given a broad mandate to deal with the allocation problem, but as he failed either to generate savings or to establish a workable system of priorities, severe local fuel shortages developed in early 1941.[34] To remedy the problem, the brilliant and rising Nazi technocrat Hans Kehrl named the general manager of the Göring works, Paul Pleiger, to head a new federal mines organization called the Reich Coal Association (Reichsvereinigung Kohle).[35] Formed in March 1941, the RVK was the first of the new streamlined industrial organizations of the type later introduced by Speer throughout the armaments economy. Its novelty should not, however, be exaggerated. Though Pleiger could impose directives on the industry through the new organization, he could only implement them with the aid of the old coal cartel, the Rheinisch-Westfälisches Kohlensyndikat. RWKS provided RVK's brain, its marketing machinery. The regime's inability to dispense with RWKS left the private owners in control of their industry. Far from challenging this state of affairs, Pleiger eventually became the mines' most prominent advocate and supporter in the ongoing struggle over fuel allocation.[36] There were nonetheless limits to RVK's effectiveness in reducing coal shortages. Since the cartel had long operated efficiently, coal and coke wastage was relatively slight and little could be recovered

[33] See John Gillingham, "Zur Vorgeschichte der Montan-Union: Westeuropas Kohle und Stahl in Depression und Krieg," Vierteljahrshefte für Zeitgeschichte 3:1986, pp. 381–405.
[34] Gillingham, Industry and Politics, pp. 62–63, 112, 114.
[35] Kehrl, Kriegsmanager, pp. 260–261.
[36] Gillingham, Industry and Politics, p. 114.

by reducing it. The savings, which amounted to only six million annual tons gained through improvements in allocation, were far too
little to have had a significant impact on the coal problem. The only
real solution was to raise production.

German reserves of male labor having become scarce even before
the outbreak of hostilities, there were only two potential new
sources of wartime mine recruitment, and one of them, women, was
unacceptable for reasons of ideology and tradition. The other was
foreigners. In April 1940 recruiting began, but the program was
never effective. Nazi racism, the basic problem, dictated making invidious administrative distinctions between Germans and foreigners,
and between the latter as well. Each so-called national group of miners had to be treated separately and unequally, thereby vexing any
attempt to develop a sensible personnel policy. Treatment of the
newcomers, who were in theory free labor, ranged from contemptuous to brutal, and few stayed on the job for long. Within two years
over two-thirds of the volunteers had run away from the mines, the
only national group remaining in any numbers being the wretched
Galicians.[37]

The use of foreign slave labor began with Speer's appointment of
Fritz Sauckel as labor plenipotentiary in March 1942. The decision
to begin forcible recruitment was taken without consulting the industry and in the face of long-standing opposition, based on World
War I experience, to the employment of foreign conscripts. The victims of this policy, mainly Soviet prisoners-of-war and civilian "East
workers" (Ostarbeiter), were housed in rude, poorly heated quarters,
clothed in rags, fed at near-starvation levels, and subjected not only
to tight discipline but to physical mistreatment at the hands of German workmates, who were themselves also being cruelly exploited.
Tens of thousands of slave workers perished from fatigue, under-
nourishment, and abuse. Their productivity was half the German
rate. By mid-1943 the forced laborers constituted nearly a quarter
of a manpower force swollen to a third larger than before the war.[38]

The slave workers nevertheless accomplished a minor miracle of
production. As late as 1943, outputs had still fallen only about 5
percent below the record year of 1939, when they were about 130
million tons. Even this did not end the coal shortages. The elimination of exports to Western Europe still failed to liberate enough fuel
to operate the armaments economy. To meet the demands for the
increased steel output planned for 1942 drastic reallocations and

[37] Ibid., pp. 118–119.   [38] Ibid., pp. 115–116.

shutdowns in many sectors of industry were necessary. In April 1943, the central planning board *(Zentrale Planung)* concluded that the coal situation was hopeless and ruled out new production programs. Though a breakdown in mine production never occurred, this was only because increasingly frequent disruptions of the economy after 1943 both reduced overall demand for coal and imposed stoppages which, in addition to providing time for repair work, gave idle miners a chance to rest.[39]

Ruhr coal was a hardship case, the sector of industry most harmed by Hitler's policies. Steel's experience was more like that of other branches of production. The value of capital goods actually increased during the war; skilled labor was plentiful after it; and the production effort catapulted a generation of dynamic young managers into positions of national and international responsibility. Important changes occurred in labor relations as well. Recognizing the legitimacy of the traditional management demand to serve as master in its own house *(Herr im Hause)*, Hitler abolished labor unions, allowed employers to institute neopatriarchal policies at the workplace, and put the powers of the law and the state at their disposal in implementing them. Industry took over the labor exchanges, the chambers of commerce assumed responsibility for labor training and credentialing, and a new pseudo-union called the German Labor Front (Deutsche Arbeitsfront or DAF) organized employee leisure with a view to increasing productivity.[40] A new system of vocational education was central to the Nazi pattern of labor relations. This was actually devised by Vereinigte Stahlwerke, or more specifically a private think tank sponsored by it named DINTA (Deutsches Institut für technische Arbeitsschulung). It developed the modern German industrial apprenticeship program. The founder of DINTA, Dr. Karl Arnhold, wanted to do more than merely impart technical skills; he sought to instill obedience and patriotism by indoctrinating young workers with an ethic of pride in achievement both for its own sake and for the good of the larger community. This was the achievement ideology *(Leistungsideologie)* featured in labor propaganda after 1933. It served as a Nazi moral equivalent to Stakhanovitism. Employers conducted the DINTA apprenticeships on the job. They were not meant to supplement the schooling of the fourteen-year-olds who entered them but to substitute for it. The Arnhold system was introduced into Ruhr heavy industry in the late 1920s, promoted nationally after 1933, and became an almost uni-

[39] Ibid., pp. 132–133.   [40] Ibid., pp. 40–41.

versal substitute for nonacademic secondary education in the years following rearmament. With some modification, it remains in effect today.[41]

The new climate of labor relations was evident at the Bochumer Verein, a large integrated steel producer belonging to Vereinigte Stahlwerke. One index of the changed mood is the factory council election, which suggests, first of all, that the depression unhinged traditional employee loyalties. In those of 1928 the free unions closely affiliated with the social democrats won three-quarters of the votes and the Nazis none at all, but in those of spring 1933, which though held after the seizure of power were still representative, they won less than a quarter. In the suspect 1935 elections conducted amid Gestapo comb-outs (Auskämmungen) of political dissidents the official slate received over 80 percent of the votes cast.[42] It is difficult to say whether and how new loyalties replaced old ones. Surely the disappearance of unemployment after 1936 counted for something in this respect, and so too did the increase in social mobility. In 1936 the number of apprenticeships at the Bochumer Verein was doubled. Instruction offered through DAF's Professional Education Program (Berufserziehungswerk) doubled yet again the number of those entering the ranks of skilled labor. Enough trained metalworkers were on hand to liberate successive classes for the military draft, which in 1943 was extended to those born in 1908, without causing serious interruptions in production. By 1940 training, upgrading, and conscription had together all but wiped out reserves of male semi- and unskilled labor. Women took over many of these jobs on an emergency basis, which over time became more or less permanent. As the war dragged on, foreigners were also brought in.[43] The mood on the shop floor at Bochumer Verein was hardly upbeat, but morale remained intact. In spite of longer workdays, backbreaking labor, reduced living standards, and the disruptions and personal losses of war, resistance was virtually nonexistent. Productivity increased until heavy bombing began in late 1943. The

[41] See Theo Wolsing, Untersuchungen zur Berufsbildung im Dritten Reich, Schriftenreihe zur Geschichte und Politischen Bildung, 24 (Kastellaun, 1977), and Rolf Seubert, Berufserziehung und Nationalsozialismus: Das Berufspädagogische Erbe und seine Betreuer (Weinheim, 1977); see also John Gillingham, "The Deproletarianization of German Society: Vocational Training in the Third Reich," Journal of Social History 9/3:1986, pp. 423–432.

[42] Gustav-Hermann Seebold, Ein Stahlkonzern im Dritten Reich: Der Bochumer Verein, 1927–1945 (Wuppertal, 1981), pp. 49, 217.

[43] Ibid., pp. 202–203.

wartime workforce of the Bochumer Verein – as well as those who had been drafted away from it – was highly skilled, efficient, hardened to sacrifice, apolitical, and obedient. Though those who labored were not necessarily enthusiastic about neopatriarchism, there is little to indicate that they rejected it.[44]

German industry, especially steel and metalworking, added substantially to, and improved the quality of, its inventories of capital goods from 1937 to 1945. Two background factors contributed to this result. German tax law made generous allowance for both depreciation and the retention of reserves, and there were disincentives either to paying out earnings or to retaining them. The latter invited a capital levy, the former the imposition of an officially imposed limit on yield, both of which were in fact eventually adopted during the war. More important still, the currency was not convertible, had no real exchange value, and could not command goods in the marketplace without some form of accompanying authorization.[45] As early as 1938 German businessmen had clearly recognized that future devaluation was inevitable. From then on they threw money at goods of all kinds, the more eagerly in that prices were held artificially low by controls. Whenever possible they added to stocks of scarce raw materials but, failing this, bought machinery, which was readily available. German investment in equipment totaled 2.9 million Reichsmarks in 1938, and in the following years increased to 3.5, 3.7, 4.1, and 4.2 million before declining to 3.7 million in 1943 and 2.7 million in 1944. The number of metalworking machines increased from 1.2 million in 1935 to 1.7 million in 1938, remained unchanged as of 1941, and rose to 2.1 million by 1944. The figure in 1944 is truly astonishing because metalworking machines are normally in demand in the *first* phase of mobilization, the retooling process, during which Germany was well supplied; inventory remained constant between 1938 and 1941. The accumulation of capital equipment late in the war was due mainly to hoarding. Outputs of machine tools provide even more striking evidence of this phenomenon; they too are normally in greatest demand during the start-up phase. Machine tool output predictably rose from some 264,230 tons in 1938 to 306,360 tons by 1941, declined slightly the following year, then, contrary to need, exploded to 772,629 in the very midst of Speer's intense campaign to eliminate nonessential

---

[44] Ibid., p. 282.
[45] Willi A. Boelcke, *Die Kosten von Hitlers Krieg: Kriegsfinanzierung und finanzielles Kriegserbe in Deutschland, 1933–1948* (Paderborn, 1985), pp. 83–161.

production.[46] But nonessential for whom? Germany's producers
were saving for a rainy day sure to come. As a result of rearmament,
the lack of serious damage to capital equipment, and an apparently
high rate of investment, Germany's industry emerged from the war
not only with more good machinery than it could put to use but in
better shape than that of any major producing nation save the
United States.

In steel as in most branches of production, the gap between the
two nations was still very wide; there were immense differences in
size and productivity of plant. The Salzgitter works of the Göring
enterprise, with two million tons of capacity, was the only new integrated facility built in the Third Reich. Old firms like the Bochumer
Verein were typically only semimodernized. The company built two
new blast furnaces in the late 1920s and had no need to invest further in them after 1933; yet it did construct three new electrosteel
furnaces after 1936, thereby tripling outputs of high-grade armaments steel. Though by 1939 the firm's rolling mills, one of which
dated from before the turn of the century, had become hopelessly
obsolete, their replacement was a long-term project and thus had to
be dropped when the war broke out. The cast products for which
Bochumer Verein was famous (train wheels, cannon barrels, and the
like) required little new investment, the bulk of which was placed
into the rapidly expanding shell manufacturing division. This was
the firm's most modern operation as well as its main profit center.[47]
However, neither the Bochumer Verein nor any other Ruhr enterprise had a large, continuous wide-strip rolling mill for producing
the huge amounts of cheap sheet metal that would be in demand
after the war.

The final years of the war brought a new generation of executives
to power in the Ruhr and provided one-of-a-kind training. It was
only partly technical, yet it did involve the use of new management
tools. Although much has been made of Speer's organizational contributions to the post-1942 increases in armaments outputs, the real
credit belongs to that most ruthless and clever of Nazi technocrats,
Hans Kehrl. He devised an equivalent to Monnet's Anglo-American
Combined Statement, a document providing the information required to channel the total resources of an economy into war production. Kehrl also organized the machinery needed to make it
work, analogues to the coal association set up in 1941, the most im-

[46] Cited in Rolf Wagenführ, Die deutsche Industrie im Kriege, 1939–1945, 2nd ed.
(Berlin, 1963), pp. 160, 162, 164.
[47] Seebold, Stahlkonzern, pp. 40–41, 119–120.

portant of which was a steel counterpart, Reichsvereiningung Eisen (RVE).[48] As directors of committees and rings, the leaders of industry also assumed the responsibility for weapons design and procurement formerly exercised by the armed forces. As such, they were able not only to eliminate waste and build better weapons but to introduce serial production on an unprecedented scale.

In August 1943, Speer empowered a few carefully selected heavy industry managers, many of them quite young, to act as a virtual collective dictatorship over the coal–steel region. The new Ruhr staff *(Ruhrstab)* was given blanket authorization to do everything required to sustain outputs. Its reach extended far beyond the factory walls. The staff men could shift production, reroute trains, appropriate food supplies, rehouse the public, and give orders to the military. Such authority as remained to the regime in the Ruhr now belonged to representatives of the dominant Konzerne: The equation of private power and public responsibility could hardly have been more complete.[49]

One of the new men was Walter Rohland of Vereinigte Stahlwerke, a youth of modest origins promoted through the ranks to captain in 1918 by age twenty. Rohland was already a recognized world expert in the field of special steels when he became general manager of a key division of Vereinigte Stahlwerke, the Deutsche Edelstahlwerke in Krefeld, in 1932. A party member prominent in the professional association for Nazi engineers, he was a natural choice to head the tank design committee set up in late 1940 by Armaments Minister Fritz Todt. Rohland soon rapidly increased outputs while improving design and performance, thus providing an object lesson in the superiority of civilian to military war management. The tank committee became a model for Speer's system and "Panzer Rohland" himself a hero of Nazi propaganda. Rohland later become deputy director of the Committee for Iron and Steel Production as well as the representative for that sector on the Ruhr staff, but his high wartime visibility made him an embarrassment after 1945; among Speer's new men he alone was unable to rejoin the heavy industry establishment.[50]

Willi Schlieker, the boy wonder of the Nazi war economy, actually chose not to enter it. Schlieker, according to Kehrl, was the "only businessman [I have] ever met who is able to blend harmonically

[48] Kehrl, *Kriegsmanager*, pp. 260–261.
[49] Milward, *German Economy at War*, passim.
[50] Walter Rohland, *Bewegte Zeiten: Erinnerungen eines Eisenhüttenmannes* (Stuttgart, 1978), pp. 68–69.

solid technical knowledge of the most precise kind with the intuition
of an aritst in order to understand complicated economic problems."[51] In *Fire in the Ashes*, an early and excellent popular American
survey of post–World War II Europe, the journalist Theodore H.
White praised Schlieker's gifts with equal extravagance, as someone
who "could talk about steel with more romance and love than any
man I have ever heard."[52] Schlieker was a supersalesman who after
a stint at hawking cement for U.S. and British firms in the West
Indies returned to Germany just before the war and soon rose me-
teorically in the raw materials division of Vereinigte Stahlwerke. At
twenty-eight the pudgy-faced young executive became Kehrl's deputy for steel production and, as White described it,

> put the industry on a punch card basis. Each item ordered by
> the army had to be accompanied by punch cards breaking down
> the end-item requirement of 1000 tanks, say into so many tons
> of light-gauge steel, so many tons of tubing. All the rolling ca-
> pacity of Europe's steel mills was similarly broken down into
> their monthly productive capacity of Europe's special categories
> of steel. Running a million punch cards a month through his
> filing machines, Willi made a balance sheet. On the one side
> were the war's requirements listed in tonnages of each type of
> steel; on the other side, Europe's capacity that month to pro-
> duce what tonnage of the needed kinds of steel.[53]

Schlieker was captured by the Russians at the end of the war and
supervised the steel industry of the Soviet Zone of Occupation until
kidnapped by the British, for whom he worked in the late 1940s. By
1950 he had made the first new postwar fortune in steel by acquiring a monopoly over the importation of American scrap. Schlieker
was not, however, to become a permanent fixture in the Ruhr: His
firm went under in 1962.

Hans-Günther Sohl had a more lasting influence. A suave, well-
connected mining engineer, Sohl started work for Krupp in the depression but shifted to Vereinigte Stahlwerke in 1941 at the
personal request of its revered chairman, Albert Vögler. The third
of the Konzern's triad of bright young men, Sohl was closer than
either Rohland or Schlieker to the industry's old guard and remained deeply involved in the firm's private affairs. Yet he too entered Speer's service, joining the administrative council of the Reich

[51] Kehrl, *Kriegsmanager*, p. 255.
[52] Theodore H. White, *Fire in the Ashes: Europe in Mid-Century* (New York, 1953), p. 169.
[53] Ibid., p. 177.

Iron Association in early 1943 and the Ruhr staff in summer
1944.[54] In the final days of the war Sohl convened another kind of
Ruhr staff, an informal collection of dignitaries who gathered to
smooth the transition to Allied control. Among them were the industrialists Paul Maulick, Walther Schwede, and Hermann Winkhaus;
the mayor of Düsseldorf, Robert Lehr; the former labor union official, Hans Böckler; and a future member of the British steel control,
Landrat Heinrich Meier. One must not exaggerate the importance
of the "Sohl Circle," an informal group that met only occasionally. It
nonetheless brought together many of the key men who instituted
co-determination *(Mitbestimmung)* after the war.[55] Sohl would eventually become managing director of the August Thyssen Hütte and
guide the development of this would-be object of Allied dismantlement to restored crown jewel of the Ruhr. The most powerful of the
postwar generation of industrialists, he would be a strong force for
continuity in German coal and steel tradition.

Those who on behalf of Reichsgruppe Industrie began making
Germany's economic preparations for peace in early 1944, among
whom Dr. Ludwig Erhard would later be most prominent, could not
have predicted this happy outcome. Compared to the postwar planning being conducted elsewhere, there was little they could do: Loss
of the war was a future certainty and political repression a present
reality. German industry did not come up with a broad program of
postwar economic reform or develop even the rudiments of a demo-
bilization plan. The studies of the Erhard group were in fact limited
to three important transitional issues: currency reform, the restoration of foreign trade, and the adaptation of the control system to
peacetime requirements. Their aim was *sauve qui peut*.[56] But who
could say how best to preserve the fundamentals of German economic strength? Erhard and his associates emphasized repeatedly
that industrial associations would have to be kept intact and foreign
trade restored rapidly in order to avert the twin evils of famine and
chaos. Beyond this, they could do little more than harp on the importance of reestablishing, maintaining, and building upon Germany's foreign business contacts. This was stating an obvious and
undeniable truth; not until such ties, frayed almost to the breaking

[54] Hans-Günther Sohl, *Notizen* (Bochum-Wattenscheid, 1985), pp. 40–41.
[55] Ibid., pp. 93–94.
[56] Ludolf Herbst, *Der Totale Krieg und die Ordnung der Wirtschaft: Die Kriegs-
    wirtschaft im Spannungsfeld von Politik, Ideologie, und Propaganda, 1939–1945*
    (Stuttgart, 1982), pp. 425–426.

point during the Allied occupation of Germany, had been restored could the former Reich be readmitted economically to Europe.

2.2 The French economy and the occupation

France's material preparations for war were skimpy by comparison to Germany's, *not* as in 1914 because the war was expected to be short but because as in Britain the prevailing strategic doctrine held that, unless defenses collapsed, Hitler could be defeated by economic means. Success was supposed to depend on keeping sea lanes open and blockading the Reich. This was an appealing, almost painless approach to waging war but an evasive one. How were the Germans to be forced to surrender? The general staff apparently had no plans for defeating the enemy militarily; none at any rate has yet been discovered. French hopes for victory hinged on little more than a naive faith that France, as Prime Minister Edouard Daladier put it at a meeting of the interministerial war council held in February 1940, could "assemble a powerful aerial fleet that will destroy the Ruhr in a sheet of fire and force Germany to capitulate" if only it were ready, as he proposed, "to spend every available franc and sell every last ounce of gold on the purchase of American aircraft."[57] French strategy required only minimal economic sacrifice. Proposals to convert factories to arms production encountered repeated objections that this would disrupt the peacetime patterns of commerce essential for keeping French trade balances healthy and strengthening national finances. If after 1938 investment in the armaments sector increased sharply, the secondary tier of the manufacturing industry continued to go about business as usual. Nor was civilian consumption cut back; rationing was considered unnecessay. Stockpiling was haphazard. There was no attempt to build up reserves of strategic material in the southern part of the country that could keep the army fighting after an initial defeat on the borders; everything depended on a single throw of the dice.

Coal supply was a special problem, and the British, who normally provided about 40 percent of France's imports, must take their share of the blame for it. In September 1939 they committed themselves to supplying France at the rate of twenty million tons per year but as of February 1940 had delivered less than one-half of the

---

[57] Alfred Sauvy, *La vie économique des Français de 1939 à 1945* (Paris, 1978), p. 48.; see also p. 53.

amounts promised. During the first four months of 1940 the French drew down coal stocks to the point that shortages threatened to imperil military operations. It is possible, as the economist Albert Sauvy has suggested, that France's insufficient mobilization of the economy assured defeat even before the military debacle unfolded; it is likely that if battlefield events had been less decisive France would have needed considerable American aid to remain in the war; and it is certain that the inadequate preparations aggravated the hardships of the occupation.[58]

Nazi policy toward the defeated nation was fully consistent with Hitler's aims and methods. The führer invaded Western Europe in order to destroy the French army and to reduce France, along with the Low Countries, to a politically innocuous rear area, thus permitting him to devote maximum attention to the defeat and destruction of the Soviet Union. Making little of the economic opportunities created by German victory, Hitler was all but unaware that a policy of "constructive exploitation" carried out by producers in the defeated nations would have added substantially to the warmaking potential of the Reich. Instead, he parcelized the occupied territories politically and let economically destructive German influences loose. Yet they did not have free play. Aided by the Reich's increasing need for supplementary production, German big business introduced a measure of order into the economic administration of occupied Western Europe. France and the Low Countries were hardly producer paradises during the war. Things would have been far worse, however, in the absence of German business influence.[59]

France's territorial integrity was one of the first casualties of defeat. Only about a third of the nation remained unoccupied. It consisted of the southern portions of the country, save the ten border departments occupied by Italy and a militarily significant strip of territory some fifty miles wide running along the Atlantic coast from the Pyrenees north to German-occupied territory. A first line of demarcation cut off the occupied from the unoccupied zones, and a second one separated most of occupied France from the heavily industrial provinces of the north and east. The coal-rich departments of Nord and Pas de Calais were attached to the military government in Brussels. Alsace-Lorraine returned to the Reich. These arrangements lasted until November 1942, when invading German troops ended Vichy's precarious independence.

[58] Court, *Coal,* p. 75; Sauvy, *La vie économique,* pp. 53–54.
[59] See Gillingham, "Zur Vorgeschichte," passim.

France's economic autonomy was another casualty of defeat. Typical New Order institutions were introduced after the armistice: financial arrangements designed to saddle the vanquished with the costs of the occupation and of exporting to the Reich; raw materials rationing imposed not only to reduce wastage of scarce commodities but to provide mechanisms for regulating production in the interests of the war effort; and wage and price controls calculated to maintain French living standards below those in the Reich.[60] Wartime France lacked an economic center of gravity. Though both the German military government in Paris and the French government in Vichy had nationwide and often overlapping responsibilities, real authority resided in Berlin and in the Ruhr. This was especially true in the domain of heavy industry, which was largely outside the jurisdiction of either Paris or Vichy.

The exercise of this administrative power depended less on official directives than on the outcomes of conflicts between Ruhr industry and groups such as the Reichswerke Hermann Göring and the Röchling interests enjoying the special patronage of the regime. At stake was the fate of producers in the occupied territories. The Reichswerke had grown during and before the war by means of expropriation, through officially ordered forcible exchange of shares of foreign producers in countries overrun by the Reich, whereas the Ruhr had increased its influence through renegotiated cartel agreements. For French coal and steel, victory of the one meant subjection, victory of the other, autonomy.

The "war aims" compiled by German heavy industry in summer 1940 at the request of the Ministry of Economics do not have much in common with those of 1914; they contain no demands for wrenching markets from competitors, forcing mergers, imposing limits on production, or removing plant, machinery, and labor. Sober to the point of dullness, the main issue dealt with by the briefs submitted by the *coal* industry was whether prewar relationships should be reinforced or simply restored. A minority in the mining syndicate wanted to organize a single cartel for all the districts of Western Europe with powers to centralize sales on international markets, eliminate tariff and quota barriers, promote mechanization, coordinate investment, and make adjustments for changes in currency parities and tax structures. The position officially adopted by the collieries was less ambitious. It emphasized the importance of strengthening national cartels in the Netherlands, France, and Bel-

[60] See Milward, *New Order.*

gium and proposed central direction of exports and more efficient routing. The main objective of Ruhr coal in the aftermath of victory in the West was actually to improve relations with Great Britain, whose government was to be advised to "take all steps to encourage an understanding with . . . the German mining industry."[61]

The war aims of the *steel* producers were similarly modest. In early June, the managing director of Vereinigte Stahlwerke, Ernst Poensgen, circularized the Konzerne in order to sound out their acquistion plans. One response, from Hoesch, was perfunctory. Two firms that had made acquistions elsewhere, Mannesmann and Krupp, did not reply. Director Kellermann of Gutehoffnungshütte actually objected to presenting a claim on behalf of his firm and did so only after being told that all the others had already submitted theirs. Only Friedrich Flick submitted plans for a foreign acquisition, the Aciéries de Rombas owned by the de Laurent interests. On 7 June 1940 Poensgen issued a directive defining Ruhr policy for the rest of the occupation. It contained a number of important points. No transfer of mines or mills should take place until the restoration of peace unless "other," presumably German, interests intervened; firms located in Lorraine, Luxemburg, and the Saar were to get priority access to minette, only the surplus being available to the Ruhr; previous German owners were to receive preference in recovering former assets once peace had returned; unexploited ore properties, finally, should be left in the hands of their owners or, in cases where this was not possible, administered by a consortium of German steel producers. In remarks appended to this document, Poensgen added that his personal plans for Lorraine and Luxemburg after "final victory" included guaranteeing minette supplies to traditional consumers; strengthening the coal–ore trade; rationalizing facilities; and integrating French and Ruhr production when justified by cost criteria. These plans presumed the maintenance of traditional relationships during the war and envisaged still closer ones after it.[62]

The inner German struggle for coal and steel hegemony in occupied Western Europe was at its most intense in summer 1940, gradually simmered down over the next year, but was never really settled; there were cease-fires but never a real peace. The first cease-fire established a boundary between spheres of direct and in-

[61] Gillingham, "Zur Vorgeschichte," pp. 390–391.
[62] BA R2/30287, "Eisenhüttenwerke in Lothringen und in Luxemburg," 7 June 1940; GHH 400101306/27, "Wg Esl, Rundschreiben, Tgb. Nr. 11788, R/Mu," 26 July 1940; Rijksinstituut voor Oorlogsdocumentatie (RO), HR13, "Circulaire, Tgb. Nr. 11788, R/Mu," 6 August 1940.

direct German rule. In late May 1940 a close associate of Friedrich
Flick's with good political credentials, the former U-boat captain
Otto Steinbrinck, became the plenipotentiary for iron and steel production in Belgium, North France, and Longwy. The ex-submariner
proved to be a loyal executor of Ruhr policy. To enable local steelmakers to act as subcontractors on large German orders, Steinbrinck
empowered the industrial associations to allocate contracts and raw
material and enforce wage and price guidelines. Otherwise, he let
the locals run their own operations. The big mills of Lorraine were
placed under Hermann Röchling. As plenipotentiary in the district
of Meurthe-et-Moselle Sud, this intensely nationalistic Saar steel dy-
nast often intervened at the plant level. There was a similar division
in coal; a trustee of Göring's administered the Dutch mines as an
adjunct to the Salzgitter works, and representatives of the Brussels
military administration supervised those of Belgium and North
France. To these mines fell the responsibility of provisioning industry in most of occupied Western Europe, but as in steel, overall direction was lacking.[63]

In summer 1940, the annexed territories of Luxemburg and Lorraine along with Meurthe-et-Moselle Sud were the scene of a no-
holds-barred struggle for influence among various contending
German factions. Motives varied. The Göring management needed
to acquire a couple of million tons of additional capcacity in order to
reach a total output potential of 14.6 million tons, at which it would
displace Vereinigte Stahlwerke as the largest overall Reich producer.
Röchling was intent on settling past grudges. The Ruhr wanted to
hold its own and, in a second cease-fire, almost managed to do so.[64]
In this one the Four-Year Plan agreed to accept Ruhr principles of
economic administration: Firms in the occupied territories were to
be put under trustees guided by "economic principles"; a new consortium was to be formed for minette allocation; and property was
not to change hands until after the war, and then only if the trustees' parent firms were given preference in making acquisitions.
These principles ruled out plunder as policy – the removal of machinery, massive displacement of personnel, and theft of assets. The
trusteeship provisions paradoxically even provided the prospective
owner with an incentive to maintain the plant. The actual assign-
ment of trusteeships, which brought about a third cease-fire, was

[63] Gillingham, *Belgian Business*, p. 50.
[64] Nürnberg Industrialists Trial, NI-322, "Usines nationales Hermann Göring:
Participation à l'industrie sidérurgique," 29 May 1941; NI-P1. 115, "Vorschlag
Pleiger zum Aufbau der Hermann Göring Werke," 9 November 1941.

Table 2.1. *Assigned trusteeships in Lorraine and Luxemburg*

Work	Previous owner	Trustee
Lorraine		
1. Karlshutte	Hout Fourneaux et Aciéries de Thionville (Laurent)	Rochlingsche Eisen- und Stahlwerke, Saar
2. Kneuttingen	Société Métallurgique Knutange	Klöcknerwerke, Duisburg
3. Uckingen	Forges et Aciéries des Nords et Lorraine	Neunkirchner Eisenwerk (Stumm), Saar
4. Rombach	S. A. des Aciéries de Rombas	Friedrich Flick, Berlin
5. Hayingen-Moevern	Les Petits-Fils de F. de Wendel	Reichswerke Hermann Göring
6. Hagendingen-SAFE	UPMI	
Luxemburg		
1. Rodingen	S. A.d'Ougrée-Marihaye	Faust/Hahl Venture
2. HADIR	Soc. Gén. – controlled	Vereinigte Stahlwerke
3. ARBED	Soc. Gén. – controlled	

Source: Archives of Gutehoffnungshütte A J, 400101306/27 RWM II EM3-31133/41 III, 21 February 1941.

less than a victory for the Ruhr but left both the Göring and the Röchling interests dissatisfied (see Table 2.1). Thus battle was resumed, this time with ARBED as bone of contention. The Reichswerke tried to take over ARBED both from without (by forcibly acquiring a controlling shareholding position) and from within (by placing its own personnel in key management positions). The Ruhr countered by protecting the firm's officers and bringing the Luxemburg steel giant into the German network of industrial associations. An officially imposed cease-fire of late 1943 finally directed both German parties to respect the status quo. This left the Ruhr on top. In most respects, ARBED was run during the war "like any other German firm."[65]

French firms also benefited from the political and economic protection of the German steel industry. Those in Röchling's area were allowed to maintain luxurious corporate headquarters in Paris, and their officers given continued access to firm records. Trustees generally maintained plant adequately, the Flick delegate at

[65] Gillingham, "Zur Vorgeschichte," p. 394.

Rombas actually investing on a large scale.[66] Even the anti-French firebrand Hermann Röchling is reported to have intervened on behalf of employees arrested by the Gestapo. He also objected to the expulsions of Lorrainers ordered in October and November 1941 by his erstwhile ally, Gauleiter Josef Bürckel. According to several postwar defense affidavits from French steel producers, Ernst Röchling, who represented cousin Hermann's interests in Paris, was a persuasive advocate of Franco-German reconciliation. The Röchling connection was demonstrably useful to France. In one notable deal, Ernst brought together both public and private interests in Germany, France, and the Netherlands in a complicated two-way swap of huge proportions that restored control of the steel firms of Meurthe-et-Moselle to their traditional owners.[67]

Hitler did little to capitalize on the comparatively congenial business atmosphere of the occupation. In summer 1940, however, the Reich Ministry of Economics *did* sponsor pleasant informal conferences of leading businessmen and officials and their counterparts in occupied Europe in an attempt to exploit the sentiments that had given rise to the economic appeasement of the immediate prewar period. Was the time not ripe, the unofficial German chairman hinted, for some updated version of Napoleon's Continental system? This seemed like a good idea to the captains of industry and finance from Western Europe. They agreed on the need to modernize industry on an American scale and to pool resources within a common framework of law and institutions, and to this end they envisaged an industrial parliament for Europe representing national industry associations and cartels.[68] Such dreaming was soon forgotten. When Hitler decided to invade the USSR, the so-called New Order discussions were abruptly terminated. In a cynical last-ditch effort to bind the occupied countries more tightly to the sinking Reich, Speer tried to revive them in September 1943. By this time it was no longer possible for Nazi authorities to stimulate speculative discussion concerning the future shape of a German-led Europe with the big businessmen of the occupied nations over thick cigars and fine cognac. Speer's intimate little "Europe Circle" had no more influence on events than its forerunner of 1940.[69]

Brainstorming by businessmen had little immediate importance because producers, though powerful within their assigned field of influence, were at the mercy of decisions made at the top. Berlin's

[66] Ibid., p. 398   [67] Ibid.   [68] Ibid., p. 402.
[69] Gillingham, *Belgian Business*, p. 92.

negligence in fact made a shambles of France. During roughly the first half of the occupation, restraints on German demand for French goods kept production down and performance anemic; depression-level unemployment prevailed, and inflation grew. Yet there were then no absolute shortages of raw material, and at the prevailing low operating levels there was even enough coal to have raised production. The Reich could have stepped up both its imports and its exports in Western Europe and encouraged the French to stimulate their economy. Revival might, however, have resulted in some form of economic partnership and thereby undermined the verdict of battle: A sensible policy was politically impossible. Things changed after about mid-1942. From then on, the surging demand and rising costs of the greater war combined with the depreciating values of currencies to result in a pell-mell pursuit of whatever goods paper could still command. Recovery still did not occur, however. The problem was structural. A decline in overall Western European coal outputs caused by a collapse of production at the long-overworked mines of the southern basin in Belgium choked off any real recovery, and the increasing share in national product appropriated by the Germans reduced French consumption still further, bringing about grave deterioration in both public health and capital goods.

What did this amount to? According to Sauvy's estimates, the Reich appropriated some 886 billion francs in financial assets between May 1940 and September 1944. This corresponds to about 400 billion 1938 francs, or fourteen to fifteen months of national income at the 1938 level of 330 billion francs. The unadorned statistic does not adequately indicate the human and capital costs of the occupation because during it national income dropped sharply, outputs averaging 32 percent below those of 1938. If one adds to this the 26 percent of national income appropriated by the Germans, the remainder left for the French was only 140 billion francs, or 42 percent of 1938 levels. Though the losses must have fallen disproportionately on stocks and capital goods – a 58 percent reduction in consumption would have resulted in famine – reliable cost estimates are not available for the occupation itself. One must instead use figures derived for the war as a whole. These shed only indirect light on Germany's responsibility for French economic hardships but do provide a good indication of the magnitude of the task facing postwar France.[70] The official French commission on

---

[70] Sauvy, *La vie économique*, pp. 97–99.

war damages estimated total losses at 1,390 billion 1938 francs and
concluded that deterioration of stocks and capital totaled 1,000 billion 1938 francs, or a sum equal to three years of prewar national
income. To reconstitute the capital stock as of 1938 would, in other
words, have taken three years even if nothing were consumed! Under even the most optimistic assumptions, it seemed unlikely in
1945 that France would be able to restore the capital structure of
industry to 1938 levels until the end of 1952.[71] Until such losses
could be made good, Franco-German reconciliation could be nothing more than a pipe dream.

The French began planning for the postwar period in 1940, when
their own war effectively ended. Defeat had a tonic effect: In both
Vichy and London agreement was nearly unanimous that there
could be no return to the pseudoliberal economy of the interwar
years. Not everyone regarded economic growth as a panacea. At
Vichy nostalgic advocates of a peasant-oriented society of virtue at
times enjoyed the patronage of Marshal Pétain. Within the resistance, Communists and reform-minded idealists demanded an economy based on social justice. These groups were on the fringes,
however. At Vichy technocorporatism was the coming thing and,
among the "Fighting French," neoliberalism. Both these schools advocated similar reforms, one of which was to create a strong central
economic directorate to guide industrial modernization. Another
was to organize producer associations to influence and implement
modernization policy. Both Vichy technocorporatists and London
neoliberals wanted to restrict state intervention to a bare minimum
in the belief that over the long run the market was the most efficient available agent of change. The few theoretical differences that
existed between the two groups had little significance in the actual
implementation of policy. Vichy created new institutions for economic modernization, the Free French preferred to live with them
rather than abolish them, and what changed over time was caused
by adaptation to the decidedly different circumstances of the postwar world.[72]

There were two main institutional features of France's political
economy during the Vichy years. One was a new Ministry of Industrial Production (MIP) powerful enough to override the Ministry of
Finance and thereby assure that economic growth received precedence over financial stability. The new administration essentially

[71] Ibid., pp. 100–110.
[72] See Richard F. Kuisel, *Capitalism and the State in Modern France: Renovation and
Economic Management in the Twentieth Century* (Cambridge, 1983).

cannibalized other economic agencies, stripping away their powers
and raiding their personnel. It also adopted an organizational
scheme similar to, though not necessarily patterned on, German
practices of subdivision into functional branches corresponding to
sectors of industry. Though MIP set up the first modern statistical
service in France, it did not live up to the hopes of its succession of
ambitious chiefs (René Belin, Pierre Pucheu, François Lehideux,
and Jean Bichelonne) and never developed into a central agency for
economic planning. It did, however, direct the system of raw materials controls set up to facilitate doing business with the Germans.
Though the ministry did not set national priorities, it could enforce
them through allocation.[73] Its instruments were the so-called COs
and something called OCPRI. Together they comprised the second
institutional feature of France's political economy during the Vichy
years. In August 1940 Minister of Industrial Production René Belin
ordered that each sector of industry set up a Comité d'Organisation
(CO). The new committees had far-reaching powers. They kept statistics, drafted price schedules, and regulated competition, enforcing these dicta through raw materials allocation. In September
OCRPI (Office Central de Répartition des Produits Industriels) was
added to this structure. It linked the COs to the Ministry of Industrial Production and in theory provided central direction to the rationing system.[74]

The CO/OCPRI system was an evolutionary rather than a revolutionary development. Although the old cartels and producer syndicates disappeared after the June 1940 débâcle, their former
managers were sometimes appointed directors of the COs. Even
when this was not the case, the CO chiefs still usually came from
within the industry they administered. In the case of steel, "the
property of the *Comité des Forges,* a trade association dissolved . . .
by decree, devolved on the CO for steel. Managers of the Schneider
steelworks and the heads of other big companies completed the staff
of this CO."[75] Distinguishing the CO/OCPRI system from the earlier
networks of producer associations was not only its rationalized structure but its allocative powers. The new organization and its branches
thus developed into real centers of business decision making,
through which each sector of industry was able within limits to plan
its own development as well as act in unison when dealing with
others with outside interests. The COs grew omnivorously. Twenty-
six were planned and more than 129 eventually created. They em-

[73] Ibid., pp. 133–134.   [74] Ibid., pp. 134–135.   [75] Ibid., p. 138.

ployed over sixteen thousand in "overstaffed, overequipped, over-heated," and overcentralized offices and worked in maddening ways reminiscent of old state bureaucracies.[76] Even their directors were thoroughly sick of them. Yet neither the postwar planners of Vichy nor their counterparts among the Free French wanted to eliminate the COs; both recognized that they had an essential role to play not only during the transition to peace but in modernization policy as well.

The most impressive French wartime industrial planning oc-curred at Vichy under the directions of Délégation Générale à l'Equipement National (DGEN). Headed by François Lehideux, who directed a small group of officials seconded from the ministries and the COs, DGEN produced two major proposals for a future planned economy, the Plan d'Equipement National of 1942 and a 1944 sup-plement to it entitled Tranche de Démarrage. The Plan d'Equipe-ment National was France's first true economic plan. Its emphases were on growth rather than stability, industry rather than agricul-ture, and openness to foreign competition and technologies rather than protection and self-sufficiency. In some 600 pages the equip-ment plan set out industry-by-industry requirements for investment, raw materials, and manpower over a ten-year period.[77] The Tranche de Démarrage dealt mainly with the economic problems expected to confront France during the two-year transitional period following liberation. Its authors were if anything even more radical moderniz-ers than those of the equipment plan. They proposed an immediate doubling of the rate of investment from 13 percent of prewar na-tional income to 25 to 30 percent, in order to make France "the manufacturing plant for the rest of the world."[78] An equipment plan such as the DGEN's was unthinkable without strong central di-rection; the reduction in consumption it envisaged would have re-quired the maintenance of strict economic controls. Once the hardships of war had been overcome, the DGEN planners hoped that the reins could be loosened, with decision-making power de-volving to industry.

The postwar planning done in London and Algiers by the Fight-ing French and in occupied France by the resistance was less ambi-tious in scope than that undertaken at Vichy but reached similar conclusions. In the most important planning effort of the London Gaullists, a 150-man commission headed by Hervé Alphand and in-

---

[76] Ibid., p. 142.   [77] Ibid., p. 147.   [78] Ibid., p. 154.

cluding many other figures prominent after 1945 (among them René Pleven, Etienne Hirsch, André Philip, Georges Boris, and Robert Marjolin) concluded that "The state, acting on nerve centers of the economy, could reach fixed goals without being obliged to intervene at every level."[79] The COs (or smaller bodies), their report proceeds, must be retained for a transitional period, and controls lifted only gradually. Alphand's group, which lacked precise information on developments in France, made few concrete policy recommendations.

The Organisation Civile et Militaire (OCM), a conservative resistance group heavily influenced by industry, had more to offer in this respect. In a detailed study of May 1943 written by CO chiefs of a technocratic persuasion, OCM condemned liberalism as a breeding ground for economic combat and exalted a form of *dirigisme* supposed to be "neither meddling nor bureaucratic" in an economy that can "direct itself . . . along a path laid out in advance."[80] Though a large role fell to the COs in OCM's thinking, its study recommended modifying the statutes of these organizations to include representation for labor and middle management in order to prevent them from becoming "closed corporations." The only additional change they had in view for the COs after liberation involved a purge of politically suspect officers.

The Comité Général d'Etudes (CGE) set up by the leader of the National Resistance Council Jean Moulin in late-1942 to formulate economic policy for the transitional period arrived at the same general conclusions as OCM. The CO/OCPRI machinery would have to be kept, "not only because it was needed" but because, though unpopular, industrialists feared that without it the state would intervene more directly into business affairs. In an article of November 1943 published in Algiers that presented his vision of the future, the CGE's leading light, René Courtin, gave voice to the planning consensus embracing the technocrats on both sides. All Maginot lines, he argued, end up by being overrun: To survive, France needed a progressive economy stimulated by competition, geared to expanding outputs, and run by well-trained workers and managers. Hoping that a new national equipment plan for industry would set France on a fresh course, he emphasized that it must not be imposed by the state but organized by its representatives in collaboration with industry acting through CO-like associations which in time would dis-

[79] Ibid., p. 162.   [80] Cited in ibid., p. 165.

charge planning responsibilities. Courtin's goal was not the classical separation of state and the economy but their virtual interpenetration as partners in modernizing France.[81]

Although an embarrassing, not to say politically *suspect*, fact, neither the Free French in exile nor the domestic non-Communist resistance developed an alternative policy to that of the Vichy technocrats. So what if this meant that the main new features of the Vichysoisse economic scene – a powerful superministry and strong producer associations – remained intact during liberation and, through various permutations, long after? A return to the debased interwar form of capitalism was unthinkable, and it would be years before a more vigorous strain of it could be developed. There was only one real option open to France short of revolutionary transformation, and planners on both sides recognized it: To modernize, state power had to increase, producers organize, and the two work intimately together.

In spite of its losses, France turned a corner during the occupation. The words of Minister of Production Jean Bichelonne, though fulsome, are true: "The Depression that began in the United States in 1929 has forced businessmen and economists to reconsider [their ideas] because . . . doctrinaire solutions have failed." Bichelonne added that, though entrenched liberals had sabotaged successive efforts of French governments to introduce effective cartelization in the 1930s, "the Germans – with their *Fachgruppen, Reichstellen,* and *Wirtschaftsgruppen,* through their Ministry of Economics, but above all as demonstrated by the incontestable successes achieved through such institutions – the Germans, I say, have provided an exemplary model of economic leadership, methodically organized with a view to the attainment of practical results."[82]

Bichelonne stood at an extreme in attributing the reform of occupied France to the good example set by Germany. An institutional convergence in the economies of the two nations is nonetheless undeniable; the war brought previously individualistic French producers abreast of their more sophisticated counterparts in the Reich. Once the cloud hanging over Western Europe after 1945 had passed, the similar institutions erected during the previous five years would

---

[81] Ibid., p. 163.
[82] AA Ha PcC IIa Frankreich, Wirtschaft 601/7976, "Vortrag von Generalsekretär Bichelonne: Die Grundzüge der Wirtschaftsführung in Frankreich" (n.d.); see also AA Ha Pol II Frankreich, Wirtschaft 6a/7976, "Vorträge der Tagung Deutscher Industrie- und Handelskammern und französischer Chambres de Commerce vom 18. bis 21, September in Paris."

make possible a much more formidable and permanent form of Franco-German collaboration than anything dreamed of in either occupied France or Nazi Germany. The agent of this change had little to do with either. He was Jean Monnet, and it was through his experience with "les Anglo-Saxons" that France would eventually cross the divide between economic Malthusianism and political nationalism on one side and industrial modernization and reconciliation with Germany on the other.

2.3   Britain's economic defeat

The British triumphed over the Nazis because they were ready to accept sacrifices above and beyond those the Germans were prepared to make for Hitler. This patriotic belief colors most histories of the United Kingdom in World War II, economic accounts included.[83] Unlike the Third Reich, the story usually goes, Britain aimed from early on to establish the armament in depth required for victory and in fact largely achieved it. Is this not proved by the war's outcome? The realities of the situation, which can be reconstructed quite easily from officially commissioned studies of wartime trade and production, are far less reassuring. They depict a half-hearted and ineffective economic response to the war's challenges owing to the reluctance or inability of those in charge to make hard decisions. The availability of U.S. arms and materiel made it relatively easy for the British not to act; the American bounty could cushion a fall, ease pain, even provide a crutch when necessary. Britain was fighting for its very existence after 1939, but at the same time it might have reversed the long-term decline of its producers. Economically, Britain never recovered from Dunkirk; its industrial and financial survival in the postwar world was at the mercy of the United States.

In the late 1930s the United Kingdom prepared for what military staff planners, borrowing terms from the world of corporate law, inaptly termed a war of limited liability. Under this doctrine, the U.K.'s military obligations did not extend beyond the empire; it simply skirted the problem of Hitler. In 1935 the Defense Requirements Committee determined that Great Britain faced two primary external threats, an immediate one from Japan and a more distant one from Germany. The navy was to be built up to counter the

[83] An exception is Correlli Barnett, *The Pride and the Fall: The Dream and Illusion of Britain as a Great Nation* (New York, 1986).

pressing Asian menace and to protect critical sea links to the overseas territories. It was also supposed to bear the main responsibility for defeating the Germans, who were thought to be highly vulnerable to blockade. Italy was perceived to be of critical importance in Britain's strategy. What would happen if in the event of war Mussolini chose not to remain neutral but instead joined the Japanese and Germans? Might the Italians not be able to control the Mediterranean?[84]

The Royal Air Force was assigned an important if vague role in Great Britain's war plans of the late 1930s. Though the RAF was to be built up to "parity" with the German Luftwaffe, it was never clear whether this was to be measured in numbers, weights, or types or even whether the air arm was to be used primarily to defend British skies or to bomb the Reich. As for the army, it was specifically *not* to be employed against the Wehrmacht in other than symbolic fashion. Plans called for the troops to continue performing their traditional functions, providing garrisoning in remote corners of the world, home defense, and a "well-equipped field force ready to proceed overseas wherever it might be wanted."[85] In 1938 force requirements were set at a paltry five divisions, only two of which were to be made available for use on the Continent.

As in contemporary France, a "doctrine of normal trade" went hand in hand with British military strategy, even more importance being attached to the maintenance of the national currency position. The Defense Requirements Committee explicitly refused to resurrect World War I's powerful Ministry of Supply. Limited liability implied that the United Kingdom was unwilling to defend Europe against Hitler if this meant the sacrifice of either too many lives or empire and tradition. The Germans were not the only perceived threat; victory would have been regarded as hollow if it brought decolonization and free trade, not to mention either Soviet egalitarianism or American productivism.

The war made no fundamental change in this respect. After Munich, and the subsequent disarming of the thirty-two operational Czech divisions, the limited-liability doctrine was an obvious political liability. Though from that point on committed to increasing force levels and armaments outputs, the British made little progress toward their ambitious projected targets. After the outbreak of war, the United Kingdom promised France that it would raise fifty-five

[84] Postan, *British War Production*, p. 83; Hancock and Gowing, *British War Economy*, p. 64.
[85] Hancock and Gowing, *British War Economy*, pp. 66, 67.

divisions, of which the Dominions, India, and prospective allies were to supply no less than twenty-three, but this scheme, an "aspiration rather than a program," was "hedged about by conditions whose fulfillment was not yet in sight and . . . destined for a time which the War Cabinet called vaguely 'as soon as possible.' " This was "no answer to the rearmament . . . achieved by Hitler."[86] Economically, muddling through continued even after hostilities had begun. Though there was a blueprint for wartime reorganization, "the new attitude was not one of hurry," and a "studied avoidance of clear cut principle" was habitual, as were "business first" and "dollar prudence."

The latter remained operative until February 1940, when a machine tools shortage became a bottleneck to increased armaments outputs and no choice remained but to purchase them in the United States. In this all-important sector of production an alarming casualness prevailed:

> Until June 1940 it was left to contractors to order machinery from abroad under individual import licenses and to pay for it under individual exchange control licenses. The orders went unlisted and unrecorded and frequently remained to all intents and purposes unknown to top production departments. Machine tools thus ordered continued to arrive in [Britain] until well into 1941, and the records of these orders and of the number of machine tools imported in 1939 and 1940 . . . remain a gaping void in official British statistics. Even machine tools purchased in the United Kingdom were [unregistered] until the introduction of licensing in December 1940. There was a corresponding ignorance of facts in the production departments.[87]

There was no economic equivalent to the new post-Dunkirk mood of political determination: "For all the fundamental [differences] in Britain's military position . . . the general aims of war production and even the separate supply plans for the three services did not undergo radical transformations. The programs of re-equipment expanded, but . . . spectacular changes were ruled out by the economic position of the country . . . Even the most essential minimum equipment of the Army turned out to be so large as to make it impossible to increase the Air Force as far as strategic plans demanded."[88] No powerful voice for escalating the productivity battle was heard. Certainly Churchill did not alter the fundamentals of

[86] Ibid., p. 96   [87] Postan, *British War Production*, p. 94.   [88] Ibid., p. 128.

British strategy, for instance by planning to open a second front in Europe. His hopes for ultimate victory rested on unleashing new technologies, whose workings he only half understood, and on the power of example. He had great faith that individual acts of derring-do – commando operations, sabotage, and assorted dirty tricks – would inspire the peoples of occupied Europe to rise against the Nazi tyrant. No less concerned than others by dollar shortages, Churchill felt strongly that Britain should manufacture as much as possible of its own weaponry. Beyond this he was disinterested in problems of industrial management and uncritically assumed that the North Americans would do those things economically necessary to win the war. Acting as if unaware of the long-range implications of a failure to modernize British production, Churchill also refused to override entrenched interests in the armaments economy. Though he appointed Lord Beaverbrook minister of supply in July 1941, a role in which the press baron was remarkably ineffective, the prime minister objected to the establishment of an overall Ministry of War Production as duplicating existing services.[89]

Only the handful of international technocrats attached to the North American Supply Commission held a different view. Monnet, Arthur B. Purvis (a Canadian later killed in an airplane crash), and Sir Arthur Salter, an associate of Monnet's both in World War I and at the League, were convinced on the basis of comparative national strengths that for two compelling reasons the United Kingdom had no choice after Dunkirk but to pool resources with the United States. The survival of Britain in the short run depended upon the availability of additional raw material and manpower. Just as surely, its long-term welfare required modernization: The real choice facing the United Kingdom was whether to enter a larger political and economic union with an ally or face domination in the future by an economically more powerful Germany. Monnet recognized earlier and more clearly than could the British themselves that they had become too weak ever to stand alone again.[90] Monnet's pleas had fallen on deaf ears back in June 1940, when he made his famous proposal for an Anglo-French union, and the subsequent attempt to persuade Britain to combine resources with the Americans was only partly successful. By "pooling," the Frenchman meant something total: Allied fighting units should all be similarly equipped and provisioned; administration should be centralized internationally; and the different national economies should be made to function as one

[89] Ibid., p. 143.   [90] Hall, *North American Supply*, pp. 293, 322–323.

on the basis of comparative advantage. The Monnet–Purvis–Salter plans would have caused upheavals within the industrial leadership, revolutionized organizational and manufacturing methods, and created an acknowledged dependence on the United States.[91]

Economically speaking, Britain lost the war after the fall of France. Fifty-five new divisions then had to be raised, and the United Kingdom, unable to supply them, needed American provisionment gratis. By July 1940 it had become evident that the dollar supply would run dry by the end of the year, as in fact happened in early December. Until the passage of the Lend-Lease Act on 11 March 1941, U.S. arms deliveries could be financed only privately and on an interim basis by American banks; to get them to advance funds, the British had to pledge most of their remaining foreign assets as collateral. It was against this background that the Allies developed their grand strategy of supply.[92]

Its instrument, the famous Purvis–Monnet combined balance sheet, never worked as intended owing to resistance of military procurement offices on both sides of the Atlantic. The existence of an American supply supremo in the person of Donald Nelson and the appointment by the British first of Lord Beaverbrook and later of Sir Oliver Lyttleton to serve as his theoretical counterpart on the Combined Boards made little difference in this respect. Instead of serving as a statistical basis for joint planning and allocation of resources, the Purvis–Monnet balance sheet merely guided American attempts to fill deficits in existing British programs.[93]

The single noteworthy common wartime production effort was the Rolls Royce–designed Merlin engine for the Spitfire fighter, but U.S. and U.K. manufacturing methods were simply too different to have made a joint program possible. British craftsmen "cut out the pieces and fitted them . . . , milling them here, machining them there, and piecing them together, until a superb engine materialized."[94] The Packard Company, which took over the Rolls Royce contract, could not build engines this way. The sketchy British blueprints befuddled the American engineers. To put the Merlin into production, they had to tear an engine down and copy it part by part, in some cases redesigning components for mass production. The result was a product which, if no better than the original, could be produced faster, cheaper, and in far greater numbers.

[91] Ibid; Postan, *British War Production*, p. 121.
[92] Hall, *North American Supply*, p. 52.   [93] See ibid., pp. 322–323, 409–410, 414.
[94] Nelson, *Arsenal* p. 226.

Apart from filling gaps in British production programs, the joint staff planners agreed that the United States should take over exclusive manufacturing responsibility for many additional items required for war on a global scale, including "nearly 100 percent of the Allied requirements for transport aircraft, nearly 100 percent of their self-propelled guns, and of 40-ton transporters, and a very high proportion of landing craft, light bombers, tanks and army transport . . . In addition, the Allied merchant shipping over and above the 80,000 to a million tons produced in British yards was to be covered by the United States, and so was a large proportion of the combatant vessels such as the auxiliary aircraft carriers, which could be made by modifying or adapting merchant vessels."[95]

Allied planners assumed that as the war continued the United States, whose rearmament started later than Britain's, would supply increasing proportions of overall Allied requirements, but even they were astonished by the speed at which this occurred. The rapidity of the shift was due partly to the lack of any real breakthroughs in the United Kingdom: "in most firms, the general character of industrial equipment was not . . . radically changed. New and special machines were often grafted on to the more ordinary equipment of firms or, as the expression goes, the equipment of the firm was 'balanced' for arms production. A large proportion of the machine tools supplied during the war years were general purpose tools: lathes and drills, or other unspecialized plant and machine tools used in engineering and metalworking industries." Further, "What [was] true of individual machines [was] truer still of entire workshop units. Much of the specialized equipment was supplied . . . in 'combined units' i.e. complements of machines making up in combination specialized and self-contained workshops . . . These units were an obvious alternative to brand new factories since they made it possible to produce munitions *en masse* by means of special machines, while making use of the facilities of existing firms: their management, their technical experience and staffs, and their tool rooms." The author of the study cited, the eminent economic historian M. M. Postan, concluded that "changes from civilian production to that of making weapons . . . were to a large extent achieved not by the complete reequipment of factories, but by the re-tooling of existing machines and the addition of a relatively few special machines to existing plant."[96] Seldom was a new plant erected. Instead, "work was taken to labor," meaning an existing factory, a policy necessitated by a lack

[95] Postan, *British War Production*, p. 246.   [96] Ibid., p. 395.

of executive talent, skilled labor, and training programs. Thus production remained atomized and inefficient.

This situation was general. Aircraft outputs, both by type and overall, failed consistently to rise according to schedule; in late 1943 the Ministry of Aircraft Production therefore intervened in the management of inefficient firms in order to shift contracts to reliable producers. The average size of U.K. plant was 255 thousand square feet as opposed to two million in the United States. Things were no different in a second supposedly modern branch of British industry, radio tubes: The bulk of wartime additions to U.K. production "was organized and tooled up on lines not radically different from those which prevailed in the industry on the eve of war. More efficient machines had become available but, in general, the processes remained dependent on supplies of skilled labor . . . The rise in military demand was not accompanied by fundamental changes in the method of assembly."[97] Imports of tubes from the United States increased from 1.4 million in 1942 to 17.4 million in 1944. The situation was even more depressing in sectors less closely related to armaments. The components industry was barely affected by wartime developments, certain traditional sectors such as shipbuilding deteriorated in spite of it, and at least one of them, coal mining, nearly collapsed.[98]

In 1936 British experts estimated that, with the existing labor force, 243 million tons of coal could be produced per year and, with additions to the labor force, possibly 285 million tons, leaving some 65 million tons available for exportation. In fact, no more than 228.4 million tons were mined in the coal year 1939–1940, and this figure slipped steadily until 1944–1945, when 182.5 million tons were mined. The declines in production not only wiped out the export surplus but reduced domestic stocks to the point at which only the importation of American coal prevented stoppages throughout the economy.[99] The fall in output occurred in spite of a substantial increase in mine mechanization, with some 61 percent being machine-cut in 1939 and 58 percent mechanically conveyed, as opposed to 72 percent and 71 percent respectively in 1945. The downward trend in production even continued in spite of a doubling of worker cash wages, which made mining the best paid form of industrial labor, and huge increases in the size of the workforce.

[97] Ibid., p. 366.
[98] Ibid., p. 395; Barnett, *Pride and Fall*, pp. 80–81; Court, *Coal*, p. 126.
[99] Court, *Coal*, pp. 29, 32–33.

The slide in fact persisted in the face of the most intense study ever devoted to a branch of British industry, and it did not end even after the government took over managerial responsibility for the mines.[100]

The root causes of the coal problem were bad management and laziness. Mine owners invested only half heartedly in mechanization. The exhaustive 1944 Reid Committee report "questioned the general conception of mechanization that had grown up in this country between the wars. [Management] regarded it as fruitless, because it neglected the important relations between face mechanization and the organization and layout of the mine as a whole, and not least because of its failure to grapple with problems of underground transport."[101] The government did not help matters by neglecting to devise a wartime fuel policy. As the official history describes the matter, "It was a defect of the pre-war plans that they made no provision to link coal control with the control of other fuel and power industries and . . . fuel and power controls with the Regional Defense System." Inadequate peacetime methods of distribution therefore did not change, ancient handling and shipping facilities were repaired but seldom replaced, and statistical services were "impoverished and . . . presented a grave handicap to any precise and comprehensive forecasting of the potential problems of coal control in wartime."[102] Technical reform faced a still greater obstacle in miner resistance to the introduction of machines, one grounded in his "memories of long years of unemployment and under-employment and upon his deep-rooted suspicion of mine managements . . . The coal industry had been struggling for many years with difficult, sometimes overwhelming, problems and the inhabitants of the mining villages bore many traces of the conflict. A long and often losing battle had left moral and intellectual scars in thousands of individuals, not less real and sometimes much longer lasting than . . . physical injuries. The experience of a generation had deeply affected the thinking and emotions of everyone connected with the mines."[103] Though an industry at its worst, coal mining presented in clear relief more general shortcomings of British production. Its own special problems bade ill not only for the foreign payments position of this import-dependent nation, but for the U.K.'s traditional customers in Western Europe as well. The collapse of the British mines was even bad news for the Germans; the Ruhr pits would soon be super-

---

[100] See Barnett, *Pride and Fall*, pp. 63–86.   [101] Court, *Coal*, p. 281.
[102] Ibid., p. 33.   [103] Ibid., p. 26.

vised by those demonstrably unable to solve their own problems. The troubles of British mining set the stage for the postwar coal crisis.

This could have troubled the Allies very little on V-E Day: The economic events of the war cannot be compared in importance to those on the field of battle. Britain nevertheless emerged from the war precariously dependent on a United States in whose global vision there was little place for the empire and still less for "tradition." Anglophilia and the sentimental bond of comradeship in arms could not conceal this fact forever. The special relationship that postwar British statesmen expected would restore some easygoing world of beneficent colonialism was merely an illusion, though one strong enough to have prevented Britain from ever becoming a permanent part of a post–World War II Europe in which its presence would temporarily loom very large.

2.4   The greater war is launched

The greater war started shortly after the fall of France and over a year and a half before the führer ordered Albert Speer to gear up the German economy for maximum effort: It was then that Roosevelt decided to commit the United States to war against Hitler. He never said this of course. The mood of the country was still isolationist. Yet there can be no doubt about his intentions. In July 1940 he sent to Congress the first of the huge appropriations bills necessary to build up the armed forces, including a new air arm of fifty thousand planes, then set up a national committee to oversee the mobilization of the economy, appointing powerful Republicans to serve as the new secretaries of war and navy.[104]

The greater war did not begin with the development of a new grand strategy – for the United States then lacked anything more than the rudiments of one – but was launched instead on a tidal wave of government purchasing. The demand created by Washington stimulated investment into vast new factories able to produce weapons and other war-related goods in far less time and at a fraction of the cost of the other belligerents. These increases in productivity did more than merely give the United States an edge in the war; they made it possible to engage in combat on a scale and in a manner not open to other nations, and, while doing so, support much of the rest of the world militarily and agriculturally without

[104] Nelson, *Arsenal*, pp. 81–82.

reducing standards at home. The growing economic might of the United States also fathered new evolving technologies and strategies that would foreshadow a far worse war in the future. Roosevelt's decision to arm doomed the Third Reich, eliminated Germany as a great power for a generation, and interrupted Europe's long period of world ascendancy. It also sowed the seeds of a future era of material abundance whose amplitude even the Continent's most ardent modernizers hardly suspected.

The person in charge of converting U.S. industry to war production after summer 1940 – Roosevelt's Speer – was Donald M. Nelson. If any one person deserves credit for the achievements of the American armaments economy, it is this rubicund, infinitely patient, once much abused, and since largely forgotten man from Hannibal, Missouri. A chemist by training, Nelson began his career at Sears, Roebuck and rose steadily through the ranks: from manager of the boys' clothing department to head of the men's working clothes department, and from thence to general merchandising and eventually to promotion as executive vice-president of the giant department store chain. This civilian background qualified Nelson perfectly to serve as Roosevelt's chief buyer. The head of purchasing at Sears had to know "where to get – and at the right price – 100,000 different articles of merchandise, and [in addition] something about the manufacture of major lines – or to be able to find out in a hurry. He and his assistants had to survey factories and, if necessary, reorganize their production methods, and even bring about the establishment of new factories in order to get the needed items in the right place at the right time.[105] His staff, Nelson added, bought "everything from the finest thread to the bulkiest heating plant, from drugs to fertilizers, and if there [was] procurement trouble in any department, he [had to] investigate and decide how to improve production methods, lower costs, and get a better and more profitable piece of merchandise."[106]

Nelson was the single person most responsible for American adoption of an expansionist policy toward war production. An "all-outer" convinced that the United States had no choice but to prepare itself for total war while also arming the nations of the world, as head of the war production board (then known as Supply, Priorities, and Allocations Board, or SPAB), he argued for the immediate conversion of industry to weapons production, a long-range policy of stockpile accumulation, and a "firm and deep" organization of the war econ-

[105] Ibid., p. 63.   [106] Ibid.

omy. As wise as this policy would later seem, it encountered intense resistance from both the War Department and private industry. The brass still had no accurate estimates of the material requirements of the war and feared losing control over weapons design if the economy grew too fast. Like their counterparts elsewhere, American big businessmen had been plagued by overcapacities for a decade and did not want to create problems in the future. Enjoying unaccustomed prosperity, they also ,vehemently objected to shutting down civilian lines of production. Why not leave things as they were? Hadn't the president specifically disavowed any intention of involving American troops in a European war?[107] Nelson's key allies in the ensuing struggle with industry were certain prominent New Dealers, labor, "the professors," and Monnet, who since his arrival in the United States had begun "preaching the gospel which was to find expression in the 'arsenal of democracy' phrase: the idea that the Axis could never be whipped unless the enormous productive resources of the United States were put to work to the fullest extent possible to produce the goods which Hitler's enemies needed."[108] Neither Nelson nor the "less than all-outers" had any real notion as to what would be needed to win the war. Nor did Monnet. His balance sheet (Combined Statement) did, however, provide a method of estimating this. In summer 1941 the head of Nelson's statistical section, Stacy May, who had become a disciple of Monnet, derived the first rates for U.K. consumption of military and naval equipment. When plugged into planned force level increases, they pointed to the need for a mobilization far greater than anything imagined possible: outlays of $150 billion by spring 1944, a doubling of planned expansion of capacity, and eventual expenditure of one-half of the national income (as opposed to the one-fifth then being spent on armaments).[109] These conclusions were not reached solely on the basis of dispassionate statistical analysis. Nelson, Monnet, and of course the British had every possible reason to inflate requirements.

There was, however, little the less than all-outers could offer as an alternative. The army and its producer allies were suspicious of economists and statisticians, one industrialist actually suggesting to Nelson that it might be a "fine idea [to] lay all the economists of the world end to end."[110] Lacking statistical data of their own to use as

[107] Ibid., p. 125.
[108] Bruce Catton, *The Warlords of Washington* (New York, 1948), p. 46.
[109] Monnet, *Memoirs*, p. 173.   [110] Nelson, *Arsenal*, p. 129.

ammunition, the less than all-outers could only listen in flabber-
gasted disbelief when shortly before Christmas 1941 Lord Beaverbrook, reciting information fed him by Monnet, told Roosevelt that
to defeat Hitler the Allies would need 45,000 tanks in 1942 rather
than 25,000, 24,000 fighters rather than 5,000, three times as many
antitank guns as thought to be necessary, and so forth. Two weeks
later, in January 1942, the president announced new official output
targets far above Monnet's. These figures sounded as if they were
pulled out of the wind: 60,000 aircraft of all types would be produced in 1942 and 125,000 in 1943; 45,000 tanks manufactured in
1942 and 75,000 in 1943; and eight million tons of new shipping
laid down over the same two years.[111] Such extravagant projections,
ridiculed by Hitler at the time, would be easily surpassed. The Victory Program thus conceived by Monnet, brought into the world by
Nelson, and baptized by Roosevelt would enable the United States
to win the war with comfortable margins to spare.

Though Nelson set the process of mobilization in motion, he
never became "arms czar" because in Roosevelt's scheme of things a
civilian, though useful for providing bipartisan consensus leadership
and bolstering public support for the war, was *not* supposed to interfere with production except when absolutely necessary.[112] The successive directing agencies of the war economy were politically
"strange mixed grills." In the person of the president of the gar-
ment workers' union, Sidney Hillman, labor was represented on
these boards as was small business, through the former brain-truster
Leon Henderson and the radical populist Maury Maverick of Texas.
Yet big businessmen ruled the roost at the central agencies of the
war economy, the Advisory Commission on National Defense, the
Office of Production Management (OPM), SPAB, and the eventual
War Production Board (WPB). They determined what got produced
where, by whom, with what, and at what prices.[113]

The ambiguity built into Roosevelt's approach to economic administration turned Washington into an almost unbearable hotbed of
bureaucratic maneuvering and manipulation, intrigue, misrepre-
sentation, defamation, double-crossing, and outright sabotage of
the war effort. Nelson lost nearly every turf battle. He failed in
late 1941 in a bid to assert overall control over raw materials allocation and had to retreat in early 1942 from a tentative attempt
to wrest purchasing authority from the armed services, something

[111] Monnet, *Memoirs*, p. 173.   [112] Catton, *Warlords*, p. 100.   [113] Ibid.

that might have created a real American Ministry of Supply. The campaign of his deputy Leon Henderson to create labor- and government-dominated supervisory councils to run the automobile industry went up in smoke, as did the intensive lobbying of Maverick and others to guarantee small business a share in government contracting.[114]

Most important of all, Nelson lacked the muscle to control his own organization, and in particular his codirector William S. ("Big Bill") Knudsen, the former president of General Motors. In addition to being immensely powerful within industry, Knudsen was politically unassailable. An oversized, rumpled-looking, horny-handed, self-educated Danish immigrant who spoke in a picturesque down-home idiom still punctuated by the accents of his native land, he neither looked, talked, nor acted like the capitalist of caricature, instead fitting conveniently into a different American stereotype, the "production man." The press thrived on invidious comparison between him and that lesser being, the "distribution man," as exemplified by Nelson. Personally decent to a fault, the Sears, Roebuck executive found it hard to dislike Knudsen in spite of the General Motors man's questionable loyalty to his boss. Lavishing praise on Big Bill and often pillorying Nelson, the media helped see to it that the automobile industry ran the U.S. war economy.[115]

The U.S. war economy took on an appearance similar to others, with alphabet agencies, prioritization, allocation schemes, rationing, production directives, and a great hustle and bustle between Washington and the field. But neither Nelson, his organization, nor for that matter anyone else

> did very much czaring as far as the actual production process was concerned. America's enormous volume of war production grew out of a very few basic actions. WPB put progressive restrictions on the [kinds of] goods industry might make and the materials it might use – it said "no more automobiles," for instance, and it kept aluminum out of the hands of producers who weren't in war work and the armed services flooded industry with enormous orders for munitions. WPB, in turn, then set up the machinery to distribute materials and component parts and did various things to make sure that there would be enough of them to meet essential needs. All the rest was up to industry . . . Even at the height of the war, government did not – except, per

[114] Ibid., p. 121.   [115] Nelson, *Arsenal*, pp. 22, 92–93.

haps, by the standards of a few Neanderthal diehards – go in
for telling the businessman how to run his business. The most
striking thing about the whole war production program was not
that there were so many controls but that all of them fell within
the established patterns of industry.[116]

It must be added that the patterns were those of the automobile
industry in particular. The wartime expansion of the motorcar producers was phenomenal. Among other things, they built most
American aircraft manufactured during the war. "Bill Knudsen,"
Nelson relates in his memoirs, "had been on the coast looking
over . . . airplane plants, and had stopped off in Detroit on his way
back to Washington. He had the idea we just couldn't get into quantity production . . . if we depended solely on the methods and manufacturing equipment of the plane industry." On 25 October 1940
he therefore called a meeting in Detroit that brought "together just
about everyone who had anything to do with automobile manufacture: the primary producers, the parts and appliance makers, the
tool and die makers [and] so many attended that the meeting had to
be held in a fancy meat market, which had closed up . . . with the
guests sitting on chairs supplied by undertakers."[117]

In these rather grim surroundings, the automobile men looked
over a few samples of structures Knudsen had brought and concluded that they could indeed make such parts in their own factories. All the big automotive firms thereupon agreed to begin
manufacturing airplane parts, except for Ford. Its representative,
Charles Sorenson, insisted on all or nothing. A couple of weeks after the meat market conference Sorenson "visited an aircraft plant
where B-24s were being assembled. Men were crawling over fuse-
lages, getting in each other's way, making scores of useless gestures,
and occasionally hammering one another over the head. The whole
procedure was a negation of mass production."[118] While gaping at
these strange antics, Sorenson made a hurried drawing for the biggest aircraft factory in the world, Willow Run. It would become a
veritable gargantua of aircraft production, larger than the combined
prewar plants of Boeing, Douglas, and Consolidated. The automobile industry produced over half (by value) of total American wartime aircraft output of 300,000, entered into the manufacture of
weaponry, both large and small, and of course also made tanks, mil-

[116] Catton, *Warlords*, p. 121.   [117] Nelson, *Arsenal*, p. 219.   [118] Ibid., p. 220.

itary trucks, and other wheeled vehicles. It was directly responsible
for more than one-third of the total American armaments output by
value.[119]

Knudsen himself largely settled "who got what," often with very
few words. He decided for instance that the contract for a recently
designed, lightweight, four-seater all-terrain vehicle should go not
to Ford as the army wished but to a smaller manufacturer named
Willys. Knudsen summoned the brass to a meeting, forced them to
agree that he knew at least "something about producing motor
cars," then told them, "Gentlemen, *this jeep is a motor car*, and I say
Willys can make it!" Willys got the job.[120] The motor men did not,
as they sometimes bragged, single-handedly win the war, but Detroit
did direct the conversion process. This was sensible, given the sheer
size of the automobile industry. It employed 500,000 persons directly and another seven million indirectly; consumed the bulk of
the nation's production of leather, glass, steel, rubber, cotton, cop-
per, chrome, aluminum, and soy beans; and provided the largest
market for the nation's second-biggest industry, petroleum. The
car manufacturers also had access to huge reservoirs of technical
and mechanical talent. Between September 1939 and 1941 Detroit
added over $1 billion in plant. To supply the motor industry or
compete with it, other branches had both to expand on a comparable scale and to imitate its methods. These included a general pooling of patents and processes, the standardization of design and
components, the upgrading of labor skills, and, paradoxically, the
promotion of competition in production methods.[121]

Once Detroit had entered the aircraft business, the small, largely
artisanal, highly dispersed producers of airframes and engines had
to organize in order to survive. National committees, with branches
on each coast, formed in 1940. By late 1942 they had evolved into
two Aircraft Production Councils, one for the East, the other for
the West. Within both regions intraindustry cooperation developed
far beyond anything known even in Germany. This included a pooling of patents and processes but also a sharing of raw materials,
parts, labor, and machine time. The industry handled in common
such policy matters as subcontracting, reduction of manufacturing
time, quality control, new technologies, substitution of materials,
training of female labor, general manpower problems, and inventory
practices.

[119] Ibid., pp. 224–225.   [120] Ibid., p. 178.   [121] Ibid., pp. 212–213.

By March 1944, "more than a million engineering man-hours had been saved by the pooling of research data and production techniques alone. More than 9000 technical reports were exchanged on the West Coast. More than 26,000 items were exchanged, plus [thousands of additional] items furnished to producers [outside of the aircraft industry.]"[122] Production was almost well enough integrated to have justified Nelson's boast that "our aircraft industry became in a very real sense a·unit."[123] It was undeniably no longer a small-scale affair. From less than fifty thousand in 1939, employment rose to over two million by November 1943, or one-twelfth of the nation's workforce; floor space expanded from roughly 13 million square feet to 167 million between January 1940 and December 1944; and the gross value of aircraft production mounted from $114 million to $3,006 billion.

Table 2.2 indicates the explosive increase in the value of American armaments outputs during the war. War outlays were the main source of the growth in gross national product (GNP) from 1939 to 1944 (from 88.6 to 135 billion or 52 percent) but not the only one. Consumer expenditures rose as well, whereas nonwar government expenditures and nonwar capital formation held remarkably steady. Overall, the nation's productive capacity grew by about 50 percent. The expansion was by no means limited to the munitions industry but included iron and steel, synthetic rubber, aviation gasoline, aluminum, magnesium, and machine tools. American wartime outputs per man hour were twice those of Germany and five times those of Japan.[124]

Combined with an abundance of resources, this superiority enabled the United States to produce about 40 percent of the world's armament in 1944 (including over a quarter of the United Kingdom's), while devoting 40 percent of GNP to the military sector as opposed to 50 percent in the U.K. and even higher percentages in Germany. The United States also increased agricultural exports during the war (which in addition occurred against a background of rising domestic consumption), providing enough to support Anglo-U.S. troops stationed throughout the world, and much of the Red army as well, and to prevent famine in both the Middle East and liberated Europe. Alan Milward, the author of the one comprehen-

[122] Ibid., p. 235.   [123] Ibid.

[124] *Industrial Mobilization for War: History of the War Production Board and Predecessor Agencies, 1940–1945*, vol. 1: *Program and Administration* (Washington DC, 1947; rpt., 1969), pp. 962, 963.

Table 2.2. *Production of munitions items by type, July 1940–August 1945*
(in millions of standard 1945 munitions dollars)

Item	1940 (July–December)	1941	1942	1943	1944	1945 (January–August)	Total	Percent of total
Munitions total[1]	2,047	8,442	30,168	51,745	57,594	83,153	183,149	100.0
Aircraft	370	1,804	5,817	12,514	16,047	8,279	44,138	24.5
Ships	391	1,852	6,957	12,498	13,420	6,011	41,138	22.5
Guns and fire control	78	355	1,794	3,180	2,926	1,471	9,804	5.3
Ammunition	87	427	2,743	4,908	5,768	4,173	18,106	9.9
Combat and motor vehicles	238	1,285	4,778	5,926	4,951	3,138	20,316	11.1
Communication and electronic equipment	27	226	1,512	3,043	3,739	2,212	10,759	5.9
Other equipment and supplies	856	2,493	6,567	9,676	10,734	7,869	38,195	20.8

[1] Excludes net increases in naval stock fund value of goods in store and stock in transit between supply offices, as follows: July–December 1940 (28); 1941 (149); 1942 (320); 1943 (613); 1944 (148); 1945 (68); cumulative, July 1940–August 1945 (1,326).

Source: Industrial Mobilization for War: History of the War Production Board and Predecessor Agencies, 1940–1945, vol. 1: *Program and Administration* (Washington DC, 1947), p. 962.

sive survey yet written on the international economy in the period 1939–1945, quite reasonably concluded that "the productive effort of the United States . . . may have been the most influential consequence of the Second World War for the post-war world."[125]

Europe's fate was now in the hands of the United States. Arriving unsought and unexpectedly, the immense increase in American power stemmed less from either the weaknesses of other nations or the dramatic economic growth of the domestic economy than from the advanced production methods and technologies developed to win the war. They had opened a new era of exponential increase in mankind's potential for good or ill. The next generation of Europeans had no choice but to meet the challenges of this age or succumb to them. The welfare of the old civilization depended in the meantime not upon *whether* but upon *how* the young one exercised its new authority.

World War II provided no cure for the Ruhr Problem but made a severe recurrence of the virus a virtual certainty. The economic verdict of the European war once again differed from the military: France was the loser, Germany the winner. The French had lost Britain as a coal supplier and were more dependent than ever on the Ruhr, most of whose mines were not fully operational. A crippling fuel shortage was in store, of which the French would be the likely victim. But even were it overcome the problem of steel competitiveness would remain. The nation's industry had languished for four years. Coming on the heels of a decade in which France's relative economic performance had steadily declined, the occupation drained and disrupted the economy and aggravated the already serious mood of national discouragement. The war affected the ex-Reich differently. Though the exhaustion of the mines, together with the wreckage of the transport system, impeded manufacturing, German industry remained the most productive in Europe. Its managers had caught a good glimpse of the greater war, were reasonably well attuned to its requirements and confident that if given a chance they would soon be able to pick things up where they had been left in 1939. The Franco-German imbalance of economic power was even greater than before.

Yet the war also created new means of treating and eventually eliminating the ailment. One was to work through the corporative tradition, whose strength had increased during the war. The nationalist zeal that after 1914 prompted the directors of German heavy

[125] Milward, *War, Economy, Society*, p. 163.

industry to disenfranchise, disperse, dismantle, destroy, or otherwise trash their counterparts in occupied Western Europe was not present during World War II. Whether through kindness, calculation, or mere disinterest, they left their former competitors largely to themselves except when, partly out of concern for their own position, protecting them against the assorted depredations of Göring and his fellow spirits in Nazi industry. This successful defense effort enabled a long-term evolution toward organized capitalism on the German model to continue in France and elsewhere. Its logical outcome was the founding of a bigger and better steel cartel. Another possibility was that France might someday become strong enough to overcome the virus that the Ruhr problem had become in the life of the nation. This must have seemed unlikely. The modernization-minded elites were islands in a Malthusian sea and managerially in limbo. They knew almost nothing of the greater war. One Frenchman *had* seen the future and understood how it would unfold. This was Jean Monnet, who returned home from wartime Washington both confident that the lessons of the American war economy could be applied to France and supported by the men with the power to shape the postwar world. Monnet faced an immense challenge, of which industrial renovation and reequipment were the lesser part. To modernize France he had to win American backing for a scheme of European political and economic reorganization that would guarantee a cheap and reliable supply of coal. Only then could France have breathed more easily. Finally, there was the chance that the very technologies and production methods developed to win the war could produce more potent new viral strains, and likewise more powerful inoculations, that would overwhelm those that had previously debilitated the patient: Coal and steel were no longer decisive in the nuclear age that had dawned.

3

From Morgenthau Plan to Schuman Plan: the Allies and the Ruhr, 1944–1950

To travel from Morgenthau Plan to Schuman Plan is to move from an American wartime proposal to destroy Ruhr industry to the actual organization of Europe around it little more than five years later. It is also to cross from the threshold of one short and unhappy era into a far longer and better one, passing from the material devastation, political chaos, intellectual confusion, and moral doubt left in the wake of the war to the recovery, reorganization, and renewal of confidence that would lead the way to more than a generation of unbroken peace, increasing prosperity, and improved social welfare. Between 1945 and 1950 changes normally requiring decades to develop and unfold took place within the space of only a few years, too rapidly to be fully recognized at the time or completely understood even now. The Germans had to give up their national ambitions, accept the military and moral verdict of the war, and somehow draw the appropriate lessons from it. The French had to force themselves out of a national mood of fatalism and complacency – a habitual shoulder shrugging – and endure the discomforts, pains, and occasional agonies of modernization. The United States had to play successfully the new historical role that World War II had thrust upon it as the only remaining great power in Europe west of the Elbe.

The United States laid out the route that led to a German return to Europe and set the pace for the voyage along it. This was the only option; no other nation could have served as guide. Diplomatic passivity was the lot of the defeated and downtrodden ex-Reich, both France and Britain were critically dependent on American aid, and the Soviets needed to rebuild their battered economy and consolidate their power in Europe. Aware also that any attempt to extend its power west of the Elbe would encounter the united opposition of the three western Allies, the USSR could at best only hope to occupy a blocking position. To influence the settlement in Western Europe, the Soviets, like France and Britain, would have to manipulate the all-powerful Americans.

In the five years after the defeat of Hitler, American policy toward Europe shifted, first tentatively and uncertainly and later more confidently and with increasing sense of purpose, from disengagement to constructive commitment. The United States entered the postwar era with a policy long on noble objectives and short on everything else. Roosevelt's Grand Design was intended to prevent a recurrence of conflicts like those that twice during the century had caused American armies to intervene on the Continent by framing a peace settlement around trade liberalization and international mediation. Enshrined in the Atlantic Charter, which the American president pressured Churchill to sign in August 1941 as precondition of American participation in the war, these principles were supposed to serve as beacons to guide Europe out of an age of economic and political nationalism without requiring the United States to abandon its traditional noninvolvement in the affairs of the Continent. Implicit in the scheme were economic alliance with Britain anchored in mutual commitment to free trade and a military counterpart with the United Kingdom and the USSR based upon respect for national self-determination.

Though animated by sound principle, the American approach had little value as a guide to action. In addition to taking scant account of the ideas and interests of the Allies whose cooperation was needed for success, it overlooked the problem of Germany, without whose economic rehabilitation European recovery and stabilization were impossible. Domestic politics eventually stood in the way of the formulation of a sound U.S. policy towards the enemy nation: though the infamous Morgenthau Plan for the destruction of industry first proposed in September 1944 overrode arrangements then being prepared in the State Department that would have set the stage for the return of a reformed Germany to Europe, it was too drastic to have been adopted full-scale. A bureaucratic compromise resulted, Joint Chiefs of Staff Directive No. 1067 (JCS 1067), which was both harsh and unworkable. Neither this constitutional charter for U.S. occupation government in Germany nor the Allied agreements regulating policy toward the ex-enemy provide the real clues to American behavior in the immediate postwar years. It resulted from impossible attempts to appease public opinion in the United States on the one hand and to make Four Power government work on the other. America left the ex-Reich divided but with Europe at the same time facing an untrammeled revival of the German economic power.

The failure of the Grand Design left a policy vacuum; no alternative approach of comparable attractiveness, comprehensiveness, and simplicity was readily available, and a substitute would take years to devise. The Marshall Plan provided the way out of the impasse. Presented on 5 June 1947 as a mere statement of intent, the policy named after the American secretary of state contained the gist of a fundamentally new departure, though one basically consistent with Roosevelt's aims. Marshall committed the United States to restoring European prosperity on the condition that the nations of the old civilization devise an overall recovery program that could serve as a stepping-stone to eventual unity. Marshall cut the United States adrift from globalism as well as the tradition of noninvolvement in the internal affairs of nations outside the hemisphere while pledging to create a new political unit, "Europe," as a transatlantic counterpart. Yet he did so in the hope that the United States could then disengage from its new responsibilities and concentrate on reform at home.

The outlines of the new aid plan were vague. No one knew who would belong to "Europe," how prominently the United States would figure in its affairs, or whether it would even develop along lines acceptable to either Washington or the American public. The "federator" would be of crucial importance. Although the Marshall Plan called for initiatives from Europeans, the United States needed the support of some nation, or other authority, that in advancing the unification process could mediate between the two continents. The Americans first turned to Britain for help. Still far from ready to abandon the imperial tradition and join the Continent but weaker than could be openly admitted, the United Kingdom could ill afford to refuse the United States outright. For two years the British successfully finessed but in so doing nearly wrecked the Marshall Plan as an integration scheme. In desperation, the United States finally turned to France which, though (thanks to Monnet) officially committed to modernization, trade liberalization, and eventual political federation, was thought to be too weak either to contain a revived Germany or to exercise leadership generally. Not until after the forced devaluation of the British pound in summer 1949 did the United States first begin to regard the French as the potential leaders of the unification movement, and only after the actual announcement of the Schuman Plan on 9 May 1950 did the United States fall behind the French scheme for reknitting Germany into the fabric of Europe. In so doing it fulfilled the promise of the Marshall Plan.

3.1   The not-so-grand design

President Roosevelt was partly responsible for the problems of American policy after 1945. Once the United States had declared war against Germany and Japan, he habitually turned away the pleas of his advisers for serious top-level consideration of a future peace settlement on the grounds that it would divert attention from the primary task of defeating the enemy.[1] Though in late 1943 the president did authorize the commencement of planning for the German treaty, the only real exception to his practice of deferring such matters until after the war was his short but ardent embrace of the Morgenthau Plan in Septmber 1944. It had serious consequences. The Morgenthau Plan provided ammunition to Goebbels, confused the British and Russians, greatly added to underlying tensions in American policy, increased the miseries arising from the occupation, and delayed European recovery.

By 1944 the United States had already developed a program with Britain, and supported at least nominally by the USSR, that either with or without the outbreak of the cold war would have reduced the vast and complex task of German and European reconstruction to more manageable proportions, resulted in more rapid recovery, and reduced Germany's alleged invidious trade advantages. The plan's two main themes, reconstruction combined with reform and German recovery in a European setting, would recur in several significant State Department initiatives of the postwar period. Standing in the way of their realization were not only the policies of the Soviets but forces at work in domestic American politics.

The reconstruction plan grew out of the activities of the so-called European Advisory Commission (EAC) set up after the October 1943 Teheran meeting at which the Big Three agreed to divide the Reich into separate zones of occupation. Beginning in early 1944 the EAC met in continuous session under the chairmanship of American ambassador in London John G. Winant in order to work out German surrender terms, allocate occupation zones, and devise long-range policy.[2] To support its work an interdepartmental Executive Committee on Foreign Economic Policy under Assistant Secretary of State Dean Acheson was formed in early summer 1944. A

---

[1] See John L. Gaddis, *The United States and the Origins of the Cold War* (New York, 1972), p. 13.

[2] Ibid., p. 105; Harry S. Truman Library (HST), Interview with James W. Riddleberger, 24 June 1971.

memorandum entitled "General Objectives of United States Economic Policy with Respect to Germany" summarized its recommendations. It called for reform followed by reintegration into a liberalized Europe. Germany was to be required to make restitution and offer reparation; its war industry would be destroyed, and its economic domination of the Continent ended by unspecified means. The German people would neither be forced to endure living standards below an "acceptable minimum" nor remain permanently impoverished, since "an indefinitely continued coercion of more than sixty million technically advanced [persons] would at best be an expensive undertaking and would afford the world little sense of real security." The Reich was rather to be "democratized," and the "organization and conduct of [its] economic life [to undergo] fundamental change," enabling it to be "integrated into the type of world economy envisaged by the Atlantic Charter."[3]

To Secretary of Treasury Henry Morgenthau, who until then had dealt only marginally with German matters, the recommendations of the Acheson committee amounted to a prescription for a third world war. In Morgenthau's view such a catastrophe could be prevented only by destroying Ruhr industry. His proposal was unblurred by ambiguity: What it demanded was quite literally the dynamiting of all factories and the flooding of all mines. Nothing was to be left standing or rebuilt. As Morgenthau saw it, no choice remained but to restore the economic conditions of 1860. Protestations that thirty million fewer Geremans had lived at that time left him unmoved. Nor was he touched by the spectacle of their being doomed indefinitely to a soup-kitchen existence, troubled by the high costs his plan would inflict upon the United States and United Kingdom, or even particularly concerned that without the Ruhr Europe would be condemned to endure another decade or two of depression.[4]

At the Quebec conference of September 1944, Roosevelt forced a reluctant Churchill, at the time deeply worried by the prospect of losing Lend-Lease aid, to append his signature to the Morgenthau Plan, and it thereupon became the official policy of the United Kingdom as well. The vehement opposition of Secretary of War

[3] Foreign Relations of the United States (FRUS), 1944/I ECEFPD-36/44, "General Objectives of United States Economic Policy with Respect to Germany," 14 August 1944, pp. 278–279.

[4] John Morton Blum (ed.), *From the Morgenthau Diaries: Years of War, 1941–1945* (Boston, 1967), pp. 343–369; Gaddis, *United States and Cold War*, p. 118; HST, Interview with James W. Riddleberger, 6–7 April 1972.

Henry Stimson, the most highly respected American statesman of the generation, soon called the policy status of the document into question, however. Roosevelt never actually repudiated the Morgenthau Plan but, rather than risk dividing the cabinet, chose not to act, allowing policy struggles to be waged at a lower level.[5] JCS 1067, which would serve for two years as the charter of the American military government of Germany (OMGUS), emerged from the fray. The directive represented a War Department effort to compromise with the Morgenthau group in the Treasury Department. Limited in theory to the military requirements of the occupation, JCS 1067 skirted the central dispute within the administration; it contained neither guidelines for the future integration of Germany into Europe nor plans for the destruction of industry but was, however, in essence punitive. The Germans were held responsible for having brought misery upon themselves by their own misbehavior and therefore were to be denazified, disarmed economically, demilitarized, and forcibly "democratized." Far from initiating general recovery from the war, the military government was to limit its economic interventions to the minimum necessary to "prevent disease and unrest."[6] Harsher in tone than either its British or its Soviet counterparts, this narrow, ill-considered, and generally unenforceable directive offered a poor guide to action and left policy subject to an unwelcome amount of domestic political pressure.

Harry S Truman had served a total of eighty-one days as vice-president before taking the oath of office on 12 April 1945. The man from Missouri had been to Europe only once, as commander of an artillery battery in 1918, and lacked any previous experience in foreign affairs. There was no one to whom he could turn for immediate help. To the extent that Roosevelt had not been his own secretary of state, he had relied on advice from his "kitchen cabinet," whose unofficial leader, Harry Hopkins, was terminally ill. The nominal secretary of state, Edward Stettinius, was photogenic but in every other respect unsuited for the office. The State Department, traditionally a small-scale affair, lacked experience in economic policy planning, much of which was conducted during the war by a

---

[5] Henry L. Stimson and McGeorge Bundy, *On Active Service in War and Peace* (New York, 1947), p. 576–577.
[6] John H. Backer, *Priming the German Economy: American Occupational Policies, 1945–1948* (Durham NC, 1971), pp. 17–18; Nicholas Balabkins, *Germany under Direct Controls: Economic Aspects of Industrial Disarmament, 1945–1948* (New Brunswick NJ, 1964), pp. 13–14.

private association of academic and business dignitaries in New York, the Council on Foreign Relations.[7]

Worse yet, the wartime political consensus that had given Roosevelt great freedom of action eroded as peacetime conditions approached; profound conflicts once held in check resurfaced in the public, Congress, the administration, and the cabinet. The new president had first of all to pacify an increasingly angry and outspoken American Left to which, heterogeneous though it was, the war presented a unique opportunity to fulfill the long-deferred promise of the New Deal.[8] That odd, mystical midwestern genius – the man who best deserves to be remembered as the father of hybrid corn – Secretary of Commerce Henry A. Wallace stood as guardian to this group. The personal relationships between the brilliant and successful Iowa agriculturalist and the bankrupt Missouri farmer and haberdasher were complicated by the fact that Roosevelt had chosen Truman as running mate in 1944 primarily in order to avoid again being saddled with the independent, unpredictable, and increasingly eccentric Iowan who had been vice-president for the four-year term beginning in 1940. On the right, Truman had to dam up a powerful and growing current of conservative opinion, whose respectable spokesmen like Senator Robert A. Taft provided timely warnings against undesirable extensions of American power overseas, but in which after 1946 also swam Senator Joseph McCarthy of Wisconsin. If the Wallace group hoped that the wartime alliance with the Soviets could serve as pathway to a brave new peacetime world, the increasing numbers on whose behalf McCarthy brayed condemned it as a road leading only to perdition.[9]

The Ruhr was but a small part of the vast world with whose immense problems an American president was for the first time about to contend and through which he, and he alone, could have begun to make headway; consideration for the special problems of the German heavy industry complex had to be subordinated to that given larger ones. Highest priority went to maintaining the wartime alliance, and only with great reluctance did the United States give up the effort to keep it intact.[10]

[7] Herbert Feis, *From Trust to Terror: The Onset of the Cold War, 1945–1950* (New York, 1970), pp. 16–17; HST, Interview with Willard Thorp, 10 July 1971.

[8] See Catton, *Warlords*, pp. 304–305.

[9] See Arthur S. Link, *American Epoch: A History of the United States since the 1890's*, vol. 3: 1938–1966, 3rd ed. (New York, 1967), pp. 665–666.

[10] Feis, *Trust to Terror*, p. 15.

The need to restore stability after World War II meant that ideological preferences could not dictate American foreign policy. Like it or not, the United States was unable to intervene in areas where order had been imposed by the Soviets, had to respect the established positions of colonialist powers and, as in much of Southeast Asia where they were collapsing, rebuild them from materials at hand. The successful use of the atomic bomb against Japan further increased the need to maintain the wartime alliance; the awesome destructive power of the new weapon, whose secret others were expected to discover, made thoughts of a divided world almost too frightening to contemplate.[11] American policy in occupied Germany was thus subordinated to maintaining cooperation with the USSR and United Kingdom. The United States would try hard to make Allied occupation government work. Occasionally this involved accepting painful compromises; more often, it meant engaging in protracted diplomatic rituals having no other real purpose than to keep alive the hope that the victor powers could someday cooperate effectively. Only after the futility of the attempt had been conclusively demonstrated did the United States take steps that led to the formation of the Federal Republic of Germany.

"Four Power cooperation" never existed, and it makes no sense to speak, as many accounts do, of its "breakdown" as having been responsible for the split of Germany and Europe. At the same time, the Americans did eventually give up on attempts to make it succeed. The shift in U.S. policy was due to disgreements over reparations, in particular removals from the Ruhr. Reparation was the drive wheel of Allied policy in Germany. Only after agreement had been reached about how much and what to remove could decisions be made about how to treat the rest of the economy. Yet reparations, as such, were really of paramount importance only to the Soviets. Recalling all too vividly the mistakes of Versailles, the British did not want to saddle Germany with payments obligations. They were also markedly unenthusiastic about taking deliveries from current production, which could have cut into export markets and aggravated the seriousness of the dollar shortages that the United Kingdom regarded as its primary foreign policy problem.[12] The only form of reparation the British wanted was dismantlement, but they were less interested in acquiring new plant than in eliminating

[11] Ibid., p. 95.
[12] Alec Cairncross, *The Price of War: British Policy on German Reparations, 1941–1949* (Oxford, 1986), pp. 15, 31, 49–50.

future competition.[13] The Americans were in a position to play honest broker. Needing no reparations for itself, the United States supported a policy of orderly removals in kind to benefit nations victimized by Hitler, something that served the additional purpose of weakening Germany industrially. Allied reparations policy, in sum, was supposed to provide the basis of Big Three economic cooperation and at the same time regulate the recovery process.[14]

The definitive failure of Four Power cooperation in Germany occurred in May 1946 when U.S. Military Governor Lucius D. Clay stopped reparations deliveries to the Soviets.[15] Though the decision was neither preceded nor accompanied by an overall policy review, its logical consequence was the buildup of a separate West German economy. Yet on the surface nothing had changed. The Allied Control Council (ACC), the central agency of occupation administration, did not disband, and the strange atmosphere of strained yet genuine cordiality well remembered by those who participated in the ACC's numerous sessions continued to prevail.[16] Though neither the president, State Department, nor Clay himself meant either to wreck the ACC or to drop the policy developed by Roosevelt during the war, none of them was in command of events. The new departure came because the American public was wary of European entanglements, increasingly suspicious of Soviet intentions, and undergoing a kind of tidal change in political sentiment. This required a purge of the American Left. It began in September 1946 with the summary dismissal of Henry Wallace and soon would include the red baiting and witch hunting that were to become its most repulsive features. The U.S. effort to reform the German economy would be an early casualty of the new ideologial climate.

The reparations dispute, which looms so large in cold war annals, was not, as often supposed, due to Russian violations of agreements reached at the famous meeting of Truman, Stalin, and Churchill (and after the Labour victory of Clement Attlee and his powerful new foreign minister Ernest Bevin) which took place from 17 July to 2 August 1945 at the huge and ornate English hunting lodge in Potsdam completed for Kaiser Wilhelm shortly before the outbreak

---

[13] Ibid., p. 231; FRUS 1944/I 862.50/11-1144, "Annex: Draft Memorandum for the President," 10 November 1944, pp. 398–399.

[14] Ibid., p. 100; Charles F. Mee, *Meeting at Potsdam* (New York, 1975), p. 188; FRUS 1944/I ECEFPD-37/44, "Report on Reparation, Restitution, and Property Rights – Germany," 12 August 1944.

[15] Gaddis, *United States and Cold War*, p. 329.

[16] James S. Martin, *All Honorable Men* (Boston, 1950), pp. 166–167.

of World War I. The most important of the Potsdam accords affecting Germany were political: The enemy nation was to be administered jointly by an Allied Control Council, and an eventual peace settlement was to be drafted by a new Council of Allied Foreign Ministers meeting periodically. The economic provisions of the conference communiqué were left deliberately vague and, though this was subsequently denied by the United States, intended to supplement rather than replace agreements reached previously at Yalta.[17]

The reparations dispute grew out of a United States withdrawal from commitments made earlier to the Soviets. They were based on a general understanding reached under the shadow of the Morgenthau Plan that the Germans were to be squeezed as hard as possible by those whom they had despoiled: Securing reparation was to be the primary object of Allied economic policy. At Yalta the United States had agreed that Germany's reparations bill would total $20 billion, which would be taken in the form of capital removals and deliveries out of current production, and also that the Soviets would get one-half of the total amount.[18] The figure, which by chance coincided with the total cost of the war to the United Kingdom, was not pulled out of the air but based upon educated guesses of Germany's ability to pay. Estimating the Reich's prewar wealth at $125 billion (a surprisingly accurate figure) and deducting 40 percent for wartime losses (a huge exaggeration as events would later prove), the Soviet delegate Maisky concluded that some 30 percent of Germany's $75 billion of "mobile capital," or $22–23 billion (roughly the $20 billion the United States endorsed) could be removed without driving living standards below an otherwise unspecified "modest and decent central European level."[19] The Allies were thus committed to making the defeated Germans bear costs equal to those of waging yet another major war. Until these were met, recovery was out of the question.

In addition to leaving open the issue of how hard Germany could be pressed before terminal collapse, the Yalta conferees failed to determine who would pay. From which zone of occupation were reparations to be removed? The subject of intense discussion and debate, the matter was left unresolved at Potsdam. What emerged from the conference, in addition to a commitment of dubious value made by

---

[17] Mee, *Meeting at Potsdam*, pp. 323–324; Michael Balfour, *Four-Power Control in Germany and Austria, 1945–1946*, Survey of International Affairs, 1939–1946, ed. Arnold J. Toynbee (London, 1956), pp. 88–89.
[18] Cairncross, *Price of War*, p. 70.   [19] Ibid., p. 71.

the United States and Great Britain to deliver 10 percent of Russia's total, yet unspecified, reparations requirements, was an agreed-upon framework in which the Allies could debate this question. It would provide the context of Four Power economic diplomacy in Germany. The three main propositions of the Potsdam protocols, to which endless appeal would be made in the future, were that Germany was to be treated as a unit, could not be deprived of more than would enable it to be self-supporting, and must be allowed to run export surpluses lest it become a net cost to the occupying powers (first charge principle). To implement these provisions, a commission was set up to determine within six months precisely how many and which capital goods could be removed without irreparably damaging the German economy. It produced the so-called Level of Industry Agreement of March 1946 which, though never operative, constitutes the most important single attempt of the United States, Britain, and the USSR to arrive at a common policy for Germany.[20]

The Potsdam protocols were a very uncertain guide to action: Tucked away within them were articles on reparations that made a mockery of the basic proposition that Germany should be governed as a single economic unit. These stipulations empowered each government of occupation to seize and remove goods independently, in effect providing licenses for separate economic policies. As expected, Soviet stripping began at once and on a huge scale. This provided the United Kingdom, backed by the United States, with a pretext for suspending deliveries of industrial goods to the Soviet Zone. Though the British had a genuine fear that these removals might place new strains on payments, these worries were without foundation: The seizures neither displaced dollar export sales nor lowered living standards in the Eastern zone. Excess capacities existed in virtually every branch of German production, and what was taken could not otherwise have been put to use.

The United States did not share Britain's foreign payments difficulties and disposed of huge reserves of food; the repeated American protests about Soviet removals, though phrased in the language of the prosecuting attorney and appealing to the logic of the dismal science, had no more basis in economic reality than in law. Yet they were understandable responses to the frustration of Four Power policy making in occupied Germany. The discussions leading to the Level of Industry Agreement of 28 March 1946, which the British more accurately called the Reparations Plan, caused many head-

[20] Ibid.

aches. Negotiations for this, the first and most serious attempt at Four Power economic planning, were supposed to conclude within six months of the Potsdam conference, their object being to determine, branch by branch, how much industrial capacity could be made available for reparations by 1949.[21]

This was an exercise in futility and madness because fundamental disagreement between the USSR and the United States/United Kingdom about what constituted an acceptable minimum standard of living undercut any satisfactory basis for compromise. In spite of a determined effort to arrive at one, the Allies produced something even worse than what the Russians might have adopted by themselves. The Reparations Plan divided industry into three groups, those to be prohibited outright (arms and munitions), those not to be restricted (of which the document makes no further mention), and those upon which output limits were to be imposed. For the latter, the plan set targets for outputs by sector and imports and exports by sector. To arrive at compromise figures took months of bickering which, given the deadline of January 1946, left no time to consider a number of essential questions: the effect of cutbacks in one sector on output in another; the relationship between allowable imports and exports on the one hand and production on the other; effects of changes in employment and consumption patterns; and, for that matter, any other structural adjustments that might have been required by drastic changes in the aggregates and components of production.[22]

One can only contemplate with horror what would have resulted from an attempt to enforce the Level of Industry Agreement: a reciprocal gearing down of production caused by shortages, a similar effect due to attempts to maintain exports at the expense of the domestic economy, an inevitable fall in living standards, and massive unemployment as well as administrative chaos resulting from the imposition of controls that could not have worked properly. What the Big Three (French participation being minimal) came up with was a complicated recipe for a dish every bit as unsavory as the Morgenthau Plan.[23] With the Level of Industry Agreement, Big Three planning for a joint economic policy toward Germany had reached a moral dead end. Clay can hardly be censured for having taken the step that would later prove to be the definitive break with it. Blame rests less with him than with his superiors and advisers: It

[21] Ibid., pp. 130–146.   [22] Ibid.   [23] Ibid., pp. 132–133.

was their job, not his, to find some basis for cooperation with the USSR beyond that of merely punishing Germans.

There was at least one serious attempt from within the State Department to locate one, the Acheson–Clayton proposal of April 1946.[24] It originated in the Office of German-Austrian Economic Affairs and was the joint effort of two former OSS (Office of Strategic Services) economists, Charles P. Kindleberger and Walt W. Rostow. They believed that a combination of dollars and federalism might well retard the drifting apart of Eastern and Western Europe and hoped that the same formula could be applied to the German problem. State's commercial and financial leaders, Assistant Secretary for Economic Affairs Will Clayton and Undersecretary Dean Acheson, supported the proposal and, through Jean Monnet's behind-the-scenes influence, managed to persuade Secretary of State James F. Byrnes to put it on the agenda of the Paris peace conference then meeting to draft treaties for the minor ex-enemy nations.

The memorandum declared that the United States had a permanent interest in safeguarding the future of Europe and to this end would seek an all-European settlement within the framework of wartime agreements and bolster it by creating economic organizations to deal with the region's problems on a federal basis under United Nations supervision. The United States promised to back the enterprise with substantial aid. Byrnes's initiative resulted in the creation of the Economic Commission for Europe (ECE), an umbrella for existing UN organizations in the fields of coal, inland transportation, and emergency planning. The Soviets paid scant attention to the proposal. It also lacked support within State. Arrayed against the Kindleberger–Rostow memorandum was a coalition ranging from old-time department area specialists to new anti-Soviet hard-liners to whom the young George Kennan, then beginning his meteoric rise, would soon give voice. Regardless of their judgments on the plan's merits, these skeptics were well attuned to the dominant mood in the American public and grasped far sooner than the two young technocrats that most Americans would have gagged on the idea of rebuilding Europe with Soviet cooperation.

The same pressures that prevented its being launched wrecked the American decartelization effort. It became a plaything of American domestic politics. As initially conceived by the group around Acheson, decartelization was to provide the means through which

[24] See W. W. Rostow, *The Division of Europe after World War II: 1946* (Austin TX, 1981), passim; also, Charles Kindleberger, *Marshall Plan Days* (Boston, 1987), pp. 108, 156.

Germany could be simultaneously reformed and disarmed industrially. It was, in other words, supposed to render superfluous the razing and wrecking subsequently proposed by Morgenthau and instead change legal and economic forms in order to prevent conspiracies such as those that had supposedly brought Hitler to power and stifled recovery in the 1930s. A Germany thus rehabilitated would be a welcome participant in a revived Europe. The plans of the Acheson-chaired Executive Committee on Foreign Policy called for disestablishing "the top financial structures of the great industrial combines and redistributing the ownership of constituent operating companies, in part through some effective form of public control exercised through a democratic regime,"[25] and committed the American government to opposing "the development of new forms of industrial combinations, whether on a German or international basis, which could contribute to renewed German economic and political aggression in Europe."[26] Though Secretary of the Treasury Morgenthau opposed decartelization as distracting from his "Program to Prevent Germany from Starting a World War III," as the plan bearing his name was officially known, once it had been shelved, his followers attempted to weaken organizationally what could not be destroyed physically. The "Morgenthau spirit" thus dwelled within anticartel policy.[27]

The conflict between the dovish and hawkish points of view carried over into the apparatus of the U.S. military government of Germany, in the process becoming so heated that its original causes were forgotten. The head of the Economics Division was a New York investment banker and "weekend warrior" named William Draper, who had been close to the War Department group under Stimson. As head of the Economics Division, General Draper was Clay's chief economic adviser. Under him as chief of the decartelization division was a Morgenthau man, Colonel Bernard Bernstein, who, however, resigned after several months of bitter opposition to Economics Division policies he thought were leading to a full-blown restoration of the Reich.[28]

Into this poisonous atmosphere arrived James S. Martin on the last day of 1945 as chief of the Decartelization Branch. For two years his office would be the center of a storm of controversy that

[25] FRUS 1944/III ECEFPP-36/44, p. 285.
[26] Ibid.; see also FRUS 1945/III, "Discussions Relating to Decartelization, Deconcentration, and Handling of the German Public Debt," pp. 1559–1577.
[27] See HST, Interview with James W. Riddleberger, 6–7 April 1972.
[28] Martin, *All Honorable Men*, p. 176.

left policy making in tatters. Martin's appointment had been arranged by Senator Harley Kilgore of West Virginia, like Morgenthau a strong advocate of a punitive settlement. Martin himself was no mere Germanophobe but a brilliant young lawyer from Thurman Arnold's Antitrust Division at the Justice Department, an unhappy Jeffersonian for whom the events occurring "in Germany before and after the occupation [seemed] in an increasing degree to be echoes of something more fundamental that was happening back in the United States. For whatever reason, the larger pattern [was] a repetition of what followed after World War I; but the pace [had] been quicker, as though greater powers were moving more rapidly to a more catastrophic result."[29]

Martin had spent much of the war tracking down cartel agreements (in artificial fuel, magnesium, plastics, aircraft engines) that had hindered the production effort, discovering in the process that the "real enemy" was something "much bigger than the popular picture of hobnailed Nazis with guns and tanks" and, more terrifying yet, something that almost supernaturally "could survive defeat because it did not need or use military weapons."[30] Martin's "real enemy" was a web of secret international agreements spun by monopolists of all countries, and Germany was its source. He attacked it with zeal, seizing records of combines wherever they could be found and compiling voluminous studies of the ownership, affiliations, finances, and political practices of trade organizations, cartels, and trusts. Martin not only left the historian with an extraordinary source of material concerning business practices in the Reich but set the stage for a U.S.-directed campaign aimed at rooting out German-style organized capitalism and replacing it with the idealized version of the American free market system envisaged by Thomas Jefferson. The trustbuster's program never got off the ground. The British successfully blocked his proposal for a law mandating the reorganization of companies above a specific size, thus confining his campaign to the U.S. Zone.[31] There he enjoyed only one clear-cut success, the break up of the IG Farben chemical complex located in Frankfurt. The decartelization law published after the formation of Bizonia in January 1947 was a banal restatement of Potsdam principles and far less ambitious than what Martin wanted.[32] He had in fact already completed the preliminaries needed for a tough bill: organized a joint control agency with the

[29] Ibid., p. vii.   [30] Ibid., pp. 13, 14.   [31] Ibid., p. 227.   [32] Ibid., p. 229.

British; drafted procedural rules for bringing cases under the law; prepared supporting legislation to eliminate large holding companies, interlocking directorates, and the issuance of bearer shares; readied draft orders requiring the reorganization of large combines; and finally compiled what would become a massive four-volume *Report on German Cartels and Combines* (the Ballwin Report), enumerating the restrictive practices adopted in the past by every branch of German industry.

Martin, whose plans got hopelessly tied up in the red tape of the Economics Division, rightly suspected that his boss, General Draper, was the source of the problem. Draper regarded decartelization as pointless, distracting, and even destructive. Was not German industry flat on its back? Had cartels not already been formally abolished? Did decartelization not interfere with the far more important matter of recovery, and should not the campaign be postponed until it was clearly underway? Whereas Draper worked behind the scenes, Martin went public. In October 1946, Senator Kilgore, who as head of the War Investigating Committee had lambasted OMGUS as a protector of Nazi war criminals, launched an investigation of reputed sabotage of U.S. policy by military government officials. Martin himself mischievously discovered "an almost pathological fear of the light of publicity in some parts of the Economics Division," particularly that provided under the byline of the sharp-penned Delbert Clark, his trusted ally, in the pages of the *New York Times.*[33] These tactics incensed the sometimes apoplectic Draper and the normally even-tempered Clay as well but did Martin little good.

The fate of his program had been sealed by the firing of Henry Wallace after his controversial 12 September 1946 speech in the cavernous New York convention hall, Madison Square Garden. Wallace's pronouncement was intended as a response to Secretary of State Byrnes's famous remarks in Stuttgart six days earlier, in which he had disclosed that the United States was ready to change direction in Germany. After ritualistic obeisance to Potsdam, Byrnes had stated that the United States had found the Level of Industry Agreement unworkable and was now prepared to promote economic reconstruction as rapidly as feasible; that if economic unification were impossible, then "maximum possible unification" would be attempted; and that German recovery was indispensable to Europe. The secretary of state added that this should occur within the framework of continued military occupation. His speech charted

[33] Ibid., p. 230.

with remarkable accuracy the future course of American policy in Germany.[34]

Byrnes's very words originated in the Central European Section of the State Department, but the extent to which they represented a confused administration policy remains unclear to this day.[35] Wallace was not alone in thinking that they did not. His purpose in delivering the Madison Square Garden speech was to keep the Rooseveltian Grand Design,intact by heading off the rising crescendo of right-wing attack on the administration that had begun with the fall electoral campaign. The secretary of commerce actually showed his text to Truman, who carelessly approved it without realizing that he had thereby lit the fuse on a powder keg. Depicting administration policy as uncharitable, as unnecessarily hostile to Russia, and as an attempt to appease hard-line Republicans as well as old-fashioned British imperialists, the Iowa mystic called for a frank American acknowledgment of Soviet predominance in Eastern Europe and proposed "healthy competition" between the different political and economic systems of the two superpowers. After Truman, in an maladroit attempt to conceal his blunder, told the press that the secretary of commerce's views represented those of the administration, Byrnes tendered his resignation as secretary of state. Two days later, on 20 September 1946, the president demanded, and received, Wallace's instead, and "thus departed from the Administration the only senior member who was deeply disturbed by the growing rift with the Soviet Union, and willing to hazard the fate of the Western democracies by trusting the Soviet Union."[36]

Subsequent events confirmed America's anti-Soviet course. Hammering away at the alleged incompetence of Truman to confront the menace of communism, the Republicans won landslide victories in the November 1946 elections. After January 1947 they would hold a majority in both houses of Congress for the first time since 1928. In addition to Taft and McCarthy, the "class of 1946" contained such stalwarts of the new Right as John Bricker of Ohio, George Malone of Nevada, and William Knowland of California. Even in far-off Berlin, "it was evident from the first that the changes in Congress more than the changed conditions in Germany were responsible for the growing confusion in policy matters."[37]

[34] James F. Byrnes, *Speaking Frankly* (New York, 1947), pp. 188–187.
[35] HST, Interview with James W. Riddleberger, 6–7 April 1972.
[36] Feis, *Trust to Terror*, p. 165; Gaddis, *United States and Cold War*, pp. 338–339.
[37] Martin, *All Honorable Men*, p. 231.

Worse was to follow. The disclosures during 1946 by a Canadian investigative commission that Communists there had passed on atomic secrets to the Soviets sparked the first of the great postwar spy scares. The dreaded and irresponsible House Un-American Activities Committee, its operations suspended during the war, was back in operations, this time under the irresponsible chairmanship of the clownish J. Parnell Thomas. At the end of the year, the demoralized intellectual Left split as the result of the founding of a new, militantly anti-Communist organization within the Democratic party, the Americans for Democratic Action. It excluded fellow travelers from the 1930s, Wallacites, and other assorted malcontents, who from thence would have no institutional defenders in either of the main parties. To contain the rising tide of anti-Communist hysteria, Truman authorized the Federal Bureau of Investigation in March 1947 to investigate the loyalty of all federal employees, a gigantic task which by early 1951 had resulted in two thousand resignations of persons with suspect, unorthodox, or merely unpopular views.[38]

The red baiting had a significant and lasting effect on policy, most spectacularly in China but carried over into occupied Germany as well. Unable to make progress in setting up enforcement machinery for Law 56, James Martin, along with several of his staff members, resigned in May 1947. Those who remained found themselves officially censured by the OMGUS personnel chief as "disloyal employees," a stigma connoting association with Bolshevists. Apparently because he participated in drafting the Progressive party platform of 1948, in which avowed Communists had a hand, Martin was later subpoenaed by the House Un-American Activities Committee, where he refused to testify under his Fifth Amendment rights. Though not included in any official statistic of resignations under pressure, his fate was in important respects similar to many others: blackballed and almost friendless, the ex-decartelization chief left Washington in 1950 for exile in Costa Rica. Fifteen years later he was still considered a security risk.[39] After Martin's departure, the OMGUS decartelization campaign, though officially still in force, degenerated into a farcical persecution of unimportant retail monopolies; the United States was left with no policy to reform or otherwise restrict the revival of Ruhr heavy industry until it was resumed in late 1949 for reasons quite different from Martin's by a

[38] Link, *American Epoch*, pp. 682–683.
[39] Interview with author, 10 January 1979.

member of the very Wall Street clique he despised, U.S. High Commissioner John J. McCloy.

The so-called Meader Report, which issued in November 1947 from the investigation launched by the Kilgore Committee of the Senate, blamed OMGUS for having failed to pursue denazification effectively, having allowed outputs to rise above levels envisioned at Potsdam, and having done little to break up the cartel system; instead of punishing the Germans, Clay was accused of turning them into allies.[40] The trend was indeed unmistakable: The end of reparations deliveries, the Stuttgart speech, and the founding of Bizonia in January 1947 were impressive evidence that Germany was being revived without being reformed. Though some on the American right applauded this development out of a conviction that the United States needed a German ally in order to counter the Soviet threat, it met with approval almost nowhere else – neither in France, Britain, the USSR, nor the Truman administration, and not even among the vast majority of Germans. The alternative to it, to imbed German recovery in that of Europe as a whole, grew out of the Marshall Plan.

The granting of aid to Europe began long before the passage of the Foreign Assistance Act setting up the European Recovery Program – the Marshall Plan agency – on 3 April 1948. By June 1947 the United States had donated three-quarters of the $3 billion in aid granted through the United Nations Relief and Rehabilitation Administration (UNRRA) to formerly occupied areas. Prior to the Marshall Plan, OMGUS dispensed several hundred million dollars in aid to Germany through the so-called GARIOA program. Additional loans of $3.75 billion and $1.4 billion were made successively to Britain and France in 1946. The latter was the so-called Blum loan, which Monnet negotiated.[41] All of these grants were intended to be one-shot emergency measures to facilitate American disengagement from Europe. The Marshall Plan would be different in both respects.

Its origins date from winter of 1946–1947, the worst in a century. From December to the end of March chilling winds, hard freezes, and repeated heavy snowfalls brought rail and barge traffic to a

[40] Martin, *All Honorable Men*, pp. 221–222, 225–226.
[41] Feis, *Trust to Terror*, pp. 227, 230; see Michel Margairaz, "Autour des Accords Blum-Byrnes: Jean Monnet entre le consensus national et le consensus atlantique," *Histoire, économie, société* 1/2: 1982, pp. 440–470, and Irwin Wall, "Les Accords Blum-Byrnes: La Modernisation de la France et la guerre froide," *Vingtième Siècle* 13:1978, pp. 45–62.

standstill and production to a halt for whole weeks at a time in much of northern and western Europe. The war, it seemed, had done more damage than previously suspected; recovery was perilously fragile, living standards in jeopardy, and the possibility imminent that Europe would be plunged into revolutionary chaos. The winter made few meaningful distinctions between victors and vanquished. Britain was a prominent casualty; a virtual collapse of the transport system brought about stoppages throughout industry, forced a drastic reduction in exports, and weakened the foreign currency position so severely that Foreign Secretary Ernest Bevin announced to Truman that the United Kingdom would no longer be able to support the Greek government's bitter campaign against guerillas from the antimonarchical wartime resistance movement, which had been armed by Tito and was suspected of being under Moscow's influence.

The American response was the Truman Doctrine. In order to "scare the hell out of the American people," whose delegates in Congress were to be called upon to provide military aid, the American president painted a lurid picture of the totalitarian threat overhanging the world, promising help to any nation menaced by it, and in particular Greece and Turkey which, though the latter faced no immediate military danger from the north, were deemed to be targets of a Soviet plan to expand into the eastern Mediterranean through the Dardanelles. By conjuring up the Communist bogeyman, Truman managed to get what was an enormous sum for the time, $350 million, from the tight-fisted, withdrawal-minded, Republican-dominated Congress.[42]

The Truman Doctrine has found few persuasive defenders and was an embarrassment even within foreign policy circles, if only on the narrow grounds that the president's sweeping commitments could easily overstretch American military resources. One of its immediate effects was to reduce whatever slim chances the Moscow Council of Foreign Ministers might have had for success. It was supposed to provide a venue for working out the terms of the long-deferred peace treaty with Germany. Instead, it was mainly remarkable, according to the French Foreign Minister Georges Bidault, for "endless festivities, frozen amusements, formal parties and sumptuous banquets, where most of the time was taken up in

---

[42] See Feis, *Trust to Terror*, pp. 191–206; Gaddis, *United States and Cold War*, pp. 348–349; also, Robert A. Pollard, *Economic Security and the Origins of the Cold War, 1945–1950* (New York, 1985), pp. 126–127.

toasts."[43] Each foreign minister presented the position of his government – and that was all: Even after sixty formal meetings no
progress could be made toward drafting a treaty to end the state of
war with the Reich. Secretary of State George Marshall, in office
since January 1947, left Moscow in April convinced that the Soviets
were utterly unwilling to cooperate on German matters.[44] The Truman Doctrine also set one particularly bad precedent: It linked the
provision of aid with defense against communism. The two could
never be separated again, not even in the Marshall Plan, which
many of its creators thought of as an alternative to the impromptu
assistance package for Greece and Turkey: not a blank check for
military hardware to anyone, anywhere, who felt threatened by
Communists but a voucher for economic support given under specified conditions and as required by the security of the United
States.[45]

Credit for devising the Marshall Plan can be claimed by no single
individual. In April 1947 the three top-ranking officials of the State
Department, General Marshall, Will Clayton (undersecretary of
state for economic affairs), and Acheson (at the time undersecretary
of state for political affairs), concluded almost simultaneously that
without immediate large-scale assistance the harsh winter would
cause the collapse of the European economy. In May the recently
formed Policy Planning Staff (PPS) headed by George Kennan
drafted the proposals that turned a foreign aid program into a plan
for European integration.[46] The 21 April 1947 report submitted by
the special study group of the State–War–Navy Coordinating Committee (SWNCC), set up a month earlier to determine what in the
way of money, food, military equipment, and "other forms of assistance" would be required to protect American security interests,
concluded on the basis of its global estimates that unprecedented
measures were necessary. SWNCC called for the United States to
take "positive, forehanded, and preventative action," without, however, specifying how this should be done.[47] The committee neither

43 Georges Bidault, *Resistance: The Political Biography of Georges Bidault* (London,
    1965), pp. 146–147.
44 See FRUS 1947/II, pp. 1–576.
45 See Paul G. Hoffman, *Peace Can Be Won* (New York, 1951).
46 See Hogan, *Marshall Plan*, pp. 40–41; Stanley Hoffmann and Charles S. Maier
    (eds.), *The Marshall Plan: A Retrospective* (Boulder CO, 1984), pp. 7–8, 19–20;
    Joseph M. Jones, *The Fifteen Weeks* (Feb. 21–June 5, 1947): *An Inside Account of
    the Genesis of the Marshall Plan* (New York, 1965).
47 FRUS 1947/III, "Report of the Special 'Ad Hoc' Committee of the State–War–
    Navy Coordinating Committee," 21 April 1947, p. 205.

linked financial assistance to an American geopolitical strategy, considered the relationship between an aid program and the long-term development of beneficiary nations and regions, nor said anything about how one could be administered.

The Policy Planning Staff worked out answers to the first two of these questions. Dean Acheson has questioned the originality and significance of its contribution to the genesis of the Marshall Plan, and it may well be that Kennan's key memorandums merely expressed what others had long been thinking.[48] The State Department's first think tank, the staff brought a new coherence to policy making; for the first time the United States could proceed, in Western Europe at least, on the basis of a well-defined, sensible, and flexible long-term strategy. Even before the PPS set to work, Kennan had defined certain essential elements of a new approach. First, Western Europe was to receive priority over all other trouble spots, a regional emphasis that constituted a clear-cut departure from Rooseveltian globalsim. Second, the area was to be defined as including Britain and the U.S. Zone of occupation in Germany as well as the nations of the Continent; breaking with traditional practices, the United States would no longer make fundamental economic distinctions between former allies and enemies; and Britain was specifically to be treated like any other European nation. Finally, Europe's political problems were viewed as being internal and economic rather than foreign and military in origin. American security was seen to rest neither, as earlier, on the operation of self-regulating mechanisms nor, as after 1950, on a military alliance but on the health of agriculture and industry and the welfare of the public.[49]

In a memorandum of 16 May 1947 Kennan spelled out a tentative list of principles similar to those upon which the future Marshall Plan would rest. European economic cooperation was the first of them: "The program for American aid should be, if possible, supplementary to a program of intramural economic collaboration among the western European countries." The assistance scheme was, secondly, to run for a period of from four to five years after which, "Europe no longer need be the object of charity from outside." The third point was that Communists should be kept out of the aid plan, and the fourth that the creation of a Western Euro-

[48] Acheson, *Present at the Creation*, pp. 302–303.
[49] George Kennan, *Memoirs, 1925–1950* (Boston, 1967), pp. 335–336; see also FRUS 1947/III 840.50 Recovery/5-2347, Director of Policy Planning Staff to Undersecretary of State, 23 May 1947, pp. 223–224.

pean federation, including the parts of the former Reich under U.S.–U.K. and French control, was to be its ultimate goal: "The program should be designed to encourage and contribute to some form of regional political association [enabling] the western zones of [Germany and Austria] to make the maximum contributions to economic restoration in Western Europe."[50]

The translation of these ideas into policy occurred rapidly. Upon returning from the Moscow conference on 28 April 1947 Secretary of State Marshall had told the nation over the radio that "the recovery of Europe has been far slower than expected. Disintegrating forces are becoming evident. The patient is sinking while the doctors deliberate."[51] By then Acheson had also begun to act. Troubled by the knowledge that Europe's need for assistance would be far greater than the $350 million Congress had appropriated for Greece and Turkey, on 7 April he secured permission from the president to deliver a speech that would shock the country into a recognition that far larger amounts of aid would soon be necessary. On 8 May 1947 Acheson told a hot, heavily perspiring audience of wealthy farmers and country lawyers from the delta region around Cleveland, Mississippi, that $8 billion in American assistance would be needed to overcome the ravages of the war and the threat of a hopeless descent into bankruptcy followed by general collapse. He added that both Germany and Japan would have vital roles to play in the reconstruction process.[52]

An alarming, even terrifying, 27 May 1947 memorandum of Undersecretary Clayton prompted Acheson to act without further hesitation. "It is now obvious," Clayton began, "that we grossly underestimated the destruction to the European economy by the war," which included, he went on, not only the physical damages but the effects of economic dislocations – nationalization, land reform, loss of capital, and the severance of commercial ties. Europe, according to Clayton, was on the verge of starvation and in need of an immediate infusion of $5 billion to prevent mass upheavals and chaos. It was also critically short of the means of payment (with Britain, France, and Italy having only enough dollars to last until the end of the year) and faced acute shortages of coal, grain, and ships with which to transport them. The American people, Clayton con-

[50] FRUS 1947/III, Memorandum by the Director of the Policy Planning Staff (Kennan), 16 May 1947, pp. 220–221.
[51] Forrest C. Pogue, *George C. Marshall–Statesman, 1945–1959* (New York, 1987), p. 200.
[52] Acheson, *Present at the Creation*, pp. 303–304.

tinued, had to be told in unembellished language that in order to
supply Europe with $6–7 billion over the next three years, belts
would have to be tightened. There could, however, be no more mul-
tilateral aid distribution à la UNRRA: In the future, Clayton itali-
cized, "The United States must run this show."[53] After intensive
discussions the following day, Acheson and Clayton persuaded General Marshall to accept a long-standing invitation from Harvard to
receive an honorary doctorate, and at commencement ceremonies
on 5 June he delivered a few momentous lines announcing the intention of the United States to support a recovery plan for
Europe.[54]

The Marshall Plan was, as once remarked, in some ways like a
flying saucer: No one knew what it looked like, how big it was,
which way it was moving, or even whether it was real.[55] Yet it was
hard not to believe that something rather strange and wonderful
had appeared on the horizon of American policy making. The future would eliminate remaining doubts concerning the object's actual existence. Even then, its outlines remained difficult to discern.
What was the relationship between Truman Doctrine and Marshall
Plan? Were the Europeans, as Kennan recommended, to administer
the aid program, or were the Americans to run the show, as Clayton
had insisted? Should the British be encouraged to act as political
leaders of Europe, as had been done since the war, or should they
be treated as economic equals of the continental nations? And
where, above all, did Germany fit into the picture, and how might
its emergence from the postwar ruins affect plans for federation?
There was, one must assert, certainty on only one fundamental issue: The USSR, its foreign satellites, and its Communist party allies
in other states were to be excluded from the aid program by entic-
ing the Soviets themselves into refusing to enter it. They obliged;
the split between the two Germanies became a division between two
Europes, in the traditionally dominant western portions of which
the direct influence of the USSR would largely cease to be felt for a
generation.

[53] FRUS 1947/III 840.50 Recovery/5-2747, Memorandum by the Undersecretary
  of State for Economic Affairs (Clayton), 27 May 1947, pp. 230–231.
[54] FRUS 1947/III Lot 54D394, Box 538, "Summary of Discussion on Problems
  of Relief, Rehabilitation, and Reconstruction of Europe," 29 May 1947, pp.
  234–235.
[55] FRUS 1947/III Lot 122, Box 13113, Assistant Chief of the Division of Commercial Policy (Moore) to Director of the Office of International Trade Policy
  (Wilcox) at Geneva, 23 July 1947, p. 239.

3.2 The not-so-special relationship

Almost as if by instinct, the United States first turned to Britain for help in transforming the promised aid package from a mere pledge to an operational program. Over the following months this effort would encounter delay, frustration, resistance, and eventually outright sabotage. The British response to the Marshall Plan finished off, so far as the United States was concerned, what remained of the special relationship so cherished by Churchill, at least to the extent that it extended to economic policy in Europe and insofar as it may be considered to have existed in the first place. The really "special" thing about the relationship was not the intimacy that grows out of deep affection and shared experience but its pretense of equality: This was no tight marriage bond but a loose liaison between a rich, rather indulgent patron and a conniving, somewhat déclassée mistress which, lacking conventional constraints, left ample room for the unexpected absence, the lapse of memory, the elaborate excuse for infidelity. The well-heeled provider could choose to tolerate this kind of thing, overlook or merely forget it until disillusionment inevitably set in, and with it an embarrassing recognition that the intimacy once felt had never really existed: It finally dawned upon "Sugar Daddy" that he had acted like a fool. The Anglo-American dalliance lasted too long and led to too little, nearly wrecking Marshall Plan integration policy and diverting Britain's attention from a choice opportunity to enter a meaningful and lasting relationship with Europe.

A basic incompatibility of outlook underlay Anglo-American relations from the war onward. The British wanted nothing less than to be thrust into the brave new technology-propelled American world of open markets and productivity gains and nothing more than to direct its vast energies into maintaining their customary way of life and upholding their traditional status as a world power. Their hopes of doing the latter rested on little more than American sympathy and their own guile. The profound gulf dividing Yanks and Brits was already evident in the discussions of postwar economic policy that began in early 1944 in London. These futile talks dragged on throughout the year, with neither side prepared either to compromise or to admit failure. Though the United States inveighed with the United Kingdom to support trade liberalization, the British were utterly uncompromising and sought through an abstruse and rather specious line of argumentation to persuade the United States to retain the wartime system of combined boards in which they occupied a privileged position.[56] Britain's survival as an imperial power depended just as surely on enjoying trade preference as America's hopes for a better world did upon eliminating it. Each party preferred not to admit that it was talking past the other.

Full employment was the official British objective. Even those businessmen and senior civil servants opposed to it in principle upheld the banner upon which it was inscribed during negotiations with the Americans. What better justification could one find for continuing to do things as they always had been done and keeping economic liberalism out of the empire and the world at large? In exploratory talks of January 1944, Lord Keynes, already regarded by his American colleagues as an economist without peer, hinted elliptically at Britain's future strategy – to place the burden of postwar adjustments on the United States. Keynes superciliously intimated that "the vast majority of American business and congressional circles had not yet grasped the fundamental principles of full employment policy and would reject the measures to apply them."[57] A couple of weeks later, on 12 February 1944, the matter was put more explicitly to Ambassador Winant. "In general," Winant reported, "government economists and permanent civil servants believe that the attitudes of the British Government, businessmen, trade unions and public have now developed to a stage which makes it politically practical to follow a domestic policy of full employment provided that external economic conditions are favorable. They are very skeptical, however, whether a corresponding development has taken place in the United States and many other countries and the question with which they are most concerned is how to meet the public demand for full employment in an economic system open to the effects of change in the rest of the world."[58]

What the British specifically wanted were policies providing the import markets needed to guarantee full employment and, where they failed to bridge deficits, loans. As put in the slightly garbled notes taken at the 12 February 1944 meeting: "In pursuing a full employment policy Britain will find itself unable to increase exports

---

[56] See FRUS 1944/II, "Informal and Exploratory Discussions Regarding Postwar Economic Policy," pp. 7–105 passim; see also D. Cameron Watt, *Succeeding John Bull: America in Britain's Place, 1900–1975* (Cambridge, 1984), pp. 90–91.

[57] FRUS 1944/II 840.50/3467, Ambassador in the United Kingdom (Winant) to Secretary of State, 19 January 1944, p. 6.

[58] FRUS 1944/III 840.50/3513, Ambassador in the United Kingdom (Winant) to Secretary of State, 12 February 1944, p. 11.

enough to offset increased imports. When adverse tendencies appear in its balance of payments, it should put up the whole problem to an international gathering, perhaps through an international commercial policy organization. Its case would be that it could not abandon a full employment policy, that the policy was increasing markets for the goods of other countries, and that those countries should undertake to seek methods of taking more British goods to avoid the necessity of restriction of British imports."[59] In addition to taking more imports and financing their deficits, the British urged the United States to maintain joint controls over world trade in the foreseeable future, participate in international commodities agreements, and endorse private international producer cartels. The United Kingdom made at least some headway on the first of these: at the Hot Springs conference of September 1944, the United States agreed to form international commodity organizations for cotton, wheat, and sugar as a necessary step toward the restoration of order on postwar markets and as a means to "promote economic expansion through trade liberalization."[60]

Cartels were another matter. They were at the heart of American concerns. U.S. negotiators tried hard to sympathize with what struck them as an excessively tolerant position, never realizing that the British feared being too weak to dispense with them. Ambassador Winant reported with some perplexity after one meeting that, though most economists in the Ministry of Supply recognized that the protectionism of the 1930s had retarded rather than accelerated the modernization of the steel industry and further admitted that the official regulatory agency was little more than an "appanage" of the Iron and Steel Federation, they had no intention of doing away with this machinery and letting the market work but thought that with a good plan and the backing of the government it could become an instrument of reform. Winant could only explain the British predilection for such joint state–industry arrangements as a consequence of the nation's unique wartime experience: "Civil servants and businessmen have become so accustomed to bulk purchasing, long term contracts, [and] planned expansion to meet guaranteed demand that some of them are reluctant to return to the uncertainties of individualist peacetime methods of trade and production." "A most important factor in Britain," he erroneously

---

[59] Ibid., p. 13.
[60] FRUS 1944/II ECEFPD-55/44, "Executive Committee on Economic Foreign Policy, Committee on Commodity Agreements, Summary. . . . " 19 September 1944, p. 95.

added, "which does not have equal force in all countries, is that the wartime methods and controls have been operated with impressive efficiency in the civilian sector of the war economy as well as in production for the armed forces with the result that on the whole distrust of the ability of the government in economic matters has diminished, especially in [official] circles."[61]

There was a further reason why Labour (whose thinking in such matters was in any case purportedly "wooly"), Liberals such as Keynes, and Tory businessmen could all agree not to dismantle the regulatory machinery: Without state trading the United Kingdom would have been precariously vulnerable not only to economic competition but to powerful political forces undermining the empire. Like Clémentel after World War I, the British sought to bolster military victory by reworking, strengthening, expanding, and perpetuating the inter-Allied machinery for allocating world supplies of raw material and production. A proposal of 17 August 1944 for appending a European Economic Committee to the Combined Boards spells out these plans.[62] The new organization was supposed to hand surplus European production over to the Anglo-Americans for reallocation and redistribute whatever was left over from the outputs of the rest of the world once U.S.–U.K. needs had been met.

The United States was no more ready to accept this self-serving scheme than the British were prepared to end empire preference, reduce quotas, or lift export limits.[63] Wartime trade controls, Secretary of State Hull wrote to President Roosevelt on 21 September 1944, should be abolished as soon as possible and the Combined Boards should restrict their future activities to eliminating immediate bottlenecks and specifically do nothing to support Britain's imperial pretensions. "As more nations become free," Hull said, "concern [will increase] about United States–British Empire domination. In the near future it will not be easy to justify United States–British decisions concerning the allocation of supplies of other nations. Furthermore, the dislocation of trade patterns, occasioned by the war and made effective through the Boards, creates opportunity for discriminatory national advantage. Within the British Empire differences of opinion, which are suppressed in wartime, will arise, and it will not be in the interests of the United States to

[61] FRUS 1944/II, 840.50/3513, Ambassador in United Kingdom to Secretary of State, 12 February 1944, p. 10.
[62] FRUS 1944/II 840.50/8-2344, British Embassy to Department of State, 28 August 1944, p. 70.
[63] FRUS 1944/II ECEFP D-54/44, "Executive Committee on Economic Foreign Policy, Committee on Wartime Trade Controls," 20 September 1944, p. 80.

become involved unnecessarily in these disputes, as it would if it remained a senior partner on the Boards."[64] As with the other subjects of the Anglo-American economic discussions of 1944, nothing came of the British proposal to transform the Combined Boards into peacetime instruments of U.S.–U.K. economic control. Their existence came to an end on 25 September 1945. Along with the abrupt termination of Lend-Lease after V-E Day and the growing American unwillingness to share atomic secrets with the United Kingdom, the shutdown of the Combined Boards placed the special relationship in grave danger.[65]

Foreign Secretary Ernest Bevin's biographer, Alan Bullock, assigns the colorful earthy former union leader credit for having managed to restore the essence of the special relationship by early 1947, and as Henry Wallace had indeed forewarned, Bevin tried to lure the United States into propping up British rule, restoring British influence, and acting as surrogate for British power where and whenever necessary. In Bullock's view, Bevin's policy represented a successful gamble that Britain would hold out until "[the] Americans were ready . . . to act the way [he] hoped and not pull out of Europe and the Middle East as they appeared to be pulling out of China with the withdrawal of the Marshall mission [in December 1946]."[66] The milestones of Bevin's accomplishment are the Truman Doctrine and the American entrance into the North Atlantic Treaty Organization (NATO) in 1949. They owe much to his success in preventing U.S.–U.K. differences in economic policy from spilling over into the political questions that were his primary concern. This feat is particularly impressive in light of Britain's consistent opposition to the Will Clayton–directed Geneva trade negotiations of 1946. These succeeded only in setting up still another toothless monitoring authority, the International Trade Organization (ITO), and thus represented a serious setback to the liberalization of commerce.[67]

In other respects Anglo-American cooperation actually increased impressively over 1946 and 1947, causing Bullock to speak of a "parallelism" of policy. In Germany the improvement found expression in the agreement to form Bizonia, which was a "joint four year plan to make West Germany self-sufficient by the end of 1949 . . . neither the US nor the UK [having] had much doubt that [its] logical conclusion was the establishment of a West German government."[68]

---

[64] Ibid., p. 82.
[65] Alan Bullock, *Ernest Bevin: Foreign Secretary, 1945–1951* (London, 1983), p. 124; see also Milward, *Reconstruction*, p. 62.
[66] Bullock, *Bevin*, p. 314.   [67] Ibid., pp. 402–403.   [68] Ibid., p. 314.

The idea for such a thing owed much to the British. Well before the United States gave serious consideration to breaking with the Potsdam framework, the United Kingdom was ready to do so. Concerned above all with reducing the drain on payments, the head of the economic division of the British element in the Allied Control Council, Sir Percy Mills, began to press for the adoption of a "tripartite approach to a variety of [economic] problems including imports and exports" as early as August 1945.[69]

Two considerations facilitated U.S.–U.K. cooperation in German matters. Apart from its impact on payments, what transpired in the British-administered area of the ex-Reich was not of great interest to London relative to other world problems. Cabinet responsibility for Germany fell to J. B. Hynd, chancellor of the duchy of Lancaster, a very junior minister "of whom little was heard either before or since"; the level of industry talks thus took place with "almost no guidance from London"; and German affairs were not even discussed in the House of Commons until March 1946.[70] With few hard and fast commitments within its zone, the United Kingdom was well positioned to trade off the exercise of power for dollars. Britain's initial moderation on the level-of-industry question – the result of an uneasy compromise between a desire to protect the payments position on the one hand and to limit export competition on the other – also made it relatively easy to accede to American demands to raise authorized outputs.

The revised Level of Industry Agreement of August 1947 provided at least one monument to joint U.S.–U.K. policy. Negotiations for this bizarre, anachronistic, yet binding pact began after Marshall's and Bevin's returns in April 1947 from the ill-fated Moscow foreign ministers' meeting. The level-of-industry discussions were a nightmare for British negotiators, who were vexed by special pleading from the Board of Trade for the imposition of limits on the German production of fine optics, watches, and cuckoo clocks while pressed hard by Bevin to reach agreement before it became too late to begin dismantling German steel mills; at the same time they had to appease recovery-minded Americans now able in a pinch to wield the heavy club of Marshall Plan aid.[71] The revised Level of Industry Agreement set targets only for steel, as the attempt to arrive at them for other sectors had to be abandoned as technically impossible.

[69] FRUS 1945/III 662.0031/8-3045, U.S. Political Advisors for Germany (Murphy) to Secretary of State, 30 August 1945, p. 152.
[70] Cairncross, *Price of War*, p. 103.   [71] Ibid., pp. 175–176.

These allowed for increases to 10.7 million tons of Ruhr output and, if the French and the Soviets later abided by the plan, to 11.5 million tons for the ex-Reich as a whole. Although the new agreement had no immediate effect on production, as even its higher limits remained substantially above actual output levels, it was again revised in 1948. The punitive occupation policies of the war's aftermath thus lived on in form if not in substance, complicating and confusing the recovery process and, contributing to institutional inertia.[72]

Bizonia made sense for the United Kingdom, not because it produced a genuine common policy with the Americans but because it reduced the costs of occupation. Churchill was mainly responsible for Britain's having bitten off more than it could chew. All through 1944 he pressured the Americans to agree that the United Kingdom be allowed to occupy and administer the northwestern quadrant of the Reich, his excessive demand seriously antagonizing the wary Roosevelt: After eventually giving in, the president remarked acidly that it had been far easier for the United States to arrive at an agreement with the Soviets to split Germany between East and West than to reach reasonable accommodation with Britain about a division of influence with the West.[73] Once installed in the Ruhr, Britain did not know how to proceed. It had no reform program beyond a vague and inconsistent encouragement of democratization. The nationalization of heavy industry, though official policy, amounted to little more than an attempt to head off domestic critics; never, certainly, would the British seriously encourage German efforts to socialize production. Far more faithful to the spirit of nonfraternization than the United States, the USSR, or the French, they established few economic or political beachheads in their zone.[74] The bizonal agreement was also useful to Britain because it kept the Soviets and the French out of the Ruhr. Like the United States and France, the British opposed every attempt of the USSR to gain influence in West German coal and steel, and they were wary of the French as well.[75]

---

[72] Ibid., pp. 186–187.   [73] Bullock, *Bevin*, p. 320.

[74] See introduction by Rolf Steininger, pp. 1–216, in Rolf Steininger (ed.), *Die Ruhrfrage und die Entstehung des Landes Nordrhein-Westfalen: Britische, französische, und amerikanische Akten* (Düsseldorf, 1988); also, Barbara Marshall, *The Origins of Post-War German Politics* (London, 1988), passim.

[75] See John Gillingham, "Die französische Ruhrpolitik und die Ursprünge des Schuman-Plans," *Vierteljahrshefte für Zeitgeschichte* 35:1987, pp. 1–24; also, Frances M. B. Lynch, "Resolving the Paradox of the Monnet Plan: National and International Planning in French Reconstruction," *Economic History Review*, 2nd ser. 2:1984, pp. 229–243, see also Steininger, *Die Ruhrfrage*, passim.

Unlike the Soviets, the French could count on receiving at least a polite hearing from the U.S. State Department regarding their thoughts on the Ruhr. In October 1945 Monnet managed to convince Clayton and Assistant Secretary of the Army John J. McCloy that a "coal czar" should be named to end the mismanagement causing, as he saw it, the shortages of combustible then crippling France. This initiative was in flat contradiction to the official French policy of impeding Four Power cooperation. Monnet was disowned by his own government before his American allies had succeeded in bringing that of the United States around to his policy.[76] Such incidents ended with the formation of Bizonia in January 1947. Until the opening of the London conference on Germany in February 1948 the French would have little voice in determining whether or when the Ruhr would be allowed to recover, who would run it, or how they would do the job. The British would let the Americans decide these things in exchange for a less burdensome financial settlement.

It resulted from decisions arrived at during the U.S.–U.K. coal talks of July and August 1947. Launched on 30 June in preparation for the Marshall Plan conference scheduled to open in Paris, they represented a major U.S. attempt to solve the production problems of the Ruhr mines and to raise coal outputs from catastrophically low prevailing levels to those needed for European recovery. The coal talks brought together technical specialists, senior diplomats, and military government officials from both countries. In the nearly unbearable summer heat of Washington they pored together over virtually every phase of mining operations: labor, supply, finances, sales, and administration. Their summary report reads, as Bevin feared it would, like an indictment of the British for mismanaging the mines.[77]

This humiliating scolding was small price to pay for the improved financial settlement the United Kingdom secured after the coal talks. It did not come at once. On 17 December 1947 the United States relieved Britain of its bi-zonal dollar obligations for the final two months of 1947 and the whole of 1948, at the same time reducing those for sterling from one-half to one-quarter of the total. Britain in turn ceded the United States a controlling voice in the Joint

---

[76] Fondation Jean Monnet pour l'Europe (JM), AMF 4/3/5, "Résumé of a Conversation between Mr. Monnet and Mr. Clayton," 24 September 1945; see also JM AMF 4/3/10, 4/3/11, 4/3/12, 4/3/15, 4/3/17, 4/3/22, 4/3/24.

[77] USNA RG260 AG47/178/1–2, "Report on the Anglo-American Talks on Ruhr Coal Production," 17 September 1947.

Export Import Agency (JEIA), which conducted bi-zonal foreign trade, and also agreed to relinquish control over the Ruhr mines as well as place them into the hands of a German trustee organization whose mandate was to end with the election of a new government.[78] Though according to the letter of the law the German people were to determine the eventual disposition of mine properties, reality dictated otherwise. Effective decision-making power rested with the occupation governments. In terminating direct control of the mines, the British forfeited their remaining chance to encourage the socialization of heavy industry and delegated the de facto authority to determine the property settlement to the Americans.

Paris rather than Washington was the scene of the developments that brought the turning point in Anglo-American relations; there spokesmen for the continental nations and the United Kingdom met to work out a joint response to the offer of American assistance. The gathering, the Conference on European Economic Cooperation (CEEC), was convoked to devise an overall aid program and establish a permanent organization to coordinate national economic policies.[79] The Paris conference gave Britain a unique opportunity to assert economic and political leadership in Europe, but the United Kingdom sabotaged the CEEC, the Organization for European Economic Cooperation (OEEC) that issued from it was stillborn, and the Marshall Plan thereby failed to become a vehicle of European economic integration.

Britain's relationship to the Marshall Plan was initially an open question, as was evident in a telegram from Secretary of State Marshall to the Paris embassy a week after the Harvard speech: "Scope and nature of program is not yet foreseeable. It might be possible for program to be somewhat along lines of Monnet plan but on much larger scale involving several countries. On the other hand it might develop that most feasible way is to concentrate on few matters of vital importance to Europe such as food, coal and transport."[80] These two options — the one requiring the thorough internal overhaul of economic policies and structures in all participating nations,

[78] Bullock, *Bevin*, pp. 428–429.
[79] See Milward, *Reconstruction*, pp. 56–90; Immanuel Wexler, *The Marshall Plan Revisited: The European Recovery Program in Economic Perspective* (Westport CT, 1983), pp. 13–14; Hogan, *Marshall Plan*, pp. 61–62; George Ball, *The Past Has Another Pattern: Memoirs* (New York, 1982), pp. 77–78.
[80] FRUS 1947/III 840.00/6-1247, Secretary of State to Embassy in France, 12 June 1947, p. 251.

the other mere "bottleneck breaking" – encompassed a vast range of possibilities.

Regardless of any past differences, most American policy makers were prepared to give special consideration to Britain's peculiar problems. On 21 July Kennan described the United Kingdom's situation to Marshall as "serious, more serious than most people know . . . it will be one year before she can be expected even to approach balance on payments," but also thought that "a program of European recovery worked out by the Europeans themselves might include provisions which would benefit Britain along with the others." Fearing that "this may not be enough . . . some special aid may have to be provided for her [lest] she have no choice but to dismantle extensively her defense and imperial commitments."[81]

Speaking bluntly but with evident sympathy, American ambassador to France Jefferson Caffery, who was also one of the American negotiators at the CEEC, described the United Kingdom's plight a month later as desperate but not hopeless:

> Britain's position today [he said] is tragic to the point that challenges description. Her problems . . . are deep-seated and grave, and require for their solution all the coolness, the realism, the energy and the unity the British people can muster [but] as a body politic Britian is . . . sick [and] incapable of viewing her own situation realistically and dealing with it capably . . . The tragedy of the Labor government lies in the fact that after waiting several decades for a chance to put certain principles into effect, it had finally come into power at precisely the moment when these principles become essentially inapplicable . . . Only the most dire practical necessity can [however] push them to recognize that [fact] . . . If we choose to hold the British Government fully responsible, as a rational body, and to treat it accordingly, we may have to despair of it – and of European recovery. If we choose to treat it as a sick man, then perhaps, by a judicious admixture of patience and pressure, we can string things out to a better state of affairs.[82]

Caffery's views were not universal in the State Department. Rather than try gently to nurse the gravely and perhaps terminally ill patient back to health, Will Clayton, another of the U.S. negotiators at the CEEC, prescribed a bracing ice-cold shower as the best

[81] FRUS 1947/III, Memorandum Prepared by the Policy Planning Staff, 21 July 1947, p. 336.
[82] FRUS 1947/III 840.50 Recovery/8-3147, Ambassador in France (Caffery) to Secretary of State, 31 August 1947, p. 399.

cure for – or at least the means of otherwise swiftly resolving – the invalid's problems. Clayton was ready to force Britain out: Weary from months spent dismantling roadblocks raised by the British at Geneva, he insisted that prior acceptance of a European custom union and of trade multilateralizaton be required for admission to the Marshall Plan talks in Paris.[83]

Few American policy makers despaired of the British so completely as Clayton. The patience of the new secretary of state, Dean Acheson, was nearly inexhaustible. Acheson was born a subject of Queen Victoria and grew up in an Anglican parsonage. This international background did not make him popular. Fitting all too well heartland America's stereotype of the affected, upper-crust eastern seaboard Anglophile and often pilloried by commentators and caricaturists, he was an easy target for McCarthyism. Acheson was the very embodiment of the special relationship. To have let it die would have been to write off a part of his own personality.[84]

Bevin understood immediately that Marshall's speech offered Britain an extraordinary opportunity, and he quickly took off for Paris in order to win Foreign Minister Bidault's approval for a policy of joint action, at the same time intimating to the American chargé d'affaires that he was contemplating an Anglo-French Monnet Plan.[85] By taking this initiative he hoped first to head off any intervention of the United Nations' Economic Commission for Europe, which he feared might serve as an opening wedge for either Soviet or minor power influence at the Paris proceedings; second, to "steal the thunder" from the French, who had been only slightly less prompt in responding to the 5 July speech; and third, to "get the Americans to accept the thesis that Britain ought to be regarded as the partner of the USA in helping Europe through the Marshall Plan rather than 'lumped in' with the other European countries." Great Britain, he asserted, still had an empire and therefore deserved to be restored to the position of influence it had occupied at the time of the Dawes Plan. In preliminary conversations of 23 and 24 June in London, Clayton unambiguously rejected the United Kingdom's claim to act as "a partner in the Marshall Plan rather than as a part of Europe, [as this] would violate the principle that

[83] FRUS 1947/III 840.50 Recovery/8-2247, Memorandum by Lt. Col. Charles H. Bonesteel, III, Special Assistant to the Undersecretary of State (Lovett), re meeting of 22 August 1947, pp. 368–369.

[84] See Acheson, *Present at the Creation*, passim; Hogan, *Marshall Plan*, pp. 262, 269–270, 276.

[85] Bullock, *Bevin*, pp. 406–407.

no piecemeal approach to the European problem would be taken."[86] Clayton spoke only for himself, however: The United States had not yet defined the United Kingdom's role in administering the aid program. This would be up to the British.

The conflict between U.S. and U.K. views first began to surface after the (27 June–2 July 1947) Paris meeting of the Big Three at which the United States and the United Kingdom euchred the Soviets into taking the fatal step of withdrawing both themselves and their satellites from the conference. Serious discussion about an organizational framework for the proposed CEEC could now begin. A British civil servant, Sir Oliver Franks, was named to head the Conference Executive Committee, in effect to direct its proceedings. Brilliant and almost universally admired, as chief of this theoretically European organization Franks would faithfully discharge his mission as a British diplomat by limiting integration to only those instances in which his country's need for dollars made this imperative, proceeding thusly on the nominal behalf of the European nations for whom as chairman he was spokesman.

The Americans received the first intimations of stormy weather in September 1947, when the CEEC embarked upon the huge task of determining overall requirements for dollar assistance. The State Department wanted considerably more from the Europeans than a mere "shopping list of aid-worthy projects"; it expected joint commitment to a binding long-term plan of development that could be presented as a report to Congress. The American negotiators left no doubt that they expected the report to set as its goal the achievement of a self-sustaining European economy within three to four years through self-help, mutual aid, and additional far-reaching measures of cooperation. Currencies were to be stabilized, trade barriers reduced, and the efficiency of plant improved. The CEEC states were also supposed to agree to periodic reviews of their progress by the United States.[87]

Bevin and Franks told the Americans that none of this was on: The Europeans would rebel at such encroachments into national sovereignty, and the conference would go down in failure.[88] To make matters worse, the shopping list grew longer, working up to a grand total far in excess of anything Congress would have been prepared to fund. On 20 September 1947 the United States had to intervene directly in the proceedings, impose a total figure on the

---

[86] Ibid., p. 413.   [87] Wexler, *Marshall Plan Revisited*, pp. 19–20.
[88] Bullock, *Bevin*, pp. 458–459.

Europeans ($22 billion), adjourn the CEEC, label everything done so far "provisional," and regroup. Only by whittling down the $22 billion figure to $17 billion and attaching numerous conditions did the Harriman Committee, which the president had set up to draft the necessary legislation, manage to come up with a bill that could clear Congress.[89]

By March 1948 the passage of the Economic Cooperation Act was in sight, and the attempt to bring the Europeans in line could resume. The task facing them this time was to transform the CEEC into a permanent body, the eventual Organization for European Economic Cooperation (OEEC), to serve as a counterpart to the Economic Cooperation Administration (ECA) enacted by Congress in the bill signed by the president on 13 April 1948. Not until the CEEC discussions went aground the following spring did the United States place full blame for the failures of September 1947 on Britain. There had been other bones of contention in the meantime, among them Britain's repeated refusal to allow the CEEC to consider bi-zonal matters and persistent opposition to trade liberalization at the Geneva negotiations. These apparently interminable talks finally ended on 30 October 1947 with the conclusion of the GATT treaty (General Agreement on Tariffs and Trade), which the United Kingdom signed only after having been threatened by Clayton with expulsion from the Marshall Plan. The British still did not commit themselves to ending empire preference, however, instead agreeing only to a phase-out of the special trade relationship in effect with Canada.[90]

If the GATT treaty added to American disappointment with the United Kingdom, the discussion over the powers and structure of the CEEC that began on 15 March 1948 brought outright disillusionment. American ambitions for the Marshall Plan had grown considerably in the half-year following the adjournment of the CEEC. "Purpose and scope of ERP [European Recovery Program] and CEEC," wrote Marshall to Clay's political adviser Robert Murphy on 5 March 1948, "[go] far beyond trade relationships. Economic cooperation sought under ERP, and of which CEEC is vehicle, has as ultimate objective closer integraton of western Europe."[91] The State Department frankly doubted that the issue was one of choice, for "unless the participating countries are

[89] Wexler, *Marshall Plan Revisited*, pp. 41–42. [90] Bullock, *Bevin*, pp. 461–462.
[91] FRUS 1948/III 840.50 Recovery/3-348, Secretary of State to U.S. Political Advisor for Germany (Murphy), 5 March 1948, p. 389.

pressed forward toward much greater cooperation . . . the continuing organization will never amount to much more than a review and discussion group . . . and the whole program may degenerate into a series of bilateral deals between the US and the individual countries, [with] only lip service being paid to true economic cooperation by continuous discussions of customs unions and the like."[92] To assure that integration went beyond this, the United States saw the need for "a strong secretariat with clearly defined powers," something "new," "dynamic," and with "specified functions."[93]

They never got it. Responsible along with France for drafting the constitution for the future OEEC, the United Kingdom refused to grant the organization any real power. The British "defended their position vigorously. Apart from objections in principle, they argued on practical grounds that more could be accomplished, even in such matters as statistics, by national representatives pooling their information in sub-committees than by an international secretariat . . . [and] they would have nothing to do with American ideas of building up the [OEEC] through the secretariat and giving it the power to review and criticize each country's economic program, including production and investment."[94] By the beginning of 12 April 1948 progress had all but come to a halt, according to Caffery, because of the British," [whose] attitudes [would] largely determine whether or not European integration will go beyond a modest level of coordinated import programming and limited cooperative measures. Key countries, such as France and Italy, [were] moving towards a more liberal economy and de-control [but] in the absence of British participation and leadership it [was] unlikely that enough other European countries [could] be brought together to create a unit of efficient economic dimensions."[95] The 16 April 1948 Convention for European Economic Cooperation establishing the OEEC made provision for a body headed by a Council of Ministers created to supervise a seven-member Executive Committee, which was to conduct continuing business through a Secretariat staffed by international civil servants. To implement the ERP, the OEEC was empowered to prepare production programs and import–export plans, an impres-

---

[92] FRUS 1948/III 840.50 Recovery/3-448, Memorandum, William T. Phillips to Assistant Secretary of State for Economic Affairs (Nitze), 4 March 1948, p. 387.

[93] Ibid.   [94] Bullock, *Bevin*, p. 535.

[95] FRUS 1948/III 840.50 Recovery/3-2848, Ambassador in France (Caffery) to Secretary of State, 28 April 1948, p. 407.

sive mandate that fooled no one: Without the British the OEEC could not work, and they "conceived of [it] as an instrument of intergovernmental consultations and negotiations, rather than as supranational machinery that could initiate and execute projects of its own."[96]

That there was more to British opposition than the sentimental preference for informal arrangements expressed by official negotiators is clear from the reports of the interdepartmental group making policy toward the CEEC, the London Committee on European Economic Cooperation chaired by Richard W. B. ("Otto") Clarke. Clarke put his incisive mind to the service of a single overriding objective, the avoidance of economic and political change, "adjustment" as he termed it. To him the Marshall Plan presented less an opportunity than a threat, something useful only insofar as it improved the payments situation. Clarke had few illusions about Britain's ability to turn its back on the assistance program but insisted that cooperation with it be minimal. Though clearly perceiving the advantages of partnership with Europe, he rejected this on the grounds that it required excessive commitments. Painfully aware that the United States would continue to demand undesirable adjustments, Clarke nonetheless recommended that Britain continue to base policy on the special relationship.[97]

Britain's tactics further weakened this fraying tie. Obfuscation increasingly became the order of the day. Bevin's famous January 1948 Western Union speech appealing to "spiritual federation of the West," had a calculated vagueness and was devoid of concrete proposals. Clarke's 27 February 1948 minute written in preparation for the forthcoming OEEC organizational conference contains "newspeak." Though counseling "economic cooperation with Europe," it stipulated that it should be restricted to the formation of joint export cartels strong enough to penetrate the American market.[98] And it was always easy to trivialize U.S.–U.K. disagreements over the OEEC. Was the controversy really anything more than the upshot of two different, but perhaps equally good, ways of looking at things, with the British, with their unwritten constitution, preferring informal, collegial approaches to diplomatic problem solving, and the Americans, like the French bound to a written docu-

[96] Wexler, *Marshall Plan Revisited*, p. 209.
[97] See Richard Clarke, *Anglo-American Economic Collaboration in War and Peace, 1942–1949*, ed. Alec Cairncross (Oxford, 1982).
[98] See Bullock, *Bevin*, pp. 498–499, 190–191.

ment, inclined to more rigorous centralized methods?[99] What could possibly be at stake in such a friendly difference of opinion?

The refusal of the United Kingdom to second "high-caliber personel" to both the Secretariat and the Council eventually wore down the resistance of even those Americans most inclined to suspend disbelief in British ill will. On 18 April Acting Secretary of State Robert A. Lovett wrote *finis* to Britain's attempt to uphold the special relationship in economic policy: "Difficulties of Brit in reconciling their position as Eur country and as center of Commonwealth," he telegraphed the Paris Embassy, "are recognized and seem inescapable. The special position they held in relation to Combined Boards allowed them great freedom of action and flexibility. Other Eur countries during much of that period were occupied by Germans or allied troops and not in same relative position as now. The special wartime position for Brit would not be consistent with concept of Western Eur integration and other objectives of ERP."[100]

Remarks made eighteen months later by Averell Harriman, the head of the European Recovery Program, leave little doubt about the legacy of bitterness left in the wake of the May 1948 meetings. He blamed the British for having "prevailed in setting the pattern of an organization whose impotency was now becoming alarming." Harriman "agreed with [U.S. Ambassador in London] Douglas in believing that the British will not cooperate in what we want in respect to European integration and, more important, what they agreed to do in signing the OEEC charter. In the last analysis the British are not facing up to the fact that they seem to be opposing the basic principles of cooperation upon which the Marshall Plan was presented to, and supported by, the US Congress, and they must be told so bluntly and immediately."[101] British hostility to American integration policy was more costly to them than to the United States. As the OEEC's first secretary-general, Robert Marjolin, would say years later, "If the British had wanted, in the years between 1947 and 1950 really to [direct] Europe towards union, they could have done it without problems. The French would have supported them, and all the others would have followed. That was

[99] FRUS 1948/III 840.50 Recovery/3-2048, Ambassador in France to Secretary of State, 20 March 1948, pp. 395–396.

[100] FRUS 1948/III 840.50 Recovery/4-848, Acting Secretary of State to Embassy in France, 18 April 1948, p. 417.

[101] USNA RG466 DC49/1, "Minutes of Conference of 1 July 1949 at Residence of Ambassador Bruce."

an opportunity that was missed."[102] As for the United States, it would eventually find France.

3.3 France turns a corner

The Franco-American understanding of the late 1940s was built on a belated discovery of common interests and similar outlooks. Neither nation wanted to rebuild Europe around a dominant Germany or turn the nations of the Continent into spokes radiating from an American hub. Added to this was a joint commitment of some importance to liberalization and political federation. Jean Monnet was the catalyst in the accommodation process. In 1946 he committed France to the renovation of industry, the restructuring of the labor process, and the economic reeducation of the public – to modernization. The United States gave him the money to begin the job and, because he could demonstrate that having American support worked to his nation's advantage, also the power to push his way through, or maneuver around, the powerful persons, groups, and interests standing in his path.[103] Thanks to Monnet, France would eventually turn a corner.

Nothing might have come of the wartime appeal by technocrats and producers for an end to Malthusianism had Monnet not created the Plan de Modernisation et d'Equipement. The Monnet Plan, as it was soon called, met few of its targets, wasted money on a grand scale, helped create runaway inflation, and had to be abandoned in all but name in 1952.[104] It has been derided both at the time and since as either an instrument of French nationalism or a sellout to American interests. Such judgments are far too harsh. Though Monnet was not always successful in creating modern, efficient, and productive new industries, he did provide the strong central direction called for by wartime planners. His interventions into the private economy also drastically curtailed producer independence. Monnet did not make a return to the old pseudoliberal methods unthinkable – far from it! But he did make it ultimately impossible.

---

[102] HST, Oral History Interview with Robert Marjolin, 2 July 1971.
[103] See HST, Interview with Gov. W. Averell Harriman (by Iroquois Institute), 10 January 1980.
[104] See Kuisel, *Capitalism*, pp. 219–248; also Harold Lubell, "The French Investment Program: A Defense of the Monnet Plan," Special Report of Mutual Security Agency, Paris, 1952; JM AMF o/1/2, "Memorandum Vergeot" (unpublished ms., 1960).

Thanks to the plan, improvements in productivity, economic growth, and living standards ceased to be merely subjects of dull academic treatises and instead became the supreme national priorities they have since remained.

Monnet stood for those things Americans applauded and after 1946 was in a position to promote them effectively from behind the scenes. His presence gave key U.S. policymakers, many of them personal friends, the welcome assurance that France was "on track" and could be counted upon not to collapse, crumble, fade away, or turn tail and join the Russians. He was also a staunch supporter of the European Recovery Program and used his influence while deputy to Franks in the CEEC to win French backing for it. France had an illiberal economy and could not have opened its markets to Europe, but for. the diplomacy of the period this was a secondary matter. Americans demanded commitment to principle rather than immediate results, and this Monnet was happy to give, as were his associates when this was coupled to the prospect of aid money. France's theoretical acceptance of the Marshall Plan was an essential preliminary to a broader European settlement.[105]

The goal of Monnet's Plan de Modernisation et d'Equipement, as evident in the name, was to update industry and make it competitive. The American war economy was his model. Three things impressed him most about it: the speed with which fundamental decisions could be made; the vast increase in wealth that could be realized through economies of scale; and the essential irrelevance of theories of any kind to the production achievement. The debate of the "isms" had no place in what he was trying to accomplish; bigger and better factories were the solution, and the rest, it was assumed, would take care of itself. In Washington Monnet had seen that when a few "key players" got together "things happened." Civil service procedures were to him annoying, obstructive, and something to be avoided at all costs.[106]

His new planning agency (Commissariat Général du Plan) was therefore no flatulent new superministry operating from behind a mighty concrete facade but a rock-hard little directorate of disciples working from cramped offices in the rue de Martignac. Monnet's men did not respect time-worn procedures but issued orders directly to subordinate agencies of their own creation. These were the modernization commissions (commissions de modernisation), one of them for each main branch of industry. Responsible for working out

[105] Ball, Another Pattern, p. 80.   [106] Monnet, Memoirs, pp. 150–177.

the investment programs needed to fulfill plan targets, their composition reflected lessons Monnet had learned in Washington.
Management was to have a voice on the modernization commissions, but they were not to represent its views. Had not most American producers been "less than all outers," men bound by
parochialism, complacency, and instinctive preference for business
as usual? Did they not need prodding? Monnet also insisted that
labor, as well as capital, be represented on the commissions. Had not
something like co-determination been effective in raising the morale
and productivity of American workers? Distinguishing Monnet's
commissions from similar organizations in wartime America and
Speer's Germany was the prominence of technocrats. Monnet
looked to the disinterested expert, the man who shared his vision of
a better world, to link capital and labor in common cause and provide the guidance necessary to make the productivity drive a
success.[107]

Monnet was not the only advocate of planning in liberated
France. Pierre Mendès-France, the first minister of national economy in the provisional government, made a bold attempt to launch
a long-term modernization program on the basis of the Tranche de
Démarrage drawn up during the occupation, but he was dismissed
by de Gaulle in April 1945 while still drafting his scheme. His successor, a Keynesian-minded socialist named Georges Boris, was also
sacked before making an impact.[108] The implementation of an actual planning program, rather than the mere theoretical formulation of policy, was bound up with American aid from the outset. In
October 1943 de Gaulle sent Monnet, who up to then had worked
for Britain, back to Washington from Algiers to run the French
Supply Council. He took with him many of the young technocrats
who would provide the nucleus of his postwar team, among them
Pierre Uri, Etienne Hirsch, and Robert Marjolin, and they too became acquainted with the latest in American statistical methods.
From Washington Monnet directed supply and distribution in conjunction with the D-Day landing and later administered the Lend-
Lease aid to which France was entitled for a few months after April
1945.[109]

Well aware of American dissatisfaction with the lack of accountability in aid programs, Monnet convinced de Gaulle in August 1945

[107] Kuisel, *Capitalism*, p. 230.   [108] Ibid., pp. 191–192.
[109] Ibid., pp. 219–220; see also confidential source, pp. 221–222.

that France would not get its share of assistance unless, on the
French side, the best-qualified persons were brought in to manage
distribution. Monnet's persuasion worked, and de Gaulle kept him
in office during the critical transitional months from war to peace.
By early 1946 Monnet had convinced the general that the restoration of *la gloire* required a national development plan. American aid,
he insisted, was needed less for current consumption than for the
long-term modernization of industry without which France could not
recover its rightful place in world affairs. Monnet laid out his case in
a letter of 5 December 1945, and on 3 January 1946 de Gaulle
signed a decree creating the Commissariat Général du Plan, an
agency attached directly to the head of the government and vested
with the authority to act autonomously under his sole direction.

The development of Monnet's plan was also intimately bound up
with American aid. It would enable him to make the plan autono-
mous de facto as well as de jure. Monnet's strong position derived
from his unique ability, at a time of dire French need, to loosen
Washington's purse strings. Monnet, rather than the once despised
but later revered head of the 1936 Popular Front government, negotiated the $900 million May 1946 Blum loan. His insistence that
the deal be done with the specific proviso that American funds be
used for modernization did not, however, gain him immediate control of this money; in 1947 the plan was funded provisionally,
through the usual inefficient parliamentary procedures.[110]

The availability of a huge tappable pool of Marshall Plan dollars
changed this; after a strenuous tug of war with the minister of finance, Monnet managed to persuade the Schuman–Mayer government that France could get a larger slice of the pie by providing
automatic budgeting of the plan's capital requirements. Its projects
thus received priority in the allocation of counterpart funds, the
franc balance equivalent of dollar aid generated by the American
assistance program. Of the one and a half trillion francs spent by
the Plan de Modernisation et d'Equipement between 1948 and
1952, no less than half derived from counterpart funds. Though
Monnet did not get the total budgetary independence he wanted, by
1948 the plan was only remotely accountable to Parliament and only
partly answerable to the inspectors of finance. Monnet enjoyed an
almost unprecedented freedom of action.[111]

The quid pro quo for the American aid was one that Monnet and,
among influential Frenchmen, possibly only Monnet was eager to

[110] Kuisel, *Capitalism*, pp. 219–220.   [111] Ibid., p. 239.

grant: trade liberalization.[112] The opening of markets encountered no less intensive resistance in France than elsewhere. Appreciating this fact and having confidence in Monnet, American policy makers were prepared to endorse *dirigiste* methods to arrive at liberal ends. Evidence of progress was nonetheless needed to keep the money flowing. This was not easy to come by. The plan was not an immediate or obvious success: It failed at times, had unpleasant economic side effects, did not immediately generate a broad new constituency of modernizers, and was under almost constant political attack. But what Monnet could not deliver domestically he could offer in foreign affairs. By constructive manipulation he moved the French toward full official endorsement of Marshall Plan integration policy while at the same time gradually winning U.S. confidence in France as a European partner. In the months of growing disenchantment with the United Kingdom, the way was smoothed for a new Franco-American alliance.

Monnet's domestic problems began with the collieries, a policy-making showpiece that became a spectacular failure. Politically, mine problems were simple. The Charbonnages de France, owned by the state, was the sole large producer, and private interests were unimportant; the people who ran the modernization commission also managed the industry. In addition, coal was scarce, dependence on foreign sources unwelcome, and the mines in terrible condition as a result of the war. The public stood behind a policy of reconstruction and expansion, paying little heed to cost. Just about everything that could go wrong in coal planning nonetheless did.[113] Projected output increases, from 50 million annual tons in 1946 to 65 million annual tons in 1950, were far greater than those that obtained. Production actually fell in 1947 to 47.3 million tons, some 2 million tons less than the previous year, and then rose by 1949 to 52 million annual tons, a level still below that of 1929. Output levels did not respond to changes in demand: In 1946–1947, when European production shrank, that of France did likewise, and after its expansion in 1949, which created a temporary coal glut, France continued to produce. Over the same years mine investment rose spectacularly from 26.5 billion francs in 1947 to 49.1 billion francs in 1948, to 65.5 billion in 1949, the following year falling only slightly to 63.5 billion francs in 1950, nearly all the increases after 1946 being due to long-term projects involving expansion of plant. Results

[112] See Ehrmann, *Organized Business*, pp. 104–158.
[113] Lubell, "French Investment Program," pp. 103–104.

were disappointing here as well. Only 12 million tons of capacity had been renewed by 1950 instead of the anticipated 23 million tons, and only 5 million tons added instead of the expected 9.5 millon tons.[114]

The shortcomings of the plan deserve far less blame for French coal shortages in the 1940s than the failure of the British to export and of the Allies to raise output in the Ruhr. Still, Monnet's experts made unrealistic assumptions about demand, prices, and costs and delivered far less than promised. Though the extent of protection needed by the mines is difficult to judge – so massive and multiform were the subsidies received by the Charbonnages de France and so highly regulated were European markets for coal – trade liberalization would have been a disaster.[115]

This was also true, French steel producers protested bitterly, with regard to their industry. Monnet encountered intense opposition, among these, the traditional and still reigning leaders of the business community. Their attitude was ungracious. As of 1945 there had been almost no new investment in French iron and steel for five years; uncompetititve generally during the 1920s and 1930s, the position of postwar French producers seemed all but hopeless, the only possible escape from it being by way of massive infusions of new capital. Yet financial markets had all but dried up.[116] Monnet provided not only money but a strategy for investing it. He intended to reorganize the industry around two gigantic new wide-band continuous-strip mills that could produce at low cost the grades of sheet metal needed to supply the key growth industries, automobiles and home appliances. One of them, SOLLAC (Société Lorraine de Laminage Continu), was to be erected in the East, the other, US-INOR (Union Sidérurgique du Nord de la France), in the North. The construction of these mammoths, the largest mills in Europe, gobbled up most of the moneys made available by the Fonds de Modernisation et d'Equipement to private industry, as well as goodly amounts of the dollar assistance provided by the Marshall Plan and the World Bank.[117]

---

[114] Ibid., p. 107.

[115] See JM AMF 9/2/14, "Conditions générales indispensables pour permettre une amélioration profonde de la production en 1948," 20 October 1947; also USNA RG260 AG45-46/103/2, "Development of French Coal Production since the Liberation," May 1946.

[116] See Lubell, "French Investment Program," pp. 127–128.   [117] Ibid., p. 129.

These reforms, though long overdue, did not please the industry, for as a result of plan-devised or -inspired policies, and their side effects, producers feared losing control over operations. To a young American economist unaware of the conflicts between Monnet and the *maîtres de forge* it appeared as if "the obvious importance of steel production has . . . led to the treatment of the steel industry in France by the government almost as if the industry were nationalized. Although ownership and 'control' of the industry is in the hands of private citizens and the managers of various private enterprises which form the industry, the government has taken active steps in promoting the growth of the industry's production and in its domestic and international development."[118] He predicted the emergence of a new type of economic system.

The gripe list of the steel producers started with prices for raw material and costs for wages and financing, each of which was set by the state and none of which was market-determined. Management had no way of knowing whether Monnet's ambitious plans made any economic sense and had legitimate reason to fear that, if they did not, the steel industry would become an unwitting vassal of the state. The raging inflation of the late 1940s, for which the plan deserves part of the blame, added to both the fears of the ironmasters and the fury of their opposition to Monnet's policies.

The industry fought back by recartelizing. The legal basis for this was a theoretically temporary but never rescinded law of 21 April 1946 authorizing the Comptoir Français des Produits Sidérurgiques (CFPS) to allocate orders to individual firms in order to "save raw material." The Comptoir, though legally a voluntary organization, represented all of France's 300-odd steel producers, its operations providing employment to no less than 500 persons. CFPS soon developed into a central clearinghouse for steel orders, allocating them according to established quotas. It also handled internal bookkeeping for the industry, apparently on the basis of unofficial settlement prices, and sometimes even set up production programs. CFPS further negotiated all public contracts on behalf of steel producers as well as those with other big buyers such as the automobile industry. The *organisation professionelle* of the steel industry, in short, "enjoyed exorbitant powers far greater than those exercised by any other [in France]."[119]

[118] Ibid., p. 193.
[119] JM AMG 8/1/3, "Note, Ripert," 14 December 1950; Ehrmann, *Organized Business*, pp. 375–376.

The Marshall Plan offered Monnet an escape from the onsetting structural paralysis but at the same time presented one great risk. Rapid German revival threatened both general French security and the specific investment programs of the plan. Monnet did not want to prevent a Ruhr recovery or even think such a thing possible, and he recognized that without one European prosperity would not return. He did, however, intend to regulate the recovery process, slowing it down when necessary to protect French security, speeding it up when necessary to threaten French Malthusians. To achieve his dual aims of modernizing the French economy by liberalizing Europe and working Germany into a peace settlement satisfactory from the standpoint of French national interests, Monnet thought that France had to take the lead in promoting economic cooperation as desired by the United States.[120] How else could the French influence the integration process and at the same time earn a fund of goodwill that could be drawn upon in dealing with the Germans?

Monnet's aims are evident in an important memorandum of 24 July 1947 written for Foreign Secretary Bidault. The crucial thing, he emphasized, was for France to seize the initiative and, as the Americans had requested, present the outlines of a sound program for building Europe.[121] A few days later he submitted a proposal to the French delegation at the Paris CEEC conference, sketching out how American assistance could contribute to creating a "lasting equilibrium." It included suggestions for the use of U.S. aid in modernizing European production and in stabilizing the economy as well as in "creating international conditions favorable to this production . . . intensifying European exchanges, facilitating the enlargement of markets, and [setting up] transfer systems both between the nations of Europe and with the rest of the world."[122] By playing such *Zukunftsmusik* France could gain brownie points without having to introduce politically risky domestic economic reforms.

[120] See JM AMF 1/6/13, "L'amélioration de la productivité, clef du Relèvement français: Reparations allemandes et concours américain," 11 February 1946; JM, "Entretien avec Robert Marjolin," 24 November 1981; Archives Nationales de France (AN), 457 AP622, "Memorandum pour M. Georges Bidault," 27 July 1947; JM AMG 20/1/7, Monnet to Bidault, 28 July 1947; JM AMF 14/1/4, Monnet to Bidault, 22 July 1947; JM AMF 14/1/6, "Memorandum remis à M. G. Bidault par M. J. Monnet," 24 July 1947.

[121] AN 457 AP622, "Memorandum [Monnet] pour M. Georges Bidault," 24 July 1947.

[122] JM AMF 14/1/3, "Projet de memorandum sur l'ensemble des travaux de la Conférence," 30 August 1947; see also JM AMF 14/1/4, 14/1/5, 14/1/6, 14/1/10, 14/8/2, 22/1/5.

During the months of struggle with the British over the organization of the future OEEC, Bidault indeed managed to impress the Americans with the constructiveness of French attitudes toward "Europe." "The closest approach to the US concept of [OEEC]," Ambassador Caffery telegraphed Acheson on 23 March 1948, "has been the original French proposals. . . . [They] accept the necessity for executive committee as method of ensuring cooperation of national governments but do not feel it should be in permanent session. They have advocated a strong secretariat with clearly defined powers [and] advocated specification of functions of organization in some detail and granting to organization of specific powers. French delegates appear to understand necessity for something new and dynamic, and indicate they believe both French public and other European peoples are also expecting something both important and different in kind from previous organizations."[123]

Though discouraged by the failure to create a strong OEEC, U.S. commitment to Europe increased as involvement deepened. What the Americans wanted was, however, never completely clear: Words like "economic cooperation," "economic integration," and even "political unification" were used almost interchangeably.[124] By the end of 1948, two concerns were nonetheless uppermost in the technocratic minds of the Economic Cooperation Administration. One was to develop a European customs union within four years. The other was to revitalize the OEEC.

For reasons of both principle and expediency, Monnet, and France, became the standard bearer for American policies of trade liberalization and political integration in 1948 and 1949. Monnet's personal convictions were a matter of record, but where apart from his devoted band of followers was he to find support for an economy of free markets in a Europe of federated states? Fully aware of the "prevalence of . . . restrictive business practices in France," ECA chief Paul Hoffman underscored in a letter to his head of mission in Paris the "central problem [of finding] within French Government . . . a group . . . that will support anti-restrictionist measures which we could endorse." Hoffman was troubled by rumors that a new international steel cartel appeared to be in the works but, to offset its influence, could suggest only lending support to the still weak non-Communist unions.[125] The growing gap between New

---

[123] FRUS 1948/III 840.50 Recovery/3-2348, Ambassador in France (Caffery) to Secretary of State, 23 March 1948, p. 401.
[124] See Milward, *Reconstruction*, pp. 57–58.   [125] See ibid., pp. 168–211.

World ambitions on the one hand and Old World protectionism and nationalism on the other gave Monnet a tailormade opportunity to win American goodwill at very little cost. He could commend trade liberalization in theory without being expected to introduce it in practice.

Certain of the foreign policy exercises conducted in Paris during these months seem to have had no other purpose than to impress the United States with the goodness of French intentions. To persuade Americans of its seriousness about lowering tariffs, for instance, France ostentatiously negotiated for two years with Italy over a customs union that never had a chance of working either politically or economically.[126] The French were also quite happy to stand back and let the British, who stood to lose far more than they, lead the opposition in the OEEC to the tariff reductions feared by all Europeans. This defense effort won Chancellor of the Exchequer Sir Stafford Cripps a stinging rebuke from Paul Hoffman. In December 1948 he pointedly told the OEEC that the ECA would pack its bags and leave Europe unless tariffs were lowered 50 percent by volume. The reduction occurred the following month.[127]

The British also took the heat for the disappointing lack of progress in the negotiations of 1948 and 1949 at OEEC for a payments union. There were no simple rights and wrongs in Europe's complicated "money muddle."[128] The American approach to settling it, as described below, clearly worked to the advantage of France and the disadvantage of Britain, however. The French were chronically in debt to their European trading partners and therefore also forever in need of credits, the more fungible the better. Anything that would make it easy for them to buy the goods of their neighbors helped, in other words; the less restrictive the terms of credit, the happier they were. Not so Britain. Its situation, though partly of its own making, was both more complicated and more difficult than that of France. Britain was the only "European" nation with a reserve currency that could be used in international trade financing. The source of its problems was that even after the 30.5 percent devaluation of 15 September 1949, the pound remained overpriced

---

[126] See Richard T. Griffiths and Frances M. B. Lynch, "The Fritalux–Finebel Negotiations, 1949–1950," European University Institute Working Paper 84/117, November 1984.

[127] HST, Interview with Paul Hoffman, 25 October 1964; see Robert Triffin, *Europe and the Money Muddle: From Bilateralism to Near-Convertibility, 1947–1956* (New Haven, 1957).

[128] Wexler, *Marshall Plan Revisited*, pp. 128–129.

because U.K. producers could not export competitively. All postwar British governments ruled out market determination of dollar–sterling parity as potentially ruinous to Britian's international standing. The pound therefore had to be protected from convertibility to the dollar.

The United Kingdom actually supported the phase of America's European payments policy in effect during 1948 as it promoted exportation by creating a mechanism for the central clearing of accounts that regulated trade bilaterally. In this scheme, Britain, as Europe's largest creditor nation, agreed to offer sterling loans to its trading partners, who could then use them to purchase goods with prices denominated in that currency.[129] The United Kingdom could not, howver, afford an arrangement in which those nations to whom it was a debtor rather than a creditor (the case, for instance, with regard to Germany in 1949) might at some point demand repayment in dollars rather than pounds because this would have weakened the hard-currency balance in the short run while in the long run providing its trading partner with the means of purchasing outside the sterling area. In 1949 the insensitive Americans nonetheless demanded that Britain make precisely this concession by accepting the introduction of so-called drawing rights enabling creditors to demand settlement in dollars. The struggle over this issue went on until April 1950, when Hoffman finally put his foot down. Protesting that Britain was trying "to secure complete freedom to pursue its domestic policies and to exploit all the benefits of its present bilateral trading relationships," he withheld $600 million of ERP appropriations until agreement had been reached.[130] Acheson "nudged" Bevin on 11 May, and on 20 June 1950 a compromise arrangement, allowing for the settlement of only 25 percent of balances by means of drawing rights, came into effect. The new European Payments Union (EPU) was born.[131]

Though the EPU would later become the most valuable institutional legacy of the Marshall Plan, it was of less immediate diplomatic importance than the controversy generated by the American attempt to install a strong director-general at the OEEC.[132] The American choice for "Mr. Europe" was Belgian Foreign Minister Paul-Henri Spaak. ERP and State, inexplicably confident that their selection would be loudly applauded, were unprepared for British foot dragging and frankly shocked by the hostility of their eventual

---

[129] Ibid., p. 137.   [130] Ibid., p. 170.   [131] Ibid., p. 171; Bullock, *Bevin*, p. 762.
[132] Hogan, *Marshall Plan*, pp. 157, 217, 218, 329.

reaction. Tired of backing away from months of Acheson's increasingly less gentle nudges, Bevin told Acheson flat out that the United Kingdom was not about to get behind either the Spaak candidacy or the kind of OEEC that the United States was trying to build, telling Ambassador Douglas bluntly that the Labour government could not be pressured into accepting the appointment of Spaak or any other Continental to a position of control in the OEEC. Britain would also never agree, he added a few days later, to any "politicization" of the OEEC, whose responsibilities should remain strictly economic. Unlike the British, Monnet, whose views were thought weighty enough to be worth communicating to the secretary of state, "had spoken wholeheartedly in favor of Spaak's appointment and shared the view that unless something were done to revivify the OEEC, it would be disastrous."[133] In January 1950 the United States dropped the Spaak nomination. In April 1950 a compromise candidate, Dirk Stikker of the Netherlands, became chairman. The United States would never again attempt to transform Marshall Plan structures into vehicles of federalization because without Britain's cooperation they could not work. The future of American integration policy was in the hands of France. There remained one enormous obstacle to its further pursuit, the Ruhr Problem, which, as Raymond Poidevin first stated in 1979, "was the main issue in all French efforts to solve the German question after 1945."[134]

3.4 The French, the Americans, and the Ruhr

The last stop on the line from Morgenthau Plan to Schuman Plan was a Franco-American agreement on German policy that centered on the Ruhr: The end of the occupation in the western rump of the former Reich would be predicated on merging its industry into Europe. This was to be assured by maintaining occupation controls over the Ruhr until, a process of federalization and liberalization once underway, the whole of Western European heavy industry could be put under the supervision of a single international authority. The arrangement harmonized French security policy and Amer-

---

[133] FRUS 1949/IV 840.00/10-1949, Secretary of State to Embassy in France (Meeting of U.S. Ambassadors at Paris, 21–22 October), pp. 469–496; FRUS 1950/III 340.240/I-550, Ambassador in Belgium (Murphy) to U.S. Special Representative in Europe (Harriman), 5 January 1950.

[134] Raymond Poidevin, "Frankreich und die Ruhrfrage, 1945–1948," *Historische Zeitschrift* 228:1979, p. 317.

ican integration policy, yet no single act or document enshrines it; the understanding was arrived at episodically, even fortuitously, and sealed only by the American reaction to the Schuman Plan announcement of 9 May 1950. Only then did the United States commit itself irrevocably to backing France's solution to the German problem.

The Schuman Plan culminated the Ruhrpolitik France had pursued consistently since the war.[135] Its history is conventionally described as having involved a simple evolution from hard to soft. The French, one is told, initially did everything in their power to cripple German industrial recovery but having enjoyed little success in the attempt wisely backed new leaders preaching a doctrine of reconciliation: Having failed to defeat the historic enemy, France thus joined him in a new cooperative effort to transform the cultural reality of Europe into something practical, institutional, and political.[136] Monnet's autobiography provides the only first-hand account of the weeks immeditely prior to the Schuman Plan announcement but unfortunately does little to demystify its origins. The circumstances are as follows. At the beginning of April 1950 Monnet took a hiking trip to Switzerland, returning from it two weeks later convinced of a need to take a bold initiative on the German question and with a general outline for a coal–steel pool. He worked out the details of the plan with the members of his team over the next several days. On 28 April he presented the completed proposal to Prime Minister Bidault, who failed to act upon it, then sent a copy to Foreign Minister Schuman, who agreed to do so. On the eve of the announcement, which was timed to precede the opening of the London conference, Monnet told Acheson and a couple of his deputies about the planned proposal, pledging them to silence until after Schuman had had a chance to act.[137] The text of Monnet's ghost–written memoirs unfortunately adopts a quasi-mosaic imagery in depicting these events: a trip to the wilderness; a return with a mental tablet inscribed with laws and commandments; a gathering of disciples and elders; and a proclamation to the assembled tribes and peoples. Though the factual accuracy of Monnet's account is not in doubt, scholar-skeptics, accusing him of self-glorification, have begun to look farther back into French diplomacy in the hopes of uncovering the "true" origins of the Schuman Plan. In their view,

---

[135] See Gillingham, "Die französische Ruhrpolitik," pp. 1–24.

[136] See Catherine de Cuttoli-Uhel, "La politique allemande de la France (1945–1948): Symbole de son impuissance?" (unpublished ms., April 1954), pp. 1–33.

[137] Monnet, Memoirs, pp. 288–289; also, Acheson, Present at the Creation, pp. 498–499.

the tripartite London conference of 1948 was when the decisive break between "hard" and "soft" phases occurred.[138] About the break itself, they have no doubts.

It is normally associated with a change of leadership at the Quai d'Orsay on 27 July 1948, when Robert Schuman replaced Foreign Minister Georges Bidault, who had held the office since 1945. Schuman was gentle, upright, pious, and celibate,[139] his predecessor, exasperating, often tedious, a whiner, and an inebriate historically suspect because of having been implicated in the 1960 *Putsch* attempt by the Organisation de Armée Sécrète.[140] Yet both Schuman and the slightly wobbly man he succeeded stood for the same policy, which indeed was not really their own but consensual, as well as continuous, clever, and even courageous. It represented the views not only of the leadership of their party, the Mouvement Républicain Populaire (MRP), but of first- and second-tier officialdom at the Quai d'Orsay as well. There were no breaks in it: French Ruhrpolitik was formulated during the war and thereafter unfolded logically step by step. As intended, it provided plenty of discussion fodder for the seemingly innumerable and interminable Allied get-togethers where the German coal–steel complex was featured on the agenda. French Ruhr policy was courageous as well, and not only because it culminated in the bold stroke known as the Schuman Plan. Bidault's habitual refrain, "I cannot afford to get too far ahead of French opinion on the German question," was more than an excuse for inaction or the application of a mild form of diplomatic blackmail; it was the truth. Both French foreign ministers deserve credit for having campaigned strenuously at great political risk to change the public opinion of their land from something anti-Boche in the manner of 1919 to something which, if hardly pro-German, at least admitted a need to reach some form of accommodation with the powerful eastern neighbor. France could not have contributed to the construction of the new Europe had not Bidault and Schuman partly succeeded in reeducating the public. Britain, where no similar effort was made, provides an instructive comparison.[141]

Most contemporary and historical commentators have understandably emphasized the problems that French Ruhr policy posed for the American integration effort. Long after the United States had initiated German revival, France remained committed to poli-

[138] See Milward, *Reconstruction*, pp. 158–159, 164–165.
[139] Kindleberger, *Marshall Plan Days*, p. 187.
[140] See Bidault, *Resistance*, passim.
[141] See Gillingham, "Die französische Ruhrpolitik," pp. 1–24.

cies in some ways reminiscent of those introduced by the Big Three during the first phase of the occupation. Though the French never advocated deindustrialization, of the powers represented at the Allied Control Council they were undeniably the most strenuous in opposing the administration of Germany as a single unit, the most consistent in advocating the territorial separation of the Ruhr from Germany, and the most restrictive in contemplating the restoration of sovereignty.[142] One must add to this bill of indictment certain incriminating realities of life in the French Zone, where self-indulgent and indolent placement of an overblown and parasitical occupation authority wallowed in viceregal luxury at the expense of the hungriest public in a badly impoverished country.[143]

Though a certain vengefulness was surely present, the source of French extremism was weakness. France was no longer a Great Power but a recently defeated nation of midsized proportions that had to return to respectability on its knees.[144] It confronted a backlog of unsolved social and economic problems exacerbated by political divisions stemming from the occupation and faced critical shortages of food and raw materials. The French were neither consulted about Yalta nor invited to Potsdam, were given a seat on the Allied Control Council thanks only to the good nature of the British foreign secretary, and were assigned a zone of occupation mainly because State Department Francophiles wanted to improve the nation's morale.[145] France had neither the strength to impose policy directly on Germany nor the standing to do so indirectly through the Allies. To be heard the French had to shout, to be noticed had to annoy, and to have any impact had to obstruct. If this unpleasantness lasted longer than it should have, it is because France unwisely did not merge its zone with those of Britain and the United States and so was deprived of influence over events in Germany in 1947 and 1948.[146] Certain French proposals undeniably would have had consequences as indefensible as those, for instance, of the first Level

[142] John Gimbel, *The Origins of the Marshall Plan* (Stanford CA, 1968), pp. 35–36, 83–84, 226–227.
[143] See Werner Abelshauser, "Wirtschaft und Besatzungspolitik in der französischen Zone, 1945–1949," in Claus Scharf and Hans–Jürgen Schröder (eds.), *Die Deutschlandpolitik Frankreichs und die französische Zone, 1945–1949* (Wiesbaden, 1983), pp. 111–140.
[144] de Cuttoli–Uhel, "La politique allemande."
[145] Bullock, *Bevin*, p. 144; Gimbel, *Origins*, pp. 35–36.
[146] See Milward, *Reconstruction*, chap. 4, "France and the Control of German Resources," passim.

of Industry Agreement.[147] Yet France was always prepared to drop, or at least postpone, its radical-sounding plans in order to meet pressing economic requirements, especially the need for Ruhr coal: Their primary use was for bargaining. There is one final proof that neither simple anti–Germanism nor old-fashioned imperialism was the primary operating force behind France's Ruhr policy. Once Monnet's modernization plans had begun to take effect, official fears of Ruhr competition abated and the readiness to strike compromises grew. France became ready, as Acheson later put it, to offer the hereditary enemy a political mortgage for the future.[148]

French Ruhrpolitik grew out of a searching investigation of the German problem begun under the auspices of the provisional government in Algiers in summer 1943 and continued at the foreign ministry in the months between liberation and V-E Day. Impelling it forward was a profound, nearly universal conviction that nothing less than national survival depended on a satisfactory outcome of the Ruhr Problem. A fundamental policy paper on Germany dated 21 August 1944 thus ominously begins, "If France should have to submit to a third assault during the next generation, it is to be feared that . . . it will succumb forever."[149] The French obsession with the German menace did not bring a return to the hard line followed after World War I but instead resulted in the adoption of the fundamentally new approaches that culminated in the formation of the European Coal and Steel Community. A 5 August 1943 "think piece" produced by Monnet and a few of his associates in Algiers foreshadowed what was to come. In the dramatic language typical of his important memorandums, the document demanded a break with the past and asserted the French right to take the lead in shaping the future. The paper began with a warning that the Clémenceau approach would encourage nationalism, stifle economic growth, and lead to war; as alternatives, it appealed to the creation of a new "European entity" that could open large markets and reduce the dangers of competitive rearmament. The document con-

[147] For instance, Ministère des Affaires Extérieures (MAE), Y 1944–1949, 396, "Memorandum de la Délégation Française au sujet de l'Allemagne" (CMAE 46/1), 1 January 1946; JM AMF 4/3/99, "Base de négotiations Franco-américaines," 15 February 1946; JM AMF 4/3/93, "Avant-projet concernant les négotiations à engager par la France au sujet de l'Allemagne," 15 February 1946; MAE Y 1944, 398, "Ministère de la Défense Nationale, Etat-major de la défense nationale . . . note pour M. le Ministère des Affaires Economiques et Financières," 22 January 1947.
[148] HST, Acheson Papers, Princeton Seminars.
[149] AN 457 AP [550 prov.], "Le Problème allemande," 21 August 1944.

cluded that among European nations only France was in a position to direct such an effort; it indeed insisted that France had a historical mandate to do so.[150]

Monnet's vision of a "new European order" was never completely absent from French Ruhr policy because its formulators, including successive military commanders, recognized that ultimately there was no realistic alternative to one. The French categorically rejected deindustrialization along the lines of the Morgenthau Plan because "its grave defects [would have deprived] other nations of a considerable store of riches, [plunged] the German people into economic chaos, and [created] a center of agitation in the midst of the continent."[151] Another military occupation like the one of 1923 was similarly ruled out, partly on humanitarian grounds. According to one study, an effective occupation would have required a minimum of 165,000 troops and the permanent maintenance of two and a half cohorts under colors. This mobilization would have aggravated manpower shortages estimated at five million. To fill them, labor drafts would have to be imposed on Germany like those experienced by France in 1942, something rejected as being as impractical as it was morally unacceptable.[152]

French policymakers faced an obvious dilemma: to break German political power while preserving, and even increasing, the Ruhr's economic importance to Europe. The only way to deal with the problem was to impose what was called "organic control" (contrôle organique), a system for the supervision and regulation of Ruhr heavy industry that could be kept intact until either the French had become strong enough or the Germans well enough behaved to render it superfluous. This was the conclusion reached after a detailed policy analysis conducted at the Quai in March 1945. Although recommending the elimination of armaments factories and the provision of restitutions to compensate France for wartime losses caused by Germany, the finding did not recommend reducing levels of economic activity in the Ruhr and rejected expropriation of the giant trusts as being too radical and leading to state control. It further objected to the displacement of industry as presenting insurmountable technical problems and opposed placing bans on the manufacture of certain products, because this could be easily dropped once

[150] JM (unclassified); see also Monnet, Memoirs, p. 293.
[151] AN 457 AP [Allemagne 60 prov.], "Desarmament économique de l'Allemagne," 19 November 1944.
[152] AN 457 AP [Allemagne 60 prov.], "Le problème rhénan-westphalien et la situation française," 22 January 1945.

the occupation was over. What, then, was *contrôle organique* to consist of? This paper recommended "effective management" of production and sales at the firm level, a campaign to break up the trusts, and a commitment to maintain a long occupation; as it correctly warned, "No matter how just the principles of collaboration between German and foreign industrialists, the European and extra-European states must retain ultimate authority over the reorganization and supervision of the German economy in order to assure that the habitual camouflaging methods do not result . . . in the reconstitution of certain . . . big concerns."[153] Though the occupation was intended to be long, it was never intended to be permanent; nor was its mechanism, organic control, to be rigid and inflexible; the finding specified rather that contrôle organique was "to take precise forms over time while being adapted as closely as possible to the reorganization of the postwar German and European economies." The ultimate objective, according to the instructions of the director of political affairs at the Quai, "[was] to integrate the productive forces of Germany into a new international order."[154]

Years would pass before France was in a position to announce this as the official goal of its policy; operating from a weak position after 1945, French diplomats had little choice but to pursue modest objectives. The first of them was merely to get a seat at the conference table. Not until 1946 was France able to participate at the meetings of the Council of Foreign Ministers (CFM), and then only because the sessions devoted to the drafting of treaties with the minor ex-enemies were to be held in Paris. The second French objective was simply to prolong the occupation. "No imaginable solution," according to an internal document of the Defense Ministry dated 8 April 1946, "[could] offer France stronger guarantees than those . . . it [enjoyed] participating in the present occupation of Germany. [This allowed] France to cross a difficult period and progressively regroup its strengths especially its army, under the protection of the other allied armies."[155] The French assiduously sabotaged every attempt at Four Power agreement on policy toward the former Reich, without, however, significantly diminishing its scant chances of success; as Military Governor Koenig wrote Bidault in December, "When real discussion starts . . . all agreement ceases. Evidence to the con-

---

[153] AN 457 AP [Allemagne 60 prov.], "Direction Générale des Affaires politiques: Contrôle de l'industrie allemande," 30 March 1945.
[154] Ibid.
[155] AN 457 AP [Allemagne 60 prov.], "Etat-major de la Défense nationale, 1ère section, no. 384 DN/IP/TS," 8 April 1946.

trary is a mere *tromp d'oeuil.*"[156] The third objective was to win Allied consideration of French proposals for the Ruhr. This was not always easy to get. In November 1945 Maurice Couve de Murville brought a team of experts to Washington to explain why France thought it necessary to separate the heavy industry district from the body of the Reich, only to be chided at their final meeting by Walt W. Rostow for harboring the illusion that control of the Ruhr was of more than negligible security value in an age of atomic warfare.[157]

The British were equally unreceptive to the major French initiative of January 1946: To them France's suggestion of transferring Ruhr assets into the public domain smacked of Rathenau's World War I "war companies."[158] But where sweet reason had failed demonstrably, might not obstructionism, brokerage, and extremism work? In 1946 the French introduced the first into the meetings of the Allied Control Council, injected the second into the sessions of the Council of Foreign Ministers, and incorporated the third into their proposals for "Ruhr-Rhénanie." These tactics were reproachable but did little real damage. Though French obstructionism at the Allied Control Council at times drove U.S. Military Governor Clay to distraction, it was only a secondary cause of the failure of Four Power rule in Germany.[159] As practiced by Foreign Minister Bidault at successive Councils of Foreign Ministers, brokerage – a kind of promiscuous trafficking between East and West – was even less consequential. A man of the Right contemptuous of both communism and Russians, Bidault himself later condemned such behavior. A domestic political necessity until 5 May 1947, when the Communists were thrown out of the French cabinet, it ended thereafter.[160]

The extremism in French proposals for Ruhr-Rhénanie cannot be so easily dismissed. Politically, they stipulated a radical and immobilizing decentralization; economically, they included measures that would have crippled recovery: the institution of international management for mines and mills, the imposition of elaborate controls throughout the economy, and the parcelization of Germany into tiny cellular customs areas. As bad as these ideas were, France never

[156] AN 457 AP [Allemagne 60 prov.], "Le General Koenig, Note personelle ridigée à la suite de la réunion du 6 November 1945, Secret absolu."
[157] AN 457 AP [Allemagne 61 prov.], "Conversations Franco–américaines au sujet de la Ruhr et de la Rhenanie," 15 November 1945.
[158] MAE Y 1944–1948, "Conversation Franco–britannique sur les problèmes économiques allemands," 9 April 1946; also, 397, "Note," 8 May 1966.
[159] See also Gimbel, *Origins,* passim.   [160] Bidault, *Resistance,* pp. 127–128.

expected to be in a position to impose them; they hardly amounted
to an equivalent to planning for the Nazi New Order. The Ruhr-
Rhénanie plans had a more limited purpose: to serve as claims
stakes, diplomatic pabulum, and bargaining chips.[161] When matters
became serious, France was quick to abandon its harebrained
schemes. Bidault admitted as much to the Americans in August
1946, when he let it be known "unofficially" that "French policy towards Germany, particularly the insistence that the Ruhr should be
detached, had been a mistake," adding that only domestic political
considerations had prevented him from dropping the demand.[162] In
an internal memorandum written a month or so earlier, the reput-
edly inflexible Hervé Alphand had already recommended the adoption of what would become the centerpiece in French Ruhr policy a
year later: inter-Allied control followed by a restoration of German
sovereignty.[163]

France's pressing need for German coal normally relegated its
long-range Ruhr schemes to the background, as for instance happened at the scene of the most important Four Power discussion of
the Ruhr Problem, the Moscow conference set in the brutal winter
of 1946–1947. France had three negotiating objectives: to secure
more coal, limit German steel output, and impose international control over the Ruhr. It used the latter two issues to gain access to fuel.
At a preconference tête-à-tête with Secretary of State George Marshall, French President Vincent Auriol had let it be known that coal
would be France's highest priority at the forthcoming sessions. Auriol told Marshall that in 1938 France had seventy-three million tons
at her disposal, of which some twenty-five million tons were from
imports, adding that in 1947 the overall figure had fallen to sixty-
one million tons, which included a substantial increase in domestic
production to over fifty million tons. Noting that imports were running at a rate of only twelve million tons, over one-half of which was
prohibitively expensive American coal, the French president announced France's intention to press at the Moscow meeting for an
immediate increase in deliveries from Germany to 500,000 tons per
month, as well as for a further increase within a year to one million

[161] MAE Y 1944–1949, 396, "Memorandum de la Délégation française au sujet de
l'Allemagne" (CMAE 46/1), 1 January 1946; AN 457 AP [60 prov.] "Note au
sujet de la politique française en Allemagne," April 1946.
[162] FRUS 1946/V 760.00119, Caffery to Secretary of State, 30 August 1946.
[163] AN 457 AP [Allemagne 60 prov.], "Note [Alphand]: Problèmes allemands," 18
July 1946.

tons per month. Without them, he warned, French recovery would fall behind Germany's.[164]

To get the coal the French introduced the so-called Alphand Plan, recommending the displacement of three million tons of basic steel capacity from the Ruhr to Lorraine, and Bidault revived dog-eared French proposals for internationalizing the Ruhr. None of this activity counted for much. After a month of acrimonious debate the Moscow conference broke up in disarray. Yet the French did not leave altogether empty-handed. By agreeing to drop the Alphand Plan, they persuaded the United States and the United Kingdom to endorse the so-called sliding-scale agreement providing for both increased gross fuel exports to France and larger German shares in overall consumption as outputs rose. Ruhr coal production failed to increase for several months, however, and French recovery continued to be jeopardized by shortages.[165]

The tactics of Ruhrpolitik changed significantly after the Marshall Plan announcement; obstruction, brokerage, and extremism gave way to initiative, cooperation, and reasonableness. What the French people wanted, Bidault told Clayton on 7 August 1947, were guarantees that German recovery would not outpace French, and once these were granted France would drop its admittedly wild proposals for control of the Ruhr in favor of international management. Thus Clayton "found little difference between the French and US view regarding the level of industry in Germany, and . . . would interpose no objections to any level of industry which we and the British might agree upon provided the French people get assurance that the resources of the Ruhr would not again be employed in war on France . . . Such assurance [Bidault added] could be had by the creation of an international board which would allocate the Ruhr production of coal, iron, and steel, and perhaps chemicals between Germany and other countries. After the peace treaty such [a] board would be composed of representatives of US, UK, France, Benelux and Germany. Prior thereto, it would be composed of the same countries minus Germany." Bidault mentioned in conclusion, according to Clayton, that "France had abandoned previous plans re-

---

[164] FRUS 1947/II 740.00119 Council/3-647, "Minutes of conversation between the Secretary of State and the President of France," 6 March 1947.

[165] Office of Military Government, U.S. (OMGUS), "Lorraine Steel instead of Ruhr Coal? A Study of the French Proposal for the Transfer of Pig Iron and Steel Production from Rheinland-Westfalen to France: Wirtschaftsvereinigung Eisen- und Stahlindustrie, Düsseldorf," August 1947; FRUS 1947/II 862.6362/4-2247, Secretary of State to French Foreign Minister (Bidault), 19 April 1947, pp. 486–487; Bidault, *Resistance*, p. 147.

garding detachment of Ruhr from Germany, suggestions for
internationalization (etc.) and is entirely willing to leave ownership
and administration with the Germans but will insist that access to
the products of the Ruhr . . . not be subject exclusively to the will of
the Germans."[166]

Auriol echoed these sentiments in a letter to his foreign minister.
Describing the Ruhr Problem as "the center of all international action and the culmination of international politics, the solution of
which is the essential condition for the success of the Marshall Plan
and the organization of Europe," the French president worried that
the United States might end up inadvertently restoring the Krupps,
the Stinneses, and the other "smoke stack barons." He therefore recommended the establishment of an "international organization for
the Ruhr like the Tennessee Valley Authority, which would be
owned by the United Nations for about thirty years and managed by
a board of directors composed of representatives of surrounding
states, a Rhenish state and German unions and supervised by a
board of surveillance constituted by the UN and the Big Four."[167]

The "Anglo-Saxons" dashed such hopes. Since 1946 France had
refused to merge its zone of occupation with those of the Americans
and the British out of fears that this would be a step toward unification and the restoration of German control, put a halt to booty
seizures, and eliminate what little leverage the French enjoyed in
international councils. France would now have to pay for this stiff-
necked attitude. The bi-zonal Allies concluded the summertime coal
discussions in Washington with a compromise that all but knocked
the wind out of French policy. Unable to agree whether to restore
Ruhr assets to private control or socialize them, they decided to let
the Germans work the matter out for themselves. The French found
themselves cut out of a role in the restoration process.[168]

This decision had two far-reaching consequences, the first of
which was the recognition that a continued refusal to work with the

[166] FRUS 1947/II 862.60/8-747, Undersecretary of State for Economic Affairs
(Clayton) to Secretary of State, 7 August 1947, and 862.60/8-647, pp. 1022–
1024; Memorandum, Director of Office of European Affairs (Hickerson) to
Undersecretary of State (Lovett), 23 August 1947, pp. 1050–1055; AN 457 AP
[622 prov.], "Memorandum de M. Georges Bidault," 24 June 1947.
[167] AN 457 AP [550 prov.], Auriol to Bidault, 7 August 1947.
[168] MAE Y 1944–1949; "Note sur le statut des mines de la Ruhr," by G. Parisot
(n.d.); also, FRUS 1948/II 740.00119 Control (Germany), POLAD to Secretary
of State, 11 January 1948, and French Desk File, Lot 53D246, Germany, Associate Chief of Division of Western European Affairs (Wallner) to Counsel or of
Embassy in France (Bonbright), 16 January 1948.

United States and United Kingdom on German problems would be a disaster. France was thus an eager participant in the London tripartite conference on Germany. Opening in early 1948 and lasting all year, it tackled a number of problems surrounding the organization of the future West German state.[169] This conference could not have been convened, however, until the diplomatic machinery of the Big Four had been effectively put out of operation. This happened at London, the CFM opening in November 1947. On 17 December, after nearly a month of predictable tedium, Secretary of State Marshall proposed to everyone's feigned astonishment an indefinite postponement of future foreign ministers' gatherings (which according to Potsdam were to have been held every six months). The Soviets thereupon walked out as expected. The deed had been done; the stage was set for the Western Allies to proceed on their own.[170]

The "Anglo-Saxon" decision to let the Germans determine the disposition of Ruhr assets caused a momentous policy shift: France set out to restore contacts with the hereditary enemy. Neither the magnetic Monnet nor the saintly, slightly sanctimonious Schuman but the often belittled, frequently besotted, and historically dismissed Bidault was the first to sound the new note in French policy toward the Germans. On 4 January 1948 he informed General Koenig that "German recovery in the cadre of Europe should take its place as rapidly as possible," emphasizing that "the first phase of the occupation is now over [and] our main concern is no longer to right past wrongs but to prepare for the future." This would rule out, he specified, both "direct administration of the Zone . . . and exploitation for our own profit."[171] Bidault ordered Koenig to stop all seizures, forced sales, and displays of conspicuous consumption at once, further directing that "everything . . . be done to develop useful contacts with Germans." Frenchmen, he said, should miss no opportunity to explain politely to them that "we [did] not intend to dominate but merely to play an honorable role in a united and cooperative Europe."[172]

[169] See Milward, *Reconstruction*, pp. 141–168; FRUS 1948/II, pp. 1–571.
[170] See FRUS 1947/II 840.6362/10-847, Memorandum of conversation by the Secretary of State, 8 October 1947, p. 684, and 740.00119 Control (Germany) 11-1847, Memorandum of conversation by the Deputy Director of the Office of European Affairs (Reber), 18 November 1947, pp. 721–722.
[171] AN 457 AP [Allemagne 15 prov.], "Les Ministère des affaires étrangères à M. le Général d'Armée Koenig," 4 January 1948.
[172] AN 457 AP [Allemagne 15 prov.], "Instructions adressées à M. Schneiter," 7 January 1948.

This desire to discover a basis for an accommodation with Germany proved highly contagious. It soon infected the Foreign Office where over the following months Roger Fabré of the Central European desk drafted numerous memorandums advocating a striking public relations gesture to emphasize France's recognition of Germany's right to equality, which he hoped would serve as a first step toward the creation of a climate of "constructive European collaboration."[173] Symptoms of the new contagion were also manifested at such intimate gatherings as a dinner held on 20 July 1948 at which the economist Henry Laufenberger expressed his "personal view" that Germany, being more malleable than the United Kingdom, Italy, or the Benelux nations, was quite capable of becoming France's best economic partner. Another guest, the MRP deputy Pierre Pflimlin, thereupon recommended setting up an "energy complex," which would include France and the Benelux nations as well as Germany and be run by a European authority. Jean Monnet interrupted this discourse to suggest that the matter be referred to his friend John Foster Dulles, to which a man named Abelin, the fourth diner, nodded gravely in silent assent.[174]

Over the coming eighteen months, as the Federal Republic moved off the drawing board and was sized, drawn, bolted together, and put into operation, such swirling brews of hope and ambition would be distilled into strategies of reconciliation; the London tripartite conference created much of the setting in which they would be put to work. Secretary of State Marshall convened it "in order to proceed at once with the economic integration of West Germany into the economy of Western Europe," something that could be achieved, he went on, "by rehabilitating the Western German economy within the framework of the program for general economic recovery in such a way as would insure a maximum contribution of the Western German economy to European recovery." "Equitable economic arrangements," he concluded without being more specific, "should enable conditions in West Germany to improve along with the economic progress of Western Europe generally."[175] The means obviously eluded Marshall for reconciling the European economic necessity of German revival with the French

---

[173] MAE Y 1944–1949, "Note [Fabré] Direction d'Europe, Sous-Direction d'Europe Centrale," 12 June 1948; MAE Y 1944–1949, Roger Fabré, "Note sur les aspects économiques du problème allemand," 1 April 1948.

[174] AN 80 AP, "Allemagne," 20 July 1948.

[175] FRUS 1948/II 740.00119 Council/2-2048, Secretary of State to Embassy in the United Kingdom, 20 February 1948.

political imperative of industrial security, his only suggestion in this respect being that Bevin's recent proposed Western European Union, itself a monument of imprecision, might magically become a vehicle of economic integration.[176]

On 27 February 1948 the French presented a proposal intended to provide an escape from the dilemma facing Marshall, to create an international authority for the Ruhr. As envisaged, the regulatory body would become the centerpiece of contrôle organique. In reality, the proposal would be discussed to, and even beyond, the point of exhaustion. Still, an International Authority for the Ruhr (IAR) did result from the tripartite negotiations in London and, though never fully operational, would serve as forerunner of the European coal–steel pool. The Ruhr Authority envisaged by the French was to include delegates from the United States, United Kingdom, France, the Benelux nations, and Germany (represented initially by the occupation authorities), who were to be empowered to appoint management, approve production plans, and inspect plant. The new directorate was also supposed to determine how much coke and coal could be consumed domestically and how much exported. The military governors were to be in charge of the proposed organization, and its powers and structure could be modified as necessary and desirable.[177]

The paper approved on 27 May 1948 by the three parties modified the French proposal in a couple of significant ways. The International Authority for the Ruhr was to assure that the "resources of the [area] shall not in the future be used for aggression but . . . in the interests of peace," guarantee "European powers operating in the common good . . . non-discriminatory access" to the Ruhr coal and steel, and encourage a general lowering of trade barriers as well as the democratization of Germany. Although the powers of the IAR were to be exercised jointly as specified in the earlier French draft, they no longer extended to management. IAR nonetheless retained the authority to prevent unfair trade practices and determine the amounts of exportable coal. The final agreement also replaced the vague initial French promise to allow future changes in structures and functions with a specific stipulation providing for a control period of defined length that was to remain in force until the Allied governments chose to end it. A Military Security Board cre-

[176] Ibid.
[177] FRUS 1948/II 740.00119 Council/2-2748, Amassador in the United Kingdom (Douglas) to Secretary of State, 27 February 1948, pp. 97–98.

ated to vet all proposals for factory construction with a view to preventing the reestablishment of an armaments industry was to
operate under IAR's authority.[178]

The IAR agreement was not universally well received within the
French government. Prime Minister Queuille, Minister of Finance
Petsche, and the inspectors of finance counseled caution. Monnet
(in "bad odor" with the current government and "generally at dag-
gers drawn with the Inspectors of Finance") nonetheless thought
the time for the "big initiative" was at hand. At a 20 August 1948
meeting of experts at the rue de Martignac he directed that long-
term planning on a European scale now begin,[179] adding that he
expected the ECA to approve such action. Though not specifying
the extent to which the "harmonization" of French and German
economic development should work to France's advantage, he obviously expected that in the short run it would offer a measure of
protection to French modernization plans and in the long run serve
as a means of economic integration.[180]

Monnet's plans got no farther, nor did the IAR. Once again unilateral action by the bi-zonal partners pulled the rug out from under the French "integrators." U.S.–U.K. Law 75 of 10 November
1948 for coal and steel reorganization, codifying the understanding
reached a year earlier, provided that "the eventual ownership of the
coal and iron and steel industries should be left to the determination of [the] freely elected German government."[181] Though the law
did not explicitly contravene the IAR agreement, it made a mockery
of the text's provisions for international management. Failing to
have the law repealed, the French lost interest in the IAR as a vehicle for reintegrating the Ruhr into Europe. Its neutering did have
one unexpected consequence, however: The French were allowed to
join the two control groups in charge of monitoring the coal and
steel industries without having to enter Bizonia. Though unable to
chair the sessions of the control groups, France would turn them

[178] FRUS 1948/II, CFM Files, Lot M-88, Box 118, File-TRI Documents, "Paper
Agreed Upon by the London Conference on Germany," 27 May 1948, pp.
285–288.

[179] MAE Z 1944–1949 [Allemagne 87], "Compte rendu de la réunion sur les affaires allemandes tenu au Commissariat du Plan," 20 August 1948; also in AN
80AJ [unnumbered–1948].

[180] MAE Z 1944–1949, Bonnet Papers, 1 August 1948–December 1949, Chauvel
to Bonnet, 7 December 1948.

[181] FRUS 1948/II 740.00119 Control (Germany) 12-1348, Acting Secretary of
State to Secretary of State, 13 November 1948, p. 492, and 740.00119/11-1448,
Clay to Department of the Army, 13 November 1948, pp. 494–495.

gradually into instruments of contrôle organique,[182] and they would figure prominently in the calculations behind the Schuman Plan proposal, the negotiations resulting from it, the compromises leading to the formation of the Coal and Steel Community, and in its operations.

Having been shown a draft version of Law 75 in August 1948, the French could not claim that its publication took them by surprise. Still, they had not expected that a bi-zonal ordinance worked out by military governors Clay and Robertson would take precedence over tripartite agreements concluded in London by senior diplomatic representatives of the United States, the United Kingdom, and themselves. The State Department was equally distressed by the precipitous publication of the law. On 19 November 1948 Ambassador Douglas cabled Secretary of State Marshall from Paris to protest that he was not informed in advance of the provisions stipulating an eventual disposition of Ruhr industrial assets by a future German government. Douglas argued that the matter was too "far-reaching" to have been handled by Clay; it concerned "the peace and security of Europe," and France as well as the other nations dependent on Ruhr coal and steel should have been consulted.[183]

The lack of coordination also troubled Averell Harriman. He complained to the secretary of state on 16 December 1948 that "there are seven or eight different subjects being currently discussed or negotiated by different groups in London, Paris, Washington, Germany and elsewhere. These include a series of subjects regarding Germany (control of Ruhr, reparations, prohibited or limited industries, occupation statute, fusion agreement and German Government) and, concurrently, ECA aid including use of counterparts, rearmament of Western Union and North Atlantic Pact." The lack of coordination, Harriman proceeded, was particularly unfortunate with regard to France where the "process of separate . . . negotiations . . . permits very sensitive issues such as reparations and control and ownership of Ruhr industries to be debated and presented to the . . . public outside of the overall economic, political, and military measures which, in fact, more basically affect French security. [Even though] we are taking and are proposing to take so many extraordinary actions favorable to France's economic well-being and

[182] FRUS 1948/II 700.00119 Control (Germany) 12-2848, Communiqué of the London Conference on the Ruhr, 18 December 1948, pp. 577–578.
[183] FRUS 1948/II 740.00119 Control (Germany) 11-1948, Ambassador in the United Kingdom (Douglas) to Secretary of State, 19 November 1948, pp. 515–516.

security, these are overshadowed by unrealistic fears engendered by our objectives in Germany considered apart from these overall concepts."[184] Diplomatic verbiage aside, Harriman was demanding that "Clay must go," Germany must be subordinated to Europe, and someone must be put in overall charge of policy.

In early January 1949 the latent conflict between the State Department and the Economic Cooperation Administration on the one hand and OMGUS and the Department of the Army on the other broke out into open warfare. The dispute had little to do with national "-philias" and "-phobias" (the men in Berlin and Frankfurt could forget the war no more easily than those in Paris) and a great deal to do with the incompatible missions of the two factions. JCS 1779 of 11 July 1947, the post–Marshall Plan replacement for the old charter of the U.S. military government, JCS 1067, merely sanctioned Clay's long-standing approach to discharging his mission: the job of OMGUS, as he undertood it, was to get the broken and defeated German nation back into normal operation in order to keep down the costs borne by the U.S. taxpayer, reduce the risks of civil disorder, and minimize the American military commitment in Europe. Clay was growing increasingly impatient with anyone or anything preventing the fulfillment of this mission. The State Department's quite different task was to get Europeans to cooperate economically and politically, to which end it was necessary to provide guarantees against the threat of renewed German aggression. This required the imposition of controls over the Ruhr; without them European integration would have been stuck in its tracks.

On 23 January 1949 Clay complained bitterly to Secretary of the Army Royall that "it looks as if the French Military Government had finally suceeded in that apparent anatomical impossibility of making the tail wag the dog. It was discovered that by protest it can stop the adoption of bizonal economic legislation, appealing if necessary for additional delay to an intergrovernmental conference. As a result, legislative progress in the bizonal economic council has almost ended. This is having an appreciable and increasingly adverse effect on morale. In fact, the French are able to stop legislation much more effectively now than they could [have] under trizonal fusion."[185] At the meeting of the interagency Policy Committee on

[184] FRUS 1948/II 711.51/12-1648, Special Representative in Europe in the Economic Cooperation Administration (Harriman) to Secretary of State, 16 December 1948, pp. 567–568.

[185] FRUS 1949/III 740.00119 Control (Germany) 1-2549, U.S. Military Governor for Germany (Clay) to Department of the Army, 23 January 1949, p. 85.

Germany in the following week the army put its foot down, with Secretary Royall accusing "the British and the French . . . of wrecking our plans for Germany with their [dilatory] tactics. Their method was to get to a certain point in piecemeal negotiations on various subjects at the Military Governors level and then raise the matter at a governmental level, thereby often gaining concessions on a piecemeal basis." To end this practice, and to "reach an agreement as a whole with respect to Germany," Royall counseled a tough line, namely, "using all forms of pressure open to us, e.g., withholding ECA aid, refusing to come into the Atlantic Pact, refusal to approve the Ruhr Agreement, etc." Secretary of the Army Royall in fact demanded a "US policy paper to be worked out through the NSC (National Security Council) and approved by the President [in which] we should examine such questions as to whether or not we should threaten withdrawal of all aid to the European countries . . . opposing our plans for Germany, and as to whether or not we should threaten to withdraw entirely from the occupation of Germany."[186]

On 18 March 1949 in an extraordinary transatlantic teletype conference with senior officials of the Department of the Army Clay went beyond anything said previously: The United States should scrap the tripartite agreements, force the French and British into line on the issue of Germany, and, following a plan developed by OMGUS, terminate the occupation as soon as possible. "What are we, Tracy," Clay demanded rhetorically of Asssistant Secretary of the Army Voorhees, "men or mice? Are we financing a Germany to be the prey of Tom, Dick, and Harry or because we want our influence felt in proportion to our aid? We hold the trump cards which need to be played but are holding them to the end where their value will be just one trick in the trump suit. If that is our German policy, then we should recommend to Congress as far as the Army is concerned – no funds."[187] Clay never made good his threat of breaking with the policy followed by the United States since the adoption of the Marshall Plan because his tenure as military governor was already over in all but name. In a review of policy toward Germany requested by the new secretary of state, Dean Acheson, the National Security Council had already come down on the side of a State Department draft from Kennan's pen establishing the first intelligent framework for handling the problems posed by German recovery.

[186] FRUS 1949/II 740.00119 Control (Germany) 1-2849, Memorandum by Geoffrey W. Lewis of the Office of Occupied Areas, 28 January 1949, p. 87.
[187] FRUS 1949/III, Department of Defense Files, Record of teletype conference between Washington and Berlin, 18 March 1949, p. 112.

The paper began by assigning primacy to the requirements of the nation's neighbors: "Any approach to the problem of Germany's future status must address itself not only to the arrangements made . . . within Germany but also [to] the conditions . . . governing [its] relationship to the remainder of the European community." Any lifting of the occupation further presupposed the existence of "an adequate framework of general European union into which Germany can be absorbed." But how was this European union to be formed? The United States, Kennan advised, should give encouragement from afar, treating "the form and pace of movement [toward union as] predominantly matters for the Europeans themselves [to determine]." The United States should welcome the seizure of initiative toward union by any democratic nation, he added, because the "framework and conditions of association offered by the other European governments" will have a critical influence on the internal development of the German political system and so must be examined carefully. Finally, Kennan wanted the United States to make three things clear to all parties: Aid could have only a marginal effect on the recovery process; the United States would never retard this in places where the population has by its own extraordinary efforts made progress; and it would never tolerate a resurgence of German militarism.[188] The message was clear enough: no solution of the German problem without some form of European federation; no federation incompatible with basic American values; and U.S. support for a federating power but no direct intervention, least of all in delaying German recovery. Kennan did not name a federator, leaving it open for Britain to play the role either alone or with France.

The first step toward the new State Department policy was to get rid of Clay, and the military occupation as well, just as soon as the Soviet blockade of Berlin imposed in June 1948 could be lifted. The French shared this aim. The chief economic adviser in the French Zone, André François-Poncet, apparently first suggested how this could be accomplished. Endorsed wholeheartedly by the United States, François-Poncet's solution was eventually adopted in the form of a new Allied High Commission. He told Kennan that while he "would not go into personalities . . . the time had passed when generals could solve [the German problem]. The . . . Military Government had exhausted its usefulness [and] was a terrible thing: irritating and discouraging for the vanquished, corrupting and demoralizing for the victors. For those who participated on the Al-

[188] FRUS 1949/III, "Principles of Basic Policy concerning Germany," 7 February 1949, pp. 90–93.

lied side, it was a schooling in totalitarian practices and administration. It was all right for the immediate tasks of the post-hostilities period; but it was incapable of leading the way to the liquidation of the war and to the tasks of psychological adjustment and reconstruction." According to Kennan, François-Poncet described the occupation statute as being overcomplicated, impractical, and politically deadening. The Frenchman also reportedly commented that "Mr. Schuman ... had no enthusiasm for continuing on this line, but felt ... the time had come for a sweeping and forward-looking solution that would give not only hope and inspiration to German political life, but respite to the Allies from their own internal differences." François-Poncet's solution was for "the Military Government to be abolished altogether, being replaced as follows: each of the three governments would have a civilian commissioner, each with a small staff of advisers, whose task it would be to control the actions of the German authority [with] total personnel [amounting to] only a fraction of the present Military Government ... It would make no effort to govern [by] itself ... but merely exercise [a] control function. The troops would remain and would act as a sanction for the ultimate power of the Allies to intervene if things went seriously wrong."[189]

The shift from military to civilian occupation government soon began. On 30 March 1949 Robert Murphy, who had been recently kicked upstairs from his position as Clay's chief political advisor to a new job as acting director of the State Department's Office of German and Austrian Affairs, recommended making an attempt during the forthcoming meeting of the three Western foreign ministers scheduled for 6 April to "bring about a radical change in the nature and operation of Military Government at the time of the establishment of a Western German government." "Future arrangements," he said, "should look toward the abolition of Military Government as such and the substitution of a small Allied control body, headed by civilian commissioners." Such a changed concept, he concluded, would simplify occupation government by limiting its scope to specific "reserved powers such as security, decartelization, and reparations."[190] On 12 May 1949, coincidentally the same day the Berlin blockade was lifted, the French joined up with the United States and United Kingdom, and the "tri-zonal" powers reached

[189] FRUS 1949/III, George F. Kennan Papers, "Notes by the Director of the Policy Planning Staff [Kennan] on a Trip to Germany," pp. 113–114.

[190] FRUS 1949/III 740.00119 Control (Germany), "Paper Prepared by the Acting Director of the Office of German and Austrian Affairs [Murphy]," 30 March 1949, pp. 140–141.

agreement on an occupation statute whose central feature was an Allied High Commission representing the three Western Allies as first proposed by François-Poncet and quickly adopted by Murphy and the State Department.

The speed with which the Allies concluded these arrangements was due to the dawning recognition that the United States would have to depend mainly upon France, and maybe even France alone, to solve the German problem. The State Department's correspondence of March and April 1949 is full of reminders that the peace of Europe would depend in the long run on a Franco-German rapprochement and that without one there could be neither European union nor even an end to the occupation of Germany.[191] Yet none of those prominent in American decision making had much confidence in the ability of the French to keep the hereditary enemy in line. The Ambassador in Paris, David Bruce, expressed the general view of senior U.S. diplomats by stating on 20 October 1949 that "no integration of Western Europe . . . is conceivable without the full participation of the UK. This is fully realized by the Continentals themselves. No Frenchman, however much of an Anglophobe he may be or however embittered he may now find himself as a result of the events of the last few months, can conceive of the construction of a viable Western world from which the UK would be absent . . . The French know that [British] disassociation [from the Continent] would be fatal to the cause of European integration . . . and [its] result can only be the reversal of the trend towards integration and a return to the worst continental type of autarchy." The State Department, he concluded, is "unrealistic in urging that France alone take [the] lead in bringing about the reintegration of Germany into Western Europe, [for] France . . . cannot take the lead without the full backing of the US and UK, accompanied by precise and binding security commitments looking far into the future."[192] Bruce recommended giving Britain a swift kick in a tender place.

France had already tried and failed to take the lead in drawing the United Kingdom into a European partnership. In April 1949 Monnet accepted a long-standing invitation from Chancellor of the Exchequer Cripps to discuss planning methods in the hope of once again raising his famous proposal for an Anglo-French economic union. Monnet's interlocutor was Cripps's deputy, Sir Edwin Plowdon, whom he had known since World War I. "If I had been the

[191] Ibid., pp. 121–122, 132–133; also FRUS 1949/III 740.00119, Memorandum of conversation by Secretary of State, 1 April 1949, pp. 158–159.
[192] USNA RG466 TS(49)24DC4931L, Paris to HICOG, 20 October 1949.

most powerful civil servant that had ever been," said Plowdon years later, "I could not have influenced the British Government at that time – they were not interested . . . Ernie Bevin especially."[193] Plowdon admits having been unable to think seriously during the discussions with Monnet about anything other than the mounting sterling crisis, which finally resulted in the devaluation of the pound from 4.10 to 2.80 to the dollar. Though the Plowdon conversations would be far from the last of the Frenchman's attempts to interest the United Kingdom in integration, they were a kind of turning point. From then on, Monnet directed his attention primarily to the Rhine.

Others had been looking in that direction long before him. Monnet had not been a main formulator of French Ruhr policy. As commissioner of the plan, he in fact had no official responsibility in foreign affairs, although of course he remained a powerful influence behind the scenes, particularly in connection with American loans. Monnet's interventions into German issues were infrequent and in general done with a view to impressing Americans with the seriousness of France's coal problems. His one independent foray into policy making, the "coal czar" proposal of late 1945, won him a knuckle rapping. Apart from this episode, there was little to distinguish his positions from those of Bidault, with whom he corresponded frequently. Monnet's relative inactivity is not surprising: He had less first-hand experience in Germany than most other top-level French foreign policy makers of his generation.[194]

At the Quai d'Orsay, as in private circles, planning for a more far-reaching Franco-German accommodation than that envisioned by Monnet had long been under way. A background paper written by the chief of the Quai's European desk shortly after the conclusion of the London accords on Germany proposed a more ambitious approach to integration than the future Schuman Plan. The paper began on an unexpected tack, criticizing previous policy for having dropped demands for the internationalization of the Ruhr before attempting to explore the possibility that it could provide the first step toward a Western union; it went on to censure French policy for neglecting to take up the suggestion made by the United States at the recent London discussions that the jurisdiction of the pro-

[193] JM AMF 22/3/3, "Notes sur les entretiens Monnet–Plowdon," 23 May 1949; JM, Interview with Lord Plowdon, 2 February 1982; JM AMF 22/3/6, Etienne Hirsch, "L'Angleterre fera-t-il antichambre?" Les Cahiers de la République, 2 January 1963, pp. 9–16; Monnet, Memoirs, pp. 250, 280.
[194] See Gillingham, "Die franzöösische Ruhrpolitik," pp. 15–16.

posed Ruhr authority be extended to the other heavy industry regions of Western Europe. France, the author insisted, should not fool itself: European integration without Germany was a myth. With it, he reassuringly added, prospects were unlimited. The paper then recommended a number of specific fields for cooperation. Private commercial arrangements between the large producer regions (the Ruhr and eastern France) could, he said, be a step toward statutory agreements at a higher level, which should be organized within the Marshall Plan. German and French political parties should also link up to promote direct association between the industries of the two countries. The author envisioned the formation of mixed Franco-German companies to exploit the large market and to serve as steps toward a customs union and industrial modernization.[195] Describing as "farcical" the pretension of French socialists and Christian Democrats of having no partners across the Rhine, he emphasized that European integration required close cooperation with them. As necessary immediate actions, the author suggested drafting political platforms, devising specific programs for political, cultural, and economic cooperation, setting up exchanges, and creating coordinating mechanisms; for the rest, he thought that the French government should encourage producers to reach understandings with German partners, form joint ventures, and prepare the way for cooperation on a larger scale.[196]

By the date of the Schuman Plan announcement, French expressions of interest in reconciliation with Germany had become quite commonplace. At the Council of Europe, ex-Prime Minister Paul Reynaud proposed a new coal–steel agreement to link the two nations. In April 1950 Prime Minister Bidault presented plans for what he hoped would be his most important contribution to foreign policy, a High Atlantic Council for Peace. Yet these initiatives were as exiguous as those of Monnet's precursors at the French plan.[197]

As with the French Plan de Modernisation et d'Equipement, U.S. support of the Schuman initiative would be critical. No guarantees were demanded in advance, and none was received. The closest thing to one came in the wake of the September 1949 pound devaluation, a unilateral move taken without consultation with either France or Germany. The French ambassador in Washington, Georges Bonnet, protested this act as an abuse of the "special rela-

---

[195] MAE Z 1944–1949 [Allemagne 83], "Direction d'Europe: Perspectives d'une politique française à l'égard de l'alliance," 30 November 1949.
[196] Ibid.   [197] Monnet, *Memoirs*, p. 300.

tionship," claiming to perceive in the move the outlines of an "historic design" in which France had no part. Foreign Minister Schuman exploited the opportunity presented by the chance to clear up Bonnet's alleged "inexplicable misunderstanding" by sounding out Acheson concerning American views regarding French initiatives on the German question. He, Schuman, the text reads, told Acheson that his understanding of the American position was that "the future of Western Europe depended upon the establishment of an understanding between the French and the Germans; that this could only be brought about by the French, and only as fast as the French were prepared to go; and that, therefore, the role of the US and UK in this matter was to advise and to assist the French and not put them in the position of being forced reluctantly to accept American or UK plans." To "be quite clear that we understood one another," Acheson responded to these leading questions by reiterating measures indicative of the "deep concern" of the United States with Europe, the Marshall Plan, the NATO pact, and the military assistance program recently passed by Congress, "steps certainly not looking towards the abandonment of France but, on the contrary, towards the increasing association of the US with the Atlantic Community." Though Schuman professed great satisfaction with this response, the lawyerly Acheson had in fact evaded the Frenchman's question.[198]

Monnet had no advance guarantee of U.S. support for the Schuman Plan proposal and could only gamble that he would succeed as previously in exercising a personal influence on American policy makers. The most critical figure was U.S. High Commissioner John J. McCloy, who had been appointed on 15 April 1949. Friendship between the two first developed in the late twenties, when McCloy, then a partner in the prestigious Wall Street law firm of Cravath, Swain, engaged Monnet, at the time Paris representative of the investment house Blair and Co., as underwriter for an issue of municipal bonds for the city of Milan. Monnet came away from the deal with little material gain but with something worth far more, his future wife and intimate co-worker, Sylvia. At the time of their courtship Sylvia was married to the Milan representative of Blair and Co. McCloy's wife nursed Monnet's future bride through a messy separation, divorce being legally unrecognized in Italy. The happy couple was united at the House of Marriages in Moscow, and the two families became close personal friends. In the early 1930's McCloy

[198] HST, Acheson Papers, Memorandum of conversation, 27 September 1949.

guided Monnet through a financial fiasco resulting from an abortive takeover of the Transamerica Corporation. The two also worked closely together during the war on supply problems.[199]

Monnet was delighted with McCloy's appointment as U.S. high commissioner in Germany, writing Schuman that according to Harriman all the key U.S. "players" were finally in position. Schuman, Harriman had said, would have a far easier time arriving at common Franco-American policy with McCloy than with Clay, intimating confidently that McCloy "is a close friend of mine and a friend of [US ambassador in Paris] David Bruce as well as the brother-in-law of Mr. Lewis Douglas, our Ambassador in London. Together we are a well-united foursome who, I suspect, will remain in close touch with French representatives, whom we have known for a long time."[200]

No evidence has ever been produced that McCloy had a direct hand in either formulating the Schuman Plan or leading France toward rapprochement with Germany. McCloy's personal views on European integration, Franco-German reconciliation, and German revival were not well formulated when he took office in June 1949 as high commissioner but developed in response to the practical problems facing him. McCloy, like Clay, soon grew restive under the restraints imposed on German economic activity and made no secret of the fact. By a remarkable coincidence, he committed his thoughts to paper on the big questions facing Europe in a memorandum prepared for the opening of the London conference on 10 May 1950. Though admitting the "inconclusiveness of our own thinking" about European federation, McCloy ruled out the United Kingdom as even a possible member of such an entity. As for France, its policy toward German revival was too restrictive and would, the U.S. high commissioner thought, have to be overridden by the United States because of the threat posed by the USSR. McCloy also recommended that the French should be offered a privileged relationship with the United States in lieu of the outmoded "special relationship" with Britain on the condition that France drop its hostility to German revival, suggesting that mixed commissions of industrialists and civil servants get together to work out the details. For the rest, McCloy thought it necessary to force the French to phase out restraints on recovery by "stressing [to them] that the continuance of controls

[199] JM, Interview with John J. McCloy for the Jean Monnet Foundation, Lausanne, 15 July 1981.
[200] AN 80 AJ (1949–1950), unnumbered, undated note of JM.

is a self-defeating process and that our estimate as to the effectiveness of such is 18–24 months; that during this period we intend to do everything in our power to support the liberal elements in Germany so that they will have the maximum of strength when the controls are withdrawn."[201] McCloy, in other words, objected to the maintenance of controls over the Ruhr. Because of Monnet's personal intervention, this would change, the United States would step forward as champion of French security policy, and the Ruhr would remain under Allied authority until long after the adoption of the Schuman Plan.

The man whose opinion would be decisive for the 9 May 1950 announcement was Washington's reputed leading Anglophile, Dean Acheson. His relationship with U.S. policymaking was almost suspiciously intimate. Not long after becoming secretary of state, he proposed to British Ambassador Sir Oliver Franks that "we talk regularly and in complete personal confidence about any international problems we saw arising. Neither would report or quote the other unless, thinking that it would be useful in promoting action, he got the other's consent... The dangers and difficulties of such an arrangement were obvious," Acheson proceeded, "but its benefits proved to be so great that we continued it for four years. We met alone, usually at his residence or mine at the end of the day before or after dinner. No one was informed even of the fact of the meeting. We discussed situations already emerging or likely to do so, the attitudes various people in both countries would be likely to take, what courses of action were possible and their merits, [and] the chief problems that were likely to arise."[202]

Franks was in Washington on 7 May 1950 when Acheson arrived in Paris after a trip to Pakistan and was informed by Schuman of the planned announcement. Initially suspecting that the French were proposing to build a huge international cartel but disabused of this notion by Monnet and McCloy, whom Monnet had brought in that very day from Bonn, he was soon won over by the proposal's "commonsense approach and its avoidance of the appearance of limitations upon sovereignty... What could be more earthy than coal and steel, or more desirable," he wondered, "than a pooling and common direction of France and Germany's coal and steel industries?"[203] Acheson asked Truman to withhold comment on the proposal until after the announcement, which he agreed to do, then

[201] USNA D(501)1299, McCloy to Secretary of State, undated memo.
[202] Acheson, *Present at the Creation*, pp. 424–425.   [203] Ibid., p. 425.

sent a second cable requesting permission to release a favorable
American reply, which to his relief was granted before prior consultation with the Antitrust Division of the Justice Department.
Acheson's hopes that, contrary to past practice, the British would
agree to join the Franco-German pool were dashed the day of the
announcement; exhausted, irritable, and suspecting U.S. complicity
in the French proposal, Bevin rejected it categorically. The coal–
steel pool was launched.

There is no reason to dwell at length on the role of either the
Soviets or the British in the events leading up to the Schuman Plan.
If the three Western Allies could agree on any single thing, it was to
deprive Moscow of any voice in the affairs of the Ruhr, and in this
they succeeded. Soviet influence on the Western European coal–
steel settlement was only indirect and important mainly as it affected American policy. Whereas, the USSR was kept outside, Great
Britain was invited in, but the British were too puffed up to pass
through the door jamb. The United Kingdom might have entered
Europe at a time when it could have led it, shaped it, and adapted
to it. Instead, the British refused to accept the war's verdict, admit
that Europe's Great Powers were no longer great and that the world
they had once dominated was no longer theirs. Bevin was in charge
of foreign policy and therefore the man most directly responsible for
its shortcomings. Still, nobody worth counting in either of the main
political parties, in any of the economic or administrative elites, or
in the public at large was prepared to pay the high price of adjustment to a world of economic competition. It would take another
twenty years of failure to begin to change this.

France, or at least a small but critical faction within the national
leadership, *was* willing to make the necessary sacrifice. Credit is
due above all to Monnet: There were other modernizers, but his
was the plan that worked. Progress was uneven and resistance to it
often heavy, yet the changes he introduced proved lasting. Monnet
understood better than anyone else in France the challenges of what
Time's publisher Henry Luce shamelessly called "the American Century," accepted them as constructive and worthwhile, and met them
eagerly. He did not, however, act alone, nor was his fresh spirit un-
representative. A desire for a new departure suffused French Ruhr
policy, whose formulators recognized that the German heavy industry complex could not be destroyed, occupied, or wished away but,
once tamed, had to be accepted; they wanted not to displace the
heavy industry complex but eventually to share power with it. If in
the short run this involved imposing restraints on German coal and

steel, in the long run it aimed at some form of partnership in an integrated Europe.

The U.S. contribution to the origins of the Schuman Plan was vital. The United States pushed Europe away from the politics and economics of the 1930s and towards federation and liberalization, advancing men like Monnet who promoted it. The plans of the State Department and the Economic Cooperation Administration were not instant successes. Still, the European Payments Union (EPU) grew out of negotiations conducted by the OEEC, and the General Agreement on Tariffs and Trade (GATT) was the end product of Clayton's long labors. Both became permanent fixtures. By giving France the backing it needed to enter a coal–steel pool with the Germans, the United States helped link the two more powerfully than any plan ever drafted by the statesmen, diplomats, and economists of the ERP could have done: This would be the most important American contribution to European integration. In making it, U.S. foreign policy had to pass through a process that was often confusing, always sensitive, and at times painful: It had to come of age. The United States emerged from the war with a Grand Design easily ridiculed as amounting to no more than a big sketch. Yet the two lodestars of American policy, free trade and self-determination, were the right ones; no other principles of international reorganization would have lifted the darkness of the interwar era more rapidly or led ultimately to a better world order. Other than those engaged in developing a radical critique of American society, critics of U.S. policy toward Europe emphasize shortcomings of execution – inconsistency, naiveté, and occasional arrogance. These qualities were abundantly evident in 1945 but appear less frequently thereafter. More striking than gaucheness or stupidity is the speed with which U.S. foreign policy makers came to grips with the new responsibilities of world power: Stettinius to Byrnes, to Marshall, to Acheson is evidence of a pretty clear trend in the caliber of American statesmanship.

The United States did not ask to be arbiter of Europe. The overriding purpose of American postwar policy was to make it possible for the United States to withdraw behind the security of its borders. The attempt to restore order by the conduct of joint policing operations with the Russians was a pathetically inadequate response to the problems facing Europe; the two could not agree on which beats to walk, not to mention how they should be walked. Four Power control of Germany was never more than a diplomatic fiction, not least of all because we lacked a coherent policy toward the beaten foe. U.S.–USSR cooperation on reconstructing Europe never existed.

Failure to restore Europe from afar forced the United States to take an active hand in reorganizing it until things could be put right. The idea that mere money, if properly allocated, could set things in order did not have a very long shelf life: The attainment of a payments equilibrium between the United States and Europe, the ERP's magical state of "viability," could not be reached within the statutory four years merely by following a timetable drawn up in Washington. Twenty-four months of struggle in, around, and over the OEEC taught the United States that federalization and trade liberalization would come only on Europe's terms. But whose, specifically? There was simply no common ground with Britain, the American first choice as bridgehead. The United Kingdom intended neither to open up world markets nor to join Europe, would not give an inch, and in the end disappointed most deeply those, like Acheson, who truly believed in the strength of the link to America's past. France, though weak and unstable, endorsed U.S. integration policy while demanding security against German revival. By supporting the 9 May 1950 proposal the United States accepted France as federator of Europe. The decision was important for another reason as well: The Schuman Plan sounded very strange to American ears, like a European supercartel: by backing it, the United States was faithful to the generous spirit of its official policy of restoring Europe to control over its own destinies.

The development of a constructive American integration policy was not easy. In the United States even more than elsewhere in the turbulent postwar years, foreign policy elites advocated ideas far more enlightened and far-seeing than those of the public whose servants they were supposed to be. As indicated by the prevailing views of the Republican-dominated Eightieth Congress, the American people's sentiments toward European policy found expression in a strident insistence on ending foreign involvement of all kinds, especially in the form of aid, and in clamorous demands for a purge of those persons, suspected Communists and their dupes, believed responsible for introducing evil foreign doctrines into American life. On the political right were those who still nostalgically wished that the United States could become a city on a hill; any more sophisticated appreciation than this of America's new world dependence, global responsibility, or international role was lacking in that quarter.

President Truman, so weak after the 1946 elections that a prominent foreign policy spokesman from his own party, Congressman William J. Fulbright, recommended from the floor of the House that he resign, found in Greece the elusive formula that could win public support for the increased European engagement his advisers warned him was necessary in the interests of American security. It was, as the witty and worldly Dean Acheson sheepishly admitted in his memoirs, demagogic appeal to crude instincts of anticommunism; the Truman Doctrine promised balm in the form of military hardware to anyone, anywhere, who felt himself afflicted with the noxious Marxist rash.

The European Recovery Program, something conceived as an emergency measure to prevent the internal collapse of the economies of the war-wrecked Continent, was thus also presented to the American public as a necessary dike against a rising Communist tide. As the anti-Red propaganda spewed out, Truman headed off congressional witch hunters by introducing his own "loyalty" campaigns. Nasty though they were, who can say that resort to such measures was not a necessary evil? How else could the representatives of a wary public have been enticed into handing over large amounts of money to a Europe they would have preferred to forget, or into taking the risk that the very nations whose conflicts had drawn Americans into two major wars within adult memory would not waste it? Under the Republican administration of Dwight D. Eisenhower guns replaced butter in U.S. policy toward Europe; its official aim was no longer economic cooperation but mutual security. Yet even by 1952 the European Cooperation Administration, which aimed at modernization, liberalization, and the organization of a sovereign Europe, had been folded into a new Mutual Security Agency set up to create a single, transatlantic U.S.-run armaments economy. The most constructive phase of American policy toward Europe had ended.

4

Neither restoration nor reform: the dark ages of German heavy industry

In 1945 Ruhr industry was a smoldering ruin, an ugly monument to the collapsed and discredited civilization for whose misdeeds it was partly responsible. Within ten years it would be rebuilt, reorganized, and integrated into a new political culture for whose impressive subsequent accomplishments it can share credit. It would be wrong to assume, as many commentators in the 1950s once did, that this remarkable transformation occurred because the management of German heavy industry was somehow "Americanized": No internal conversion from evil old ways to good new ones took place after the war.[1] Nor should one have been expected. Punishment and retribution were the main themes of Allied policy toward the Ruhr trusts; by comparison to them, the missionary impulse felt by a few U.S. policy makers was unimportant. The captains of German industry were unimpressed by such do-gooding. To them Allied policy was confused, senseless, arbitrary, and destructive: They wanted above all else to get out from under it. The coal and steel leaders had only limited means of defense: The Allies alone could decide when the Ruhr would recover its freedom of action. Yet they needed the cooperation of management and labor in order to keep the economy going and, after August 1949, had to be careful not to discredit the new West German government. Ruhr heavy industry could also count on receiving at least some support from foreign producers and governments whose economic welfare required the revival of the German economy.

The histories of West German coal and steel, so closely intertwined in other decades, are quite distinct for the years of occupation. The immediate economic needs of Europe stamped the history of the one, its future political requirements that of the other. In coal the threat of famine and unrest caused the Allies to concentrate single-mindedly on increasing outputs. The production problem was

[1] See, most recently, Volker R. Berghahn, *The Americanization of West German Industry, 1945–1973* (Cambridge, 1980).

not solved in the 1940s, however. Shortages would remain acute into
the following decade, creating a psychosis of scarcity that was difficult to dislodge. Fears of further unheated winters and disruptions
caused by fuel shortages governed West European coal policies long
after they had ceased to be an actual threat. The conditioning and
motivation of the labor force were the primary concerns of Allied
coal policy, and in an effort to bolster worker morale the occupation
authorities encouraged the political ambitions of the unions. The
need for increased production also ruled out attacks on mine management as well as the industry's special organizations for marketing, provisioning, and technology. Questions concerning the future
ownership and control of the collieries were deliberately left in abey-
ance. Outcomes in steel would determine the future organization of
the mines. Allied diplomacy rather than economic necessity drove
policy toward this sector of industry. Shortages of ferrous product,
unlike the case of coal, were not an imminent threat to recovery,
and several European postwar governments launched impressive
mill-building programs after the war that could have been expected
to prevent scarcity in the future. At the same time, the widely held
belief that he who controlled steel could dominate Europe made
this industry the focus of the Allied attempt to disarm Germany
economically. Decartelization and dismantlement did not die out as
occupation policies after 1945 but, contrary to fears and expectations, lived on and even intensified, remaining under way after the
creation of the new West German government. On 1 January 1950
the future of the Ruhr was a question mark. Events in the international sphere would be decisive.

4.1 Life at the pits: the mines, the miners, and the
coal supply

At the end of the war the European coal situation was grave indeed.
The Continent faced deficits of upward of seventy million tons.
About one-half of them were due to the shift of Britain from exporter to importer and another tenth to the diversion of Silesian
coal to the Soviet Union. The remainder stemmed from lost outputs
in Western Europe. High shipping costs and dollar shortages prohibited any large-scale substitution of overseas coal for that normally
produced in Germany except in an emergency.[2] The recommenda-

[2] See USNA RG260 AGTS25/3, "Memorandum for Mr. Lewis W. Douglas," 21
May 1945.

tions of the joint technical committee known as the Potter–Hyndley commission governed Allied coal policy. Ten million tons of coal were to be made available for export in 1945, according to its report, and another fifteen million by the end of April 1946; Germany's own consumption requirements were to be held to a bare minimum by use of armed force if necessary.[3] The restoration of Ruhr production depended above all on the miner. Though an estimated 10 to 15 percent of the pits were out of operation on V-E Day, physical structures and machinery were essentially intact.[4] Not so the labor force. Overaged, overworked, and reduced by three-quarters, it had to be reconstituted if outputs were to increase. Manpower thus became a constant preoccupation of Allied coal policy. Once enough healthy men were at work, supply became the bottleneck to increased outputs. The deepest problem facing the mines went beyond either the human or the material factors of production: It was the occupation itself.

For the twenty months from May 1945 to January 1947 the British ran the collieries on their own. Their record was unimpressive. To step up production they relied mainly on forcible recruitment. All adult males registering at labor exchanges were ordered to the pits on the grounds that work was unavailable elsewhere. Through the use of such methods mine employment increased to prewar levels but productivity remained at about 1,200 kilograms per day, little more than half as much as during the 1930s. Compulsion merely aggravated the already severe morale problem.[5]

The wretchedness of those drafted for mine labor is difficult to exaggerate: Few were in any sense fit for work. A typical survey of miner recruits reports that one lad examined had two bullets lodged in his chest, and another suffered from malarial attacks and had an enlarged spleen and liver. Widespread dermatitis covered the extremities of a third, who had been classified earlier as unsuitable for employment. Few recruits in the survey had been trained for mine labor and they, like the others, resented the brutal tone used by

[3] USNA RG260 1/165-1/4, "Combined Resources and Allocation Board, Coal Report," 12 July 1945, and RG260 AG45–46/103/1, "The European Coal Situation," 11 July 1945.
[4] USNA RG260 AG47/178/1-2, "Review of Hard Coal Problem in Ruhr–Aachen Area since the Capitulation" (Enclosure to Report on the Anglo-American Talks on Ruhr Coal Production), 17 September 1947.
[5] Abelshauser, *Der Ruhrkohlenbergbau*, p. 34; USNA RG260/2/102-1/8, "Report of the Manpower Directorate on Compulsory Direction of Labor in the Coal Industry," 13 January 1947.

work party supervisors.[6] These so-called green miners as well as those veterans who had not fled the pits at the end of the war were, if not quite on the verge of starvation, certainly too ill nourished to work productively in 1946. Average miner weight was 120 pounds, average daily caloric intake only 2,000, or less than one-half the amount normally required. Accident rates increased after every hour worked. Sickness affected between 11 and 12 percent of the workforce on the average. Absenteeism – much of it caused by food foraying – stood at the astronomical rate of 30 percent. Of those forcibly recruited for the mines, 80 percent later deserted them.[7]

The only defense that can be offered of this policy is the lack of an available alternative.[8] The usual technical approaches to raising outputs were unfeasible: The best seams were already being worked; decreasing the number of operating mines would have reduced overall seam quality; mechanization beyond that introduced during the war was unpromising; and support workers could not have been shifted to the seams without an improvement in the supply situation. The British also lacked a reserve of foodstuffs that could have been diverted to those doing heavy labor and, facing a severe coal crisis of their own, could not afford to second supervisory personnel to Germany.[9] A policy of intervention was thus not adopted. During the first few weeks after the onset of occupation only a handful of military officers was on hand to keep the mines operating. In July the North German Coal Control (NGCC) was set up as a permanent organization, which, with American concurrence, made no attempt to interfere with the existing industrial associations for provisioning, marketing, and engineering. The Bergbau Verband, Rheinisch-Westfälisches Kohlensyndikat, and Bergbau Verein continued to operate.[10] British power in the field of labor relations was also lim-

[6] USNA RG260 1-165-1-23 GPRB, Intelligence Division, Social Survey Rep. No. 5, "The Life and Working Conditions of the Miner Trainee," 31 December 1946.
[7] USNA RG260 45-46/2, "Report on Inspection Trip to the Ruhr Coal Mines," 10 January 1946.
[8] For critical view, see USNA, Department of State 862.6362/6-1046, C. P. Kindleberger to W. W. Rostow, 10 June 1946, and 862.6362/6-846, Rostow and Galbraith to Clay, "Subject: Coal" (enclosure Murphy to Riddleberger, 8 June 1946).
[9] Wilhelm Salewski Papers (SP), "Tagung des deutschen Wirtschaftsrates," 1–2 April 1946 (speech by Dr. Regul).
[10] USNA RG260 AG47/178/1-2, "Review of Hard Coal Problems"; USNA RG84 739/13, Charles Willwright, U.S. Bureau of Mines, to Chief of Production Control Agencies, U.S. Group C.C., 20 July 1944; USNA RG84 739/14, Murphy to Secretary of State, 16 July 1945; USNA RG260 17/257-117, "Labor Situation

ited. In this field they could not fall back on existing institutions, as the Nazi Labor Front had been disbanded. The British had little luck in setting up new organizations. In summer 1945 NGCC held elections for works councils, which it expected to become nuclei of a future union. Instead, the Germans founded one of their own, Industrie Verband Bergbau, which soon became a force for moderation whereas the works councils unexpectedly became loci of radicalism.[11]

Even without the formation of Bizonia in January 1947, British failures would have obliged the United States to step into comanaging the mines, although it initially appeared as if the NGCC was getting results. Average daily mine output, optimally about 400,000 tons, had fallen to about 20,000 tons in the last weeks of the war, and from this low point recovery was rapid. By the end of summer daily outputs were at 100,000 tons, and these rose to 150,000 in November. The gains were due solely to the forcible recruitment of 50,000 new miners, who either would not or could not work very hard. In December 1945 discipline was reported to have broken down as a result of refusals by foremen to enforce punishment from fear of coming before de-Nazification tribunals.[12] Additional drafting expanded the labor force to 290,000 in March and raised daily tonnage to 170,000, but thereafter it plummeted because of a 500-calorie reduction in rations. Not until October 1946 did levels return to those of March, and if they rose thereafter it was because American food supplements added 2,000 calories to the daily diet.[13]

The U.S. presence made a substantial difference. The availability of additional resources enabled the new mining authority, the Bipartite Coal Control (BICO), to abandon forcible recruitment, to which American officials strongly objected, and offer incentives in the form of foodstuffs for employment at the collieries.[14] Yet progress was slow, uneven, and difficult to sustain. Bad administration of the new CARE package program, as well as those supple-

in the Rhine-Westphalian Coal Mining Industry"; Klöckner Archives, Günter Henle Papers (KA/Henle), Industrie Kreise, Bergbau-Kreis/Eisen-Kreis, "Gegenwärtiger Stand von Kohle und Eisen," 30 May 1945.
[11] USNA RG260 AGC17/Gen 13, "History of Coal Production since the Occupation," July 1947.
[12] Ibid., p. 46.
[13] USNA RG260 3/45-2/51, "Situation Chart as of 1 July 1948, Coal Production"; Abelshauser, Der Ruhrkohlenbergbau, pp. 33, 34, 43.
[14] See USNA RG260 45-46/2, "Report on Inspection Trip to Ruhr Mines," 10 January 1946; USNA RG260 2/102-1/8, "Report of the Manpower Directorate," 13 January 1947.

menting it, was one explanation for the persistence of production problems. More important yet was the poor physical condition of mine workers; not until early 1948 had consumption been brought back to peacetime levels long enough to have raised individual productivity to normal levels.

The wealth of documentation produced by the batteries of nutritional scientists, psychologists, sociologists, social workers, manpower experts, and other diverse intelligence gatherers who examined the mines makes the production problems of 1947 readily understandable. According to "Social Survey No. 6, Part II" of British intelligence, the green miner lived under appalling conditions. Twelve to fourteen persons per barracks was the rule, with neither linen nor bedding provided. The air in such rooms was fouled by odors of drying clothing and perspiring men. Recreational facilities were nonexistent. A miner normally possessed only one pair of shoes, a single suit of clothing, and either tattered underwear or none at all. Many green miners were lame, even more of them ill, and all of them anxiety-ridden.[15]

Lack of food was the main problem. To understand its psychological as well as physical effects the University of Minnesota performed a study called "Experimental Starvation in Man." This research shed new light on the effect of inadequate nourishment on productivity and the problems encountered in overcoming it. For six months thirty-four young volunteers did heavy labor on a diet of 1,850 calories, the average Western European consumption in 1946. Their weight dropped at a rate of 14.7 percent, and endurance even more rapidly: "At first running on the treadmill averaged 245 seconds to complete exhaustion, [but] the final average was only 51.9 percent." No less distressing than these findings was the conclusion that "body and mind will [not] respond at once or even proportionately to the amount of food intake." In the short run "rehabilitation feeding" drove individuals into "new depths of depression," and many became "overtly irritable," "argumentative over trifles," and suspicious. Personal doubts, rumors, and expressions of discontent were pervasive. Even after the period of "physical rehabilitation" was concluded, the control group given a "normal" diet regained only 13 percent of its strength, and the one fed only marginally better than before remained in deteriorated condition.[16]

[15] USNA RG260 5/172-2/12, PORO Rep. No. 6, Part 2, "The Mining Trainee, 1947," March, 1948.
[16] USNA RG260 7/43-2/38, Klingenfeld to Beckman, "Food and Productivity," 25 March 1948.

American policy makers would have been well advised to study the Minnesota experiment before adopting the only partially successful "point system" in 1947. By turning the miner into a "calorie aristocrat," it drew additional labor into the industry, thus raising production, but did not increase productivity proportionately. Actually the brainchild of Industrie Verband Bergbau, the point system, as originally devised, rewarded mere attendance, those who appeared regularly for work receiving food bonuses. As later amended, workers exceeding their quotas were issued huge CARE packages containing 40,000 calories worth of typical miner fare such as bacon as well as such otherwise all but obtainable goods as schnapps and cigarettes. Yet the hewer had to exhaust himself to qualify for the premiums. As a result, productivity tapered off no less rapidly after the conclusion of the successive CARE package actions than it had risen before them.[17]

In 1948 inadequate supply replaced food shortages as the main bottleneck to production. So long as the miner had no energy to spare, breakdowns in machinery or lack of essential production materials often provided a welcome respite from work. Thus "when an air hose splits on a pneumatic pick [used for hewing], it not only affects the pick being operated on the broken air hose. [Rather] the air pressure is reduced by the resultant leakage . . . and it practically shuts down every tool operated from the compressor [and] when a conveyor belt or chain breaks every man working along it is forced to remain idle for many hours while the damage is repaired."[18] By mid-1947 this enforced time off was no longer necessary. Increases in man-shift output at the face had begun to outpace overall productivity gains, an indication that abnormal amounts of time (estimated at 25 percent of total hours worked) were being spent handling defective materials and repairing broken-down equipment.[19]

The condition of mine plant improved little between V-E Day and 1947. There was still neither enough timbering nor the right profiles of steel for roof construction. Air hosing was old and often split. Cars for conveyancing were battered and full of holes. Machin-

[17] See Abelshauser Der Ruhrkohlenbergbau, p. 36–37; USNA RG84 809/27, Sanford to Estill, 8 April 1948; USNA, Department of State 862.6362/1-2948, "Report on Coal Production in the Ruhr by Mr. Hugo Stinnes," 29 January 1948.

[18] USNA RG260 AG48/163/5, "A Report Showing the Effects of the Lack of Mine Supplies on Production and Safety," by A. F. Marshal, U.S. Control Group, May 1948.

[19] Ibid.

ery was badly lubricated and subject to a high rate of breakdown. Replacement parts were lacking. Tunnel collapses were frequent, dustiness was worse than ever, lighting poor, and safety precautions totally inadequate. What H. E. Sanford found at the Julia Mine in Herne was typical of the condition of nearly all the collieries visited by U.S.–U.K. inspection teams in early 1948. Roofing was "weak and required extensive timbering. There were 6000 too few steel props, and inferior wooden substitutes permitted mainfalls within the working face along several of the longwalls. The ropes on the hoist near the loading point on one seam contained several 'dog-knot' splices and frayed ends. Four of the twenty-one compressed air locomotives and seven of the fourteen diesel locomotives in stock had been dismantled for replacement parts, and men often idled while waiting for empty cars. Air hoses on pneumatic picks were too short and had to be discarded after only eight days' use . . . The pipe fittings on most compressed air lines [were] so loose that most of the pressure was lost before it reached the working places."[20]

To restore the mines to normal operation the whole gamut of economic controls imposed during the occupation had to be lifted. The currency reform of June 1948 was only one step in this direction. It neither unblocked coal prices nor freed distribution and even diverted available private capital into less-handicapped industries. The greatest immediate impact of the currency reform was to push the mining industry into a liquidity crisis caused by the increased cost of essential supplies.[21] The measure *did* have a very positive influence on miner attitudes, although its effects were not fully evident at once. The first three weeks after the reform daily outputs increased impressively in response to the appearance in the markets of the first imported fresh vegetables seen since the war, Dutch tomatoes. Yet they fell again in July and August, with absenteeism rising alarmingly to over 18 percent. This proved to be a blessing in disguise, a case merely of miners, with a bit of money in their pockets and the exceptionally clear blue skies of that summer overhead, taking their first few real days of vacation since the war. This was a physical and emotional necessity: By the end of the year absenteeism had declined to 10 percent, man-shift production was up 120 kilograms over the previous year, and total daily output rose to

[20] USNA RG260 AG48/163/5, "Excerpts from Mine Inspection Reports," by H. E. Sanford, 27–31 March 1948.
[21] USNA RG260 7/25-2/8 BIB/B(48)113, "Financial Problems of the Coal Industry," September 1948.

320,000, a 50,000-ton improvement over the first six months of the war.[22]

More than hard work was needed to return production to the daily tonnage level of 400,000 prevailing before the war: A massive infusion of capital, its precise amount hard to specify, was absolutely essential.[23] The huge losses at which the mines operated prior to the currency reform meant little, since the circulating medium had no ascertainable value. Nor has it been possible to establish the value of mine investment during this period, though clearly it was inadequate for renovation of plant and machinery. In 1949 and 1950 the Marshall Plan provided 450 million Deutschmarks. Not until after German industry, through the so-called Investitionshilfegesetz (investment aid bill) of 1952, had provided additional billions in capital did Ruhr coal outputs again equal those of Hitler's Germany.

Political dissatisfaction was relatively unimportant as a cause of the coal production problem. Germans understandably anxious at the time to reclaim as much authority as possible tended to exaggerate the extent of it. As the improvement in diet outpaced that of productivity in 1947, discussion often turned to mobilizing the supposed "achievement reserve" (Leistungsreserve), the amount of surplus energy theoretically remaining after the miner had done his job. The old owners had no doubts about why man-shift productivity lagged: With senior management suspended from its functions, no one could order foremen to maintain discipline. Chaos and communism were the only alternatives to restoring employer prerogatives, to their way of thinking.[24] The union leadership was sometimes only slightly less crass in its claims. The president of Industrie Verband Bergbau, August Schmidt, told Allied intelligence, which valued his judgment, that productivity could be increased by at least 10 percent simply by eliminating Communist influence at the mines. He recommended socialization as the surest and best approach to attaining this result.[25]

[22] See USNA RG84 809-30, "Ruhr Coal Situation after Currency Reform," 21 July 1948; USNA RG260 11/104-1/39 OMGUS, Office of the Director of Intelligence, "Some German Views of the Political, Economic, and Sociological Aspects of Ruhr Coal Production," 15 June 1948; USNA RG84 809-32, "Cases of Insufficient Production in Coal Mining Industry" 3 September 1948.

[23] See "Finance for Ruhr Coal," The Economist, 26 March 1949.

[24] USNA RG260 11/104-1/39 OMGUS, Office of the Director of Intelligence, "Some German Views."

[25] Ibid.

Schmidt confounded a commitment to political doctrine with the radical discontent born of suffering. Though Communists dominated 630 of the 1,732 work councils elected in 1946 and 460 of the 1,926 elected the following year, American intelligence officials found that this prominence was due solely to their effectiveness as bargaining agents.[26] One judged it "very misleading to conclude . . . that one third or more of the miners would be willing to follow the implications of Communist propaganda and adopt slowdown tactics on the job."[27] Nor did miners, contrary to Schmidt's belief, care deeply about the socialization issue. According to a British intelligence report, "the result of interviews running into the hundreds [is that] one can state without hesitation [that the miner cares] 'almost not at all' about the matter."[28] Nor was it frequently discussed. When a group of miners being surveyed were asked how to best raise coal outputs, none of the 239 respondents mentioned socialization. Asked what effects it might have on production, only 20 percent anticipated big increases. Another 23 percent did not think it would have an appreciable impact, 27 percent thought it would have none whatsoever, 20 percent had no opinion on the subject, and 8 percent actually thought it would cause declines.[29]

In November 1947 BICO attempted to coopt the Germans into management responsibility by forming the Deutsche Kohlen Bergbau Leitung (DKBL) to serve as trustee for the mining industry. Its job was to draft a reorganization scheme. DKBL, which at first had seemed a serious threat to the French, became nothing of the kind.[30] Even after months of discussion DKBL's board of directors, which was divided equally between labor and capital, could not agree on a plan for restoring the mining industry to German control. By June 1948 the leading officials of the entity were working only part time.[31]

The poor performance of the Ruhr mines not only disrupted France's economy but threatened its long-term growth. An increase in French coal consumption requirements was the inescapable con-

[26] USNA RG260 7/23-1/21, "Election Results, Mining Works Councils" (n.d.).
[27] USNA RG260 11/104-1/39 OMGUS, Office of the Director of Intelligence, "Some German Views."
[28] Cited in ibid.
[29] Ibid.; also USNA RG89 829-830, Murphy to Secretary of State, 19 July 1948 (enclosure: "Communist Influence in Ruhr–Aachen Coal Production").
[30] MAE Y 1944–1949, "Note sur le statut des mines de la Ruhr," by G. Parisot (n.d.).
[31] USNA RG260 7-23/21, "Review of Hard Coal Production in the Ruhr," 10 June 1948.

sequence of the attempt to reequip industry. Even with an expansion in domestic outputs to sixty-five million annual tons projected by the Monnet Plan, France expected to face deficits of from twenty to thirty million annual tons by 1952. American coal, which could have been substituted for previous imports from Britain, was prohibitively expensive and a heavy drain on the tight supply of dollars. As a former Ally, furthermore, France had a privileged claim on German coal resources. It was thus expected that the Ruhr would deliver about twelve million tons per year, a rise in imports over the prewar period roughly proportional to that projected for domestic outputs. Reality dictated otherwise. The increase of Ruhr coke and coal outputs consistently fell behind the growth of the West German economy as a whole, shrinking the export surplus in relative and sometimes absolute terms. Actual coal deliveries fell pathetically short of French aims, never amounting to more than a fraction of requirements. By 1950 France was meeting a higher proportion of its coal and coke needs through domestic production than before the war, the Ruhr provided slightly less of them, and the United States had replaced Britain as the main source of supply.[32]

Successive failures to deal effectively with the Ruhr coal problem helped force France into Europe. The French wanted above all to raise mine production but, refusing to join with Bizonia, had no voice in management. Nor, in spite of Monnet's frequent entreaties and Bidault's numerous threats and blandishments, did the Allies cede France preferential access to available supplies;[33] responsibility for allocation fell to the nationally neutral authority attached to the United Nations, the European Coal Organization (ECO).[34] A first attempt to secure coke supplies by restraining the growth of German steel – the Alphand Plan introduced at the abortive Moscow CFM – failed to gain the support of Britain or the United States, though as a sop France gained the so-called sliding-scale agreement guaranteeing a minimum supply of coal that would increase, though at a diminishing rate, with incremental rises in output. The sliding scale had minimal operative importance; France probably received

[32] See Milward, *Reconstruction*, pp. 130–138.
[33] See USNA, Department of State 862.6362/5-146, "Memorandum of conversation: Coal," and, 862.6362/4-2446 (attachment), "Memorandum on French Import Coal Requirements from Germany," April 1946; FRUS 1946/VI 840.6362/5-1546, Acting Secretary of State to Secretary of State at Paris, 15 May 1946, p. 779.
[34] Public Record Office (PRO), CAB134/295, "The ECE Coal Committee," 5 June 1950.

no more coal than would have been the case without it.[35] Unable to play a direct role, receive preference, or win the support of the occupation powers to a one-sided bilateral agreement, France joined with Anglo-Americans in 1948 to work out a tripartite settlement of the Ruhr Problem, at the same time also turning to the Germans.

4.2 Steel and the future of German industry

The steel industry of the Ruhr was in as much peril at the beginning of 1950 as it had been on V-E Day. The harsh policies adopted at Potsdam proved to be unenforceable because of division among the Allies and the chaotic condition of the German economy. A strictly punitive policy (the elimination of trusts, cartels, and business associations, combined with seizures of stocks and factory dismantlements) would have opened the floodgates of revolution and chaos. Cooperation of some kind with both management and labor was an inescapable necessity. Only after the Soviets had effectively been eliminated as an influence in the Ruhr and the reconstruction process had begun were the Western Allies able to take decisive action against the heavy industry complex. Spurring it on was the need to strengthen the Western alliance: Without the support of France and the United Kingdom, the United States could not have contemplated promoting German revival. The price demanded by the French and the British was some form of economic disarmament. The dismantlement of Ruhr steel factories began at the insistence of the United Kingdom after over two years of occupation had already passed. First getting under way at the end of 1947, it proceeded at a steady pace until November 1949, notwithstanding successive reductions in the list of factories slated for removal. Though officially scheduled to end, the campaign was actually still in progress when the Schuman Plan was announced.

Though Ruhr industry was neither reformed nor restored during the occupation, three avenues for its future development were paved. One of them pointed to restoration. Rather than abolish cartels, the British relied on them to allocate scarce product. This set the stage for an eventual re-creation of the machinery of industrial self-government. Another avenue led in the direction of reform. Unable to agree on whether to socialize Ruhr industry or return it

---

[35] USNA, Department of State 862.6362/1-1249, "Aide-mémoire," 23 March 1949.

to private ownership, the United States and the United Kingdom placed the matter in the hands of the Germans themselves, creating a situation in which management and labor were obliged to work together. This opened the way to a new kind of system championed by the unions based on the principle of co-determination (*Mitbestimmung*). The third avenue followed a course headed toward what can be termed recombination. Faced with the threat of dismantlements, German producers desperately tried to restore wartime business contacts, actually seeking a friendly takeover from the *maîtres de forges* at one point. This was the first step toward organizing a European steel industry under private control.

At Potsdam the Big Four agreed that to eliminate Germany's ability to wage aggressive warfare the big Ruhr trusts should be dismembered; steel plant in excess of that needed to maintain standards of living at minimally acceptable levels was to be dismantled and used for reparations. Steel was not the sole sector of industry of special concern to the Allies; war plant and certain branches of synthetic production were to be eliminated altogether. They nonetheless regarded ferrous metals as the "lead industry" and by regulating its operations, meant to exercise far-reaching control over German economic development. Official policy was draconian. In July 1945, the Labour cabinet endorsed the nationalization of Ruhr industry. In March 1946 the permissible level of steel output was set at 7.2 million tons, far below the 22.6 million tons produced in 1938. (In August 1947 the Anglo-Americans, with French concurrence, raised this figure to 10.7 million tons.) With the establishment of the Steel Trusteeship as an arm of the North German Iron and Steel Control in November 1945, moreover, the decartelization program was launched. In August 1946, the British officially seized the assets of the mills, applying a measure used in coal eight months earlier.[36] Yet policy did not unfold as intended because of Four Power conflict and the near-total collapse of the German economy. In the second half of 1945, German industrial production averaged less than 20 percent of 1936 levels. At 2.5 million tons, 1946 steel output was actually below that of 1894. Recovery from these low levels was slow and uneven until 1948, during which year industrial production soared from 60 to 80 percent of the last "normal" pre-

---

[36] Isabel Warner, "Allied-German Negotiations on the Deconcentration of the West German Steel Industry," in Ian D. Turner (ed.), *Reconstruction in Post-War Germany: British Occupation Policy and the Western Zones* (Leamington Spa, 1989), p. 159.

war year.[37] Until 1948 the main task facing the British in the Ruhr was not, as defined by Potsdam, to restrain production but to maintain it while preventing a breakdown of distribution. This was no easy matter in a country suffering from hidden inflation. To conceal huge increases in the volume of the circulating medium, the Nazis had imposed strict price controls with the result that the currency had become worthless.[38] The determination of market prices being impossible, the British could run the economy of their zone only by administrative decree. Enforcement depended upon the cooperation of the Germans themselves: Everyone would have to take a hand in the rationing of penury.

The occupation actually began fairly well for Ruhr steel in spite of the severity of official Allied policy. Meeting informally in the so-called Sohl Circle, the leaders of industry had planned for the post-surrender period in advance and were prepared to collaborate with the new master.[39] This required the acceptance of certain changes in customary ways of doing business. Technical Instruction No. 49, issued in May 1945 by the First British Army Group, ordered the abolition of all cartels and agencies of the war economy. On 30 June 1945 the six leading Konzerne of the region renounced their memberships in the main steel industry cartels and the Crude Steel Union (Rohstahlgemeinschaft), the Pig Iron Association (Roheisenverband), and the Tube Rolling Association (Röhrenverband) as well as less important syndicates therewith formally ceased to exist.[40]

The syndicates' disappearance was never more than official. On 30 May 1945 the British military government began the issuance of permits to steel product cartels (Verbände) supplying goods to occupation troops. On 20 June the pipe cartel contracted to supply one thousand tons of rolled product to the Twenty-first Army Group stationed in Oberhausen. This deal indicated, according to one

[37] See Werner Abelshauser, "Probleme des Wiederaufbaus der westdeutschen Wirtschaft, 1945–1953," in H. A. Winkler (ed.), *Politische Weichenstellungen im Nachkriegsdeutschland, 1945–1953* (Göttingen, 1979), p. 226; SP, "Anlage 2. Jahresbericht der Geschäftsführung erstattet von Herrn Dr. Wilhelm Salewski," 20 February 1947.
[38] HST, Papers of J. Anthony Panuch, "Price Control, Compensation Trade, and Inflation" (Special Report of the Military Governor), June 1947.
[39] Sohl, *Notizen*, p. 93; Wirtschaftsvereinigung der Eisen-und Stahlindustrie (WVESI), Sitzungsberichte, "Eisenkreis," 30 May 1945, 6 June 1945, 13 June 1945, 14 June 1945, 28 June 1945, 11 July 1945, 13 July 1945.
[40] KA/Henle Verbände, Wiederaufnahme der Tätigkeit der Verbände nach Kriegsende (5-10.45), "Auszug aus der technischen Vorschriften Nr. 49," May 1945.

tongue-in-cheek account, that it would be "inadvisable to [eliminate] cartels at present, [since] the English would interpret this as resistance and replace our organization with their own."[41]

On 21 June discussion began within steel circles about the future of cartels. No one was fully satisfied with the way they had worked in the past. The steel industry was weary of being "overorganized" and subject to the vexations of price controls, rationing, and currency restrictions, as well as the petty official contracting practices and burdensome, not to say bizarre labor regulations that were features of the Nazi economy. When in 1935 and again in 1940 the steel cartels were up for renewal, Wilhelm Zangen of Mannesmann, Paul Reusch of Gutehoffnungshütte, and Peter Klöckner had each presented eloquent pleas for the need to become more market-sensitive. Not even at the darkest hour of German industry, however, was any steel spokesmen prepared to break with a fifty-year tradition and abandon cartels as a business way of life. The issue facing producers after 1945 was how best to adapt them to new circumstances.[42] At a long meeting of the leading firms on 21 June 1945 Walther Schwede of Vereinigte Stahlwerke, a leading light of the interwar ISC, vigorously advocated a de facto restoration of the prewar structures. He wanted a new cartel to conduct sales, allocate orders, manage accounts, keep customers in line, collect statistics, and provide unified political representation. This proposal was too extreme for most of the representatives present; for the time being the majority declared itself in favor of setting up price and production cartels but against restoring marketing agreements until conditions had returned to normal and quotas could be properly assigned.[43]

On 27 June 1945 the six leading Ruhr firms dissolved and then re-formed the main syndicates for crude steel, rolled products, pipe, and forged pieces. Under the new arrangements each firm retained control over a portion of its own output (the exact amounts of which would be the subject of much disagreement), but the new organizations were to have control for overall sales, order placement, ship-

[41] KA/Henle Verbände, Wiederaufnahme der Tätigkeit nach Kriegsende (5-10.45), "Aktenvermerk," 20 June 1945; and "Einschaltung der Verbände in Aufträge für den Truppenbedarf . . . ," 7 May 1945.
[42] KA/Henle Verbände, Wiederaufnahme der Tätigkeit nach Kriegsende (5-10.45), "Aktennotiz 25.6.45"; also, SP, "Eisenschaffende Industrie in Zeitpunkt des Zusammenbruchs," 7 October 1947.
[43] KA/Henle Verbände, Wiederaufnahme der Tätigkeit nach Kriegsende (5-10.45), "Aktennotiz 25.6.45."

ment, and bookkeeping.[44] The decision to create new cartels was not universally popular. The second in command at Klöckner warned that "to regroup so soon after dissolving the cartels [*Verbände*] while virtually maintaining the old leadership" was unnecessarily risky. He need not have worried.[45] In September the British asked the steel industry to set up a "steering mechanism" consisting of staffs for planning and administration, order placement, raw materials, technical assistance, and relations with foundries. Thereupon six leading firms, reviving the "Kleiner Kreis" of the 1930s, set up WVESI, the Wirtschaftsvereinigung der Eisen- und Stahlindustrie. From then to now, the "Economic Association of the Iron and Steel Industry" would serve as official spokesman for producer interests, act as the Ruhr's brain trust, and oversee numerous syndicates for specialized products.[46]

The British initially intended to maintain a close relationship with the Ruhr steel industry. On 5 October 1945 the organization later known as the Administrative Office for Steel and Iron (Verwaltungsamt Stahl und Eisen, or VSE) opened its doors. As set up, VSE could do little more than act as agent for steel industry policy, its competence being limited to "planning and market organization in the strictest sense of the word."[47] The steel producers' association, in other words, was to retain control of substantive matters: keep statistics, place orders, be responsible for provisioning, make recommendations regarding pricing, and handle transportation and social policy. As if to reinforce a close relationship with industry, VSE was also to exercise responsibility for implementing recommendations of an advisory board attached to the Economic Council in Minden. Included as candidates to the advisory board were delegates from Klöckner, Otto Wolff, and the Gutehoffnungshütte. Abraham

---

44 KA/Henle Verbände, Wiederaufnahme der Tätigkeit nach Kriegsende (5-10.45), "Aktennotiz Baum 27.6.45."
45 KA/Henle Verbände, Wiederaufnahme der Tätigkeit nach Kriegsende (5-10.45), Kunke to Henle, 4 July 1945.
46 KA/Henle Bergbaukreis/Eisenkreis, "Sitzung Kleiner Kreis am 7.9.45"; also, SP, "Niederschrift über die Mitgliederversammlung der Nordwestgruppe," 31 August 1945; also KA/Henle Industriekreise, Besprechungen der Ruhrwerke, (9.1945–9.1946), "Besprechung am 10.9-1945 .... über die Bildung eines Steuerungsorgans auf Wunsch der MG."
47 SP, "Aktenvermerk über eine Besprechung am 1.11.1945 in Dortmund"; SP, "Besprechung bei Herrn Abraham Frowein," 18 December 1945; KA/Henle Industriekreise, Besprechungen der Ruhrwerke (9.1945–9.1946), "Aktennotiz: Betr. Eisenkontingentierung," 5 October 1945; SP, "Eisenkontingentierungen zwischen Verwaltungsämtern und Wirtschaftsvereinigungen"; SP, "Aufgabenbegrenzung zwischen VSE und Wirtschaftsvereinigung."

Frowein headed the Economic Council. Though a "non-Aryan" forced into exile after 1933, he retained a strong sense of loyalty to the Ruhr.

Ruhr steel's honeymoon with the British came to an abrupt end. Late in the night of 30 November 1945 and in the early hours of the following morning seventy-five leading Ruhr industrialists were arrested, then chained together and trucked to the British concentration camp at Bad Nenndorf near Hannover. For six months, in some cases much longer, most of them would be hermetically sealed off from the outside world. The inmates were never charged with specific crimes; interrogation was perfunctory; reeducation attempts minimal. Though the motives behind the *Razzia* remain obscure, as do the reasons for ending the action, the victims would not soon forget it.

Hans-Günther Sohl, later chairman at Thyssen, learned the real meaning of defeat and degradation while taking the "cure" at Bad Nenndorf, a famous region for spas. He recalled in his memoirs having shared cell 44 with the three other arrested industrialists whose names were alphabetically closest to his, Hans Reuter (DEMAG), Walter Schwede (Vereinigte Stahlwerke), and Rudolf Siedersleben (Vereinigte Stahlwerke), who would later die in camp. Their quarters, a former bathroom, were large enough only to contain two rude wood-frame bunk beds separated by a corridor so narrow that the four inmates were able to share two blankets between them, assuming of course that no one tugged too hard. At one end of the room was a slit window, at the other an iron door. Sohl vividly remembered suffering through long and painful waits in line for the toilet and finally being "driven frantically to the accomplishment of one's purpose" by insistent shouts of "Hurry Up!" accompanied by furious pounding on the door.[48] The daily routine was monotonous and humiliating, Sohl recalled. After three months spent in total incarceration, the inmates were allowed to work in the garden for a few hours each day, following which they were stripped naked and their clothes searched for illicit food and cigarette butts. The diet was sufficiently meager to have caused Sohl to lose fifty pounds, which, his memoirs sardonically noted, is something that the prosperous *Bundesbürger* of the 1980s might well have envied, especially considering the outrageous fees charged at the better weight-loss clinics.

Günter Henle also lived with hateful memories of Bad Nenndorf. Himself politically irreproachable, the young Klöckner director

[48] Sohl, *Notizen*, pp. 98–99.

claims to have met real Nazis for the first time while imprisoned there. Though Henle found them personally as well as politically disagreeable, he departed the camp "less ashamed to be a German."[49] The experience of Bad Nenndorf left a legacy of fear, mistrust, and contempt for the victors that men like Henle would never quite overcome.

The mass incarcerations did not result in a clear-cut break with policy so much as leave it dangling. The British did not shut down WVESI, which continued to work with VSE, but left its legal status obscure. Though German efforts to rebuild the industrial associations continued, even extending to the national level in spite of the absence of key men, it must have seemed doubtful that anything permanent would result from this: Under the rubric of decartelization the British and the Americans had by then begun their reform program. Though its outcome could not have been predicted with accuracy, the intent behind it was unmistakably to eliminate the old elites. The British military government, though it had approved the formation of WVESI, had moreover never formally recognized its statutes.[50] It fell to the steel association's business manager, Wilhelm Salewski, to navigate the industry through the perilous shoals of occupation governance and administration. The trick was to act in an official capacity only when discharging orders from the occupation authorities, while claiming that all other industry decisions resulted from voluntary cooperation between producers. In reality, no such distinction existed. VSE and WVESI were interdependent; VSE was the sole source of demand for steel products, and WVESI controlled the apparatus needed for their manufacture. The one could not have existed without the other.[51]

An 18 May 1946 note written by Salewski in response to mounting criticism that WVESI was becoming power-hungry provides a glimpse into the legal and functional ambiguities of the situation facing Ruhr producers. WVESI, he said, was not an official organization; its membership was voluntary and in this respect represented a break with the principle of compulsion enforced between 1933 and 1945. While not denying that every firm in the steel industry belonged to it, Salewski insisted that this was by their own choice. The business manager did not discuss how it might have been possible for a maverick firm to have resumed operations with-

---

[49] Günter Henle, *Three Spheres* (Chicago, 1970), pp. 57–58.

[50] WVESI Sitzungsberichte, "Sitzung des Eisenkreises am 14. November 1946."

[51] WVESI Sitzungsberichte, Meetings of "Eisenkreis," 7 March 1946, 9 May 1946, 23 May 1946, 27 May 1946.

out the consent of other producers or how after the end of the occupation it could have coexisted with them. Salewski further denied the charge that WVESI engaged in production planning, which, he emphasized, was the official responsibility of VSE. Yet Salewski did admit that his organization allocated orders among producers, justifying this on the grounds that it alone had the statistical information needed to undertake this task on an efficient basis, and further acknowledged that though WVESI had never sought the power to allocate raw materials it had assumed responsibilities for administering the "maintenance quota" (*Instandhaltungskontingent*) and that it gave advice on export matters. The *Syndikus*'s note concluded that WVESI was a private, nonmonopolistic association useful for relieving the occupation authorities of burdensome administrative detail as well as for delegating essential economic tasks to producers. It is, in other words, evident from Salewski's remarks that Ruhr steel producers performed many of the same functions during the occupation that they had done earlier under Hitler. This was less wrong than unavoidable.[52]

The revival of business organizations advanced another step on 30 August 1946, when representatives of the twenty-three main West German producer associations meeting in Wuppertal endorsed the bylaws of a new national association for industry. The Federal Association of German Industry (Bundesverband der Deutschen Industrie or BDI) bore unmistakable resemblance to both the Reichsgruppe Industrie of the Third Reich and the pre-1933 Reichsverband der Deutschen Industrie.[53] Each of these associations was responsible for formulating overall policy for industry and acting as a lobby for its interests, and none of them dealt with labor relations. This restriction was a way to divert potential union demands for a voice in management, which were to be shunted on to a separate lower-level body, if possible one at the local level.[54] The unions immediately challenged BDI's legitimacy, demanding that its status be revised to include responsibility for labor relations. This would be but one early skirmish in the long battle over co-determination (*Mitbestimmung*).

[52] SP, "Anlage II: Stellungnahme zu den Notizen über die ministerielle Verwaltung der Eisen-und Stahlindustrie Deutschlands," 18 May 1946.
[53] SP, "Aktennotiz über eine Besprechung der Leiter der Wirtschaftsvereinigungen in der britischen Zone am 30. August 1946."
[54] SP, "Stellungnahme zu Fragen der Wirtschaftsorganisation," 17 July 1946; also, SP, "Besprechung von Organisationsfragen . . . ," 2 August 1946.

Its main theater was Ruhr steel. On 20 August 1946 the North German Iron and Steel Control named Heinrich Dinkelbach trustee for steel, empowering him to plan the decartelization and reorganization (*Entflechtung und Neuordung*) of the industry. In theory, Dinkelbach's authority extended only to interim measures, as the British never officially abandoned their ultimate aim of expropriating the shareholders of the trusts and socializing industry. In practice, the British stood aside, allowing the steel trustee a free hand to formulate policy, giving him a chance to create faits accomplis difficult to undo. Dinkelbach was perhaps the only leading figure in the Ruhr who looked upon defeat as an opportunity to break with the cartel tradition, the one potential Americanizer.[55] Yet he had remarkably little influence on events, in small part because of failures of leadership, in larger measure because the kinds of changes he championed lacked support within German society. In 1946 the unions captured the trusteeship from within and took over its program, and in 1947 and 1948 the old owners fought it to a standoff, leaving the future organizational form of industry in suspense. There it would remain until the Schuman Plan negotiations for the coal–steel pool reopened the ownership question.

Dinkelbach had become convinced as a member of the board of directors at Vereinigte Stahlwerke (VS) in the 1930s that the trust was too large and unwieldy to be managed properly.[56] What was true of VS, he declared in 1945, held good for the other big steel firms as well: They were backward, uncompetitive, and generally in as sorry a state as the broken cities of western Germany. Dinkelbach attacked the quota systems for sales as being unfair and inefficient; blamed "mixed pricing" (whereby profits at one level of production subsidized losses at another) for distorting cost structures; and condemned the tax breaks (*Schachtelprivileg*) accruing to vertically integrated combinations for further skewing supply. In his view these evils arose from "unnatural concentrations" of the 1930s that were neither necessary nor inevitable. Dinkelbach rejected the customary defense made by the trusts that the efficient exchange of fuels and supplies (*Verbundwirtschaft*) would inevitably be impaired by firm re-

[55] KA/Henle Umgestaltung des deutschen Kohlenbergbaues und der deutschen Stahlindustrie, Entflechtung, Schriftwechsel 1.1.46–30.9.47, "Der Mensch in der Wirtschaft" (radio address), 9 February 1948.
[56] KA/Henle Umgestaltung, des deutschen Kohlenbergbaues und der deutschen Stahlindustrie, Entflechtung, Schriftwechsel 1.1.46–30.9.-47, "Niederschrift über den einstündigen Vortrag von Dr. Klein (STV) über Konzernenflechtung," 13 May 1948.

organization, arguing instead that the laws of the market would assure the most efficient distribution of energy and resources among the firms of the region. To save the Ruhr industry from its own traditions Dinkelbach wanted to separate steel mills of optimum size from the trusts, which would then become low-cost operating units. Ownership of the new companies would be widely dispersed and their managements "transparent" and thus easy to supervise.[57]

The union movement provided the only real basis of support within German society for a program of detrustification.[58] In 1945 and 1946 its positions on this and other key issues were still at the formative stage. The policy of *Mitbestimmung* with which it would be so closely associated owed its beginnings to the spontaneous demands of shop-floor radicals that began to crop up in the steel industry at the end of 1945. According to the veteran lord mayor of Duisburg, Karl Jarres, "On 7 January [1946] a meeting of shop stewards from all the Klöckner mills ... took place ... at which equal representation on the supervisory and management boards and the appointment of [worker representatives by factory councils] were demanded. The seats on the board of management reserved for social and personnel questions were [also] insisted upon, and in addition a voice in all economic questions."[59] The unions, Jarres added, *opposed* these demands. This would soon change because Ruhr management, shaken by the recent imprisonments and facing disorder in the factory, was ready to accept co-determination with the unions as the least of possible evils threatening them.

To do nothing in the present crisis, Jarres told industry representatives on 18 January 1946, could only aggravate an already "very grave" situation.[60] Power sharing was simply "unavoidable" in his view, the only option being whether to grant representation on the board of management or on the board of supervisors. Everyone present preferred the latter alternative, as the board of supervisors had little authority over the conduct of business, and agreed that discussions on this basis should be opened immediately with the union chief, Hans Böckler. They took place over the following six weeks. The mood of the industry representatives was grim, even abject. On 6 March 1946 Jarres reported to WVESI that although

[57] KA/Henle Umgestaltung, Gesetz Nr. 27, Schriftwechsel 1.10.50–31.12.50, "Die Neuordnung der Eisen-und Stahlindustrie (Vortrag gehalten am 1.3.1948)."
[58] USNA RG260 7/25-2/9-15, "Developments in the Ruhr Steel Industry," 21 November 1947.
[59] WVESI Sitzungsberichte, "Sitzung des Eisenkreises," 18 January 1946.
[60] Ibid.

the recent meeting on 19 February had gone well management negotiators had been forced to accept a proposed law concerning the composition of the board of supervisors that would "destroy all authority at the factory." Worse than there being "no real chance to get rid of the [offensive] regulations" was the fact that co-determination in the chambers of commerce (*Industrie- und Handelskammern*) was on the agenda of the next meeting.[61] Thus began the union campaign to force employer acceptance of the co-determination principle for the general organization of the economy.

The most critical negotiations over *Mitbestimmung* took place under the aegis of the Steel Trusteeship. The first phase of its activities had begun in October 1946 with the commencement of "Operation Severance." This action was patterned on something called the Dinkelbach Plan of August 1945, originally an internal draft for the reorganization of Vereinigte Stahlwerke. Whether or not it represents the prevailing views of the board of supervisors at that sprawling and loosely coordinated production complex, or some minor subcurrent of dissidence, must remain moot. The rest of the steel industry construed the plan as being little more than a propaganda document put together to placate the occupation authorities.[62] Operation Severance neither transferred titles nor changed internal firm organization and was in the strict sense a deconcentration rather than a decartelization measure, being limited to defining the complicated transitional arrangements for new steel enterprises that were to be carved out of the old trusts. By 1 March 1947 four new so-called unit firms had been "severed" from the parent firms, and another twelve were soon due to be sundered.[63]

Co-determination did not originate with the Steel Trusteeship; rather, pressure from the unions caused *Mitbestimmung* to be introduced into the new unit firms. The official history of the Steel Trusteeship stated the matter somewhat awkwardly as follows: "In the initial discussion of 15 October 1946 between the North German Steel Control, the Trusteeship ... and the union federation (*Einheitsgewerkschaft*) of North Rhine-Westphalia, Hans Böckler first demanded the complete inclusion of labor in the reorganization of the steel industry by means of occupying positions on the board of su-

[61] WVESI Sitzungsberichte, "Sitzung des Eisenkreises," 6 March 1946.
[62] KA/Henle Umgestaltung des deutschen Kohlenbergbaues und der deutschen Stahlindustrie, Dinkelbach Plan 1945, "Neuordnung der Wirtschaft," 19 October 1945.
[63] SP, "Die deutsche Eisenindustrie unter Kontrolle," 16 April 1947.

pervisors and in the firm leadership."[64] In January 1947 the managements of Gutehoffnungshütte, Otto Wolff, and Klöckner offered to include labor representatives on their boards of supervisors. Industry spokesmen touted the concession as the dawning of a new era of economic democracy. At a meeting of Klöckner's board of supervisors on 12 February 1947 Henle proclaimed "the creation of a new social structure, [and] the introduction of what one commonly calls economic democracy [in which] employees should have a voice in running the company." "We are," he added, "convinced [that] to create a sound social structure in the German economy employers and employees must share responsibility equally."[65]

This statement was intended for the press; in reality, the three big trusts would try to head off socialization and decartelization whenever possible. The threat of the former was apparently acute but short-lived. Shortly after the conclusion of secret negotiations between Dinkelbach and the unions taking place between Christmas of 1946 and New Year's Day of 1947 North German Radio quoted "well-informed sources" to the effect that socialization of the steel industry was at hand. Dinkelbach refused either to confirm or to deny the rumor at a meeting of WVESI on 9 January.[66] Paul Reusch of Gutehoffnungshütte, then chairman of the steel producers' association, was still not too worried; he suggested at a meeting of 23 January 1947 that a few concessions might be all that was necessary to weather the difficult period until the impending *Landtag* elections, which he, like most others, expected to produce a nonsocialist majority. "The CDU," he said, "is against socialization and the unions . . . and even the SPD have reservations about it. Perhaps a combination of profit-sharing and co-determination on the board of supervisors will do the job."[67]

Well under way by early 1947, decartelization was a particularly annoying burr under the saddle of the steel producers. Hermann Reusch described the Dinkelbach Plan as having been "immoral," having "burned his fingers," and something deserving to be overturned at once. He resigned as chairman of WVESI rather than be responsible for helping execute it.[68] The most pressing problem facing pro-

[64] BA B109/1310, "History of the Reorganization of the German Steel Industry" (1953), p. 243.
[65] KA/Henle Umgestaltung des deutschen Kohlenbergbaues und der deutschen Stahlindustrie, Entflechtung, Schriftwechsel 1.1.46–30.9.47, Henle speech, 12 February 1947; compare with Henle, *Three Spheres*, pp. 80–81.
[66] WVESI Sitzungsberichte, "Sitzung des Eisenkreises," 9 January 1947.
[67] WVESI Sitzungberichte, "Sitzung des Vorstandes," 23 January 1947.
[68] WVESI Sitzungsberichte, "Sitzung des Eisenkreises," 20 June 1947.

ducers, however, was that the decree for Operation Severance stipulated that the new companies be allowed to acquire raw materials from the old trusts at official prices presumed to be grossly below actual value. The working capital of a firm forced to deliver on these terms would, barring some unforeseen piece of compensatory legislation, be wiped out once market pricing had been restored. Though Dinkelbach had merely failed to anticipate this problem, the industry accused the steel trustee of having deliberately plotted its destruction.[69] Forced to the wall but hoping that the entrepreneurial Americans (bi-zonal partners in North Rhine–Westphalia after 1 January) would view their problem more sympathetically than the socialist-minded British, the steel producers refused further concessions to the trusteeship and other reformers. On 3 April WVESI's Salewski begged off of general discussions of codetermination with the union representative Hans Potthoff on the grounds that the business associations were not yet completely organized.[70] There matters would remain for the rest of the year.

By midsummer 1947 battle lines were clearly drawn. The industry presented its position in a 102-page position paper whose title was crudely translated into English as "The Measures for a Dispersal and for a New Order of the Iron and Steel Producing Industry," a study endorsed by each of the trusts – Gutehoffnungshütte, Ilseder Hütte, Krupp, Hoesch, Klöckner, Mannesmann, Vereinigte Stahlwerke, and Otto Wolff. Though the paper's main premise was that the old owners were falling victim to an unjust expropriation, it revealed the industry as having been unrepentant and unreformed, opposed to any form of decartelization, and reluctant to share power with labor, the state, or other economic interest.[71] The document took the standard form of Ruhr apologia, denying first of all the validity of the customary Allied political reproaches: that coal and steel financed Hitler's rise to power, enthusiastically supported the regime, and sought to benefit from war and aggression. It also refused to acknowledge anything wrong with the trustification process in Germany, positing that industrial concentration had resulted from meager resource endowments, high capital requirements, and

[69] SP, "Zur Problematik der eisenindustriellen Entflechtung," 13 September 1947.
[70] WVESI Sitzungsberichte, "Eisenkreis: Engerer Vorstand," 3 April 1947, and "Sitzung des Eisenkreises am 28. August 1947."
[71] USNA RG260 7/25-1/2-3, "The Measures for a Dispersal and for a New Order of the Iron and Steel Producing Industry" (submitted to the Department of Economic Administration at the Economic Council of Frankfurt), 1948, 102 pp.

the heavy investment needed to capture new technologies. The document further noted that by comparison to American equivalents Ruhr firms were not large. The remainder of the paper contained more questionable assertions. Reorganization, stigmatized in the document as "dispersal," was dismissed as economically and ethically unthinkable. The current decartelization policy received predictably rough treatment, being attacked as illegal, procedurally opaque, and financially disastrous. No less than thirty pages were devoted to describing the dire consequences that would follow from its full implementation. They include tearing asunder well-established business relationships, cutting off efficient exchanges of fuel and power, and terminating production planning within the individual enterprise.

The document revealed the difficulty of the trusts' leadership in even admitting the very possibility of constructive change. Rejecting outright the decartelization process already under way, the authors of the text asserted that socialization, which by then seemed unlikely, was a "matter for Germans to decide" but in any case something that had to include compensation for all assets removed from private ownership. *Mitbestimmung* was also held to be something best reserved for future German legislation. As for possible alternatives to restoring the status quo ante bellum, the document did contemplate the possible formation of ten "minitrusts," each outfitted with optimum capacities of four to six million tons of coal, one million tons of steel, and "corresponding" manufacturing components. This was hardly a radical proposal. The new enterprises could all have been put together from production units of Vereinigte Stahlwerke. Since the parent firm was to have been compensated with stock in the successor companies, the old trust could have been reassembled later without great difficulty.

The steel trustee responded to the blast from the old firms with a countercritique of his own, which put the situation facing the industry in a historically more realistic context. Customary rights of ownership, Dinkelbach argued, had become meaningless as a result of defeat: The trusts were already bankrupt in everything but name, having lost outright 40.5 percent of real total assets plus another 30.3 percent in less tangible values. Further, the separation of the mines had reduced their net worth by another 11 percent, leaving as "disposable remaining assets" only 17.3 percent of prewar capital. Depressing though these facts were, they excluded a number of additional debits, according to Dinkelbach, among them losses on foreign properties, overdue payment on dollar bonds, and various fees owed cartels and trade associations. An even more immediate problem was that the low official steel prices in effect since 1933 had drained the firms of liquidity and the rising demand of 1947, which required increased provisioning, threatened to eliminate it altogether. To put themselves back in business, Dinkelbach concluded, the trusts would need not merely the restoration of ownership but the active support of the government.[72]

Though Dinkelbach's diagnosis of the Ruhr malady was sound, his remedy was not the one that brought about eventual recovery. What he prescribed was a healthy dose of *Mitbestimmung:* "Only after the working man has been convinced that everything is being done to improve the hard conditions under which he has operated since defeat will he be able to live with them. He must therefore be given a second chance to grow into leadership responsibility."[73] For this reason, he added, employee representatives had been included on the boards of supervisors of the successor steel firms. Yet this was, in his view, only the first step toward introducing economic democracy; the next one would have to be the establishment of codetermination at the level of the industry associations.

It never came to this. In the second half of 1947 the steel producers waged a delaying action that postponed a reckoning with codetermination for another three years. Though aided by external forces, this was a considerable feat. In spring 1947 the prognosis for WVESI was not good. The military government had still not recognized the producer association, nor did it apparently intend to do so until reforms were introduced.[74] On 6 May 1947 *Landrat* Maier appeared before WVESI to demand that the association admit successor firms to full membership. He talked tough. The supervisory boards of the new unit companies included 50 percent representation for employees, he explained, and decartelization, as then scheduled, was to extend to 90 percent of the industry. Would this mean an eventual takeover of the association? A mere matter of justice, Maier responded: Had the trusts not already written off their capital investment three times, in 1924–1926, in 1934, and now again after the war? And was it not true that once again the public would be called upon to bear the cost? Noting that fourteen companies were

---

[72] "Die Entflechtung und Neuordnung der Eisenschaffenden Industrie: Stellungnahme der Treuhandverwaltung im Auftrag der North German Iron and Steel Control" (Düsseldorf, April 1948), 72 pp.

[73] Ibid., p. 47.

[74] KA/Henle Verbände, WVESI (1.1.46–31.12.47), "Zur Frage der freien Wirtschaftsorganisation in den Grundindustrien, Salewski," 29 August 1947; KA/Henle Industriekreis, Bergbau-Kreis/Eisen-Kreis, "Sitzung des Eisenkreises am 9.10.1947."

scheduled to be created over the following few months and thirty-five within the next year, Maier warned that the efficient new production units would force the old firms out of business. "Perhaps," he speculated in conclusion, "we'll have to build everything from the ground up."[75]

This would not prove necessary. WVESI managed to stonewall until October 1947, when it entered negotiations with the Steel Trusteeship over the terms of new firms' membership in the association. The talks ended on 28 November 1947. On 18 February of the following year WVESI was formally reorganized, with the board of directors increasing from twenty-five to forty-five seats, half of which were reserved for the decartelized firms, three additional directors (Bruns, Barish, and Sendler) being named to the board of management. This was no coup d'état in the Ruhr. Each of the new appointments had been senior executives in the parent companies, and none of them was happy about Dinkelbach's reforms.[76]

Nor was anybody else outside of the trusteeship. Too small to be viable, the new companies soon started piling up unacceptable losses.[77] The unions, which were interested in decartelization mainly as a means of introducing co-determination, thereupon began reconsidering their political commitment to the Dinkelbach program, and the United States decided for its part to relaunch the entire reorganization process.[78] With the enactment of Law 75 in November 1948 the trusteeship lost its independence and was placed under the direction of a supervisory council consisting of four delegates from the unions and an equal number from industry. Though two theoretically independent delegates, Dinkelbach and Hermann J.

[75] SP, "Besprechung am 6. Mai 1947 mit Landrat Maier, Treuhandverwaltung der North German Iron and Steel Control," 10 May 1947.

[76] SP, "Sitzung des Eisenkreises am 3. Juli 1947"; KA/Henle Verbände, WVESI (1.1.46–31.12.47), Schröder to Henle, 29 August 1947; SP, "Besprechung vom 8. September 1947 bei der Treuhandverwaltung"; KA/Henle Industriekreis, Bergbaukreis, "Sitzung am 9.10.1947," and "Sitzung des Eisenkreises am 28.11.1947"; SP, "Pläne für eine Neuordnung in der deutschen Eisenindustrie."

[77] USNA, Department of State 862.6511/7-947, Fassberg to de Wilde, "Reorganization of the Iron and Steel Industries of the British Zone," 9 July 1947; 862.6511/7-2048, U.S. Consul-General in Bremen (Altaffer) to Secretary of State, "'Cold' Socialization of the Iron and Steel Industry in the British Zone of Occupation," 29 July 1947; and 862.6511/9-2947, Consul-General in Bremen (Altaffer) to Secretary of State, "Approaching Crisis in Decarteliztion Proceedings," 29 September 1947.

[78] USNA RG260 7/25-2/9-15, "Developments in the Ruhr Steel Industry," 23 November 1947.

Abs, were supposed to hold the balance of power, Law 75 was a formula for institutional gridlock. Union demands for co-determination could not be reconciled with management insistence upon a restoration of private control. The British remained doctrinally committed to socialization, the Americans ideologically attached to decartelization. Both feared that the reform process had gotten out of control, and neither knew where it was heading. The Steel Trusteeship, which would waste another two years drafting reorganization plans, was a mess.[79]

4.3 The dismantlement of Ruhr steel

During the very months that Marshall Plan aid began to pour into Europe the Western Allies destroyed much of the Ruhr steel industry. Dismantlement was not, as one might first suspect, caused by the left hand not knowing what the right was doing: The reduction of the material plant and the reconstruction of the European economy were inseparably linked. The British, seconded by the French, conditioned their acceptance of German participation in the European Recovery Program on the elimination and removal of the mills. Rather than risk weakening the Western alliance the United States reluctantly went along until the dismantlement policy, which, by arousing massive public resistance, threatened to discredit and destabilize the new Adenauer government. According to the official explanation as represented by Acting Secretary of State Lovett, the breakup of plant actually contributed to economic reconstruction. "The only way Germany could use its total industrial capacity," he said, "would be to grant her absolute priority over other European countries in the allocation of scarce raw materials. Rather than serving as a drain upon the United States ... the removal program will ... lessen this burden [by fulfilling] capital requirements [and]

[79] USNA, Department of State 862.6362/4-749, Foreign Minister Bevin to Secretary of State Acheson (with enclosures), 7 April 1949; 862.6362/4-749, Memorandum of Department of State, 31 May 1949; 862.6511/4-748, Consul-General in Bremen (Hildebrand) to Secretary of State, "Draft Proposal of the German Union Federation of the British Zone for Reorganization of the Iron and Steel Industry," 7 April 1948; and 862.6511/3-2847, Consul-General in Bremen (Altaffer) to Secretary of State, "Further Details Concerning Steps Taken for Separation of Firms from the German Iron and Steel Trust," 28 March 1947; also, "Ein Jahr Gesetz 75," *Der Volkswirt* 47:1949, p. 13.

through the product manufactured by this equipment."[80] The Petersberg Agreements, concluded in November 1949 between the Allied high commissioners and the government of the Federal Republic, finally ended demolition work at all factories where it had not yet already begun, and after two years the discreditable campaign wound down.

Dismantlement was the most serious threat faced by the Ruhr trusts during the occupation but also their greatest opportunity. Between the end of 1947 and the beginning of 1950 some 700 million Deutschmarks' worth of plant were unbolted, disassembled, cut apart with blowtorches, and carted away.[81] By weight and value, the removals amounted to slightly more than a third of the original program, which if executed in full would have crippled recovery. As it happened, its effects were relatively unimportant.[82] Many of the junked installations were outmoded, inefficient, genuinely surplus to peacetime need, and had only scrap value. Further, the Adenauer government was committed to compensating owners for losses. The real threat faced by the Ruhr steel industry was the replacement of the demolished plant by something better: Investment might have been channeled into new sectors of production in other locations rather than being plowed back into reconstructing the traditional type of vertically integrated trust. Dismantlement might in this case have accelerated the modernization process already under way in the West German economy: captured new technologies, moved production closer to new markets and sources of supply, and introduced new management techniques.

The dismantlement campaign made it politically impossible for an elected German government even to consider such an alternative investment strategy. As factory after factory came down, tens of thousands were thrown out of work. For months on end German heavy industry appeared to be doomed. The common threat faced by employers and employees renewed the sense of national solidarity shattered by the war, blunted the sharpness of socialist attacks on private ownership, and restored a measure of public respectability to the old combines. In ending the *Demontagen*, the shaky Adenauer government won its first great victory, its increasing prestige adding

---

[80] MAE Y 372, U.S. Department of State, "Dismantling and Removal of Plant from Germany," 26 November 1947.

[81] Walter Först "Die Politik der Demontage," in Walter Först (ed.), *Entscheidungen im Westen* (Cologne, 1979), p. 138.

[82] Wolfgang Benz, *Die Gründung der Bundesrepublik: Von der Bizone zum souveränen Staat* (Munich, 1984), p. 83.

to the likelihood of the eventual restoration of private ownership and control in the Ruhr. The dismantlement campaign had important international repercussions as well. The ERP objected strenuously to it from the outset. The factory removals also aroused the ire of tight-fisted congressmen troubled by the high costs of the occupation and wary of Marshall Plan expenditures, angered union leaders in both the United States and the United Kingdom, and fueled the opposition of broad sections of the public in both countries.[83] It also helped smooth the way for a West German entrance into Europe: Isolated internationally and facing the prospect of extinction, the Ruhr steel industry would seek a friendly takeover from the French. An overture to a theme with many variations, this was the first post–World War II attempt at a businessmen's solution to the Ruhr Problem.

Dismantlement represented a distinctly British solution to the Ruhr Problem and must be clearly distinguished from French security policy. France sought to protect itself by imposing institutional controls (*contrôle organique*) over production, allocation, and investment, and during the two years following the 5 June 1947 Marshall Plan announcement ceased to play the role of carping critic to U.S.–U.K. German policy and instead drew closer to the Anglo-Saxons by joining the control groups for coal and steel, shaping and participating in the new International Authority for the Ruhr, and winning U.S.–U.K. agreement to establish military security boards with veto powers over new investment in heavy industry. Though the precise use of such controls would depend upon circumstance, their purpose was to protect France during the recovery process and strengthen its long-term position vis-à-vis the powerful neighbor to the east. Though contrôle organique implied gearing German recovery down to French speed, it did not require destruction of plant. France supported the British position on dismantlement at

[83] FRUS 1948/II 740.00119 EW/2-2348, Memorandum by the Secretary of State, 13 February 1948, p. 721; 740.00119 EW/2-1448, Memorandum by the Assistant Secretary of State for Occupied Areas (Saltzman) to the Undersecretary of State (Lovett), 14 February 1948, p. 726; and 740.00119 EW/3-2248, Memorandum of conversation, by Frank G. Wisner, Deputy to the Assistant Secretary of State for Occupied Areas, 22 March 1948, p. 738; USNA RG260 17/257-2/9, "The Significance for the German and European Economy of the Factories Remaining on the Dismantlement List," September 1949 (enclosure to Consul-General Bremen to Secretary of State, "Transmitting Memorandum on Dismantling Issued by the Social Democratic Party of Germany," 8 September 1949); MAE Y 372, "Marshall's Letter [of 4 November] on German Plants," *New York Times*, 9 February 1948; Cairncross, *Price of War*, p. 180.

gatherings of the Big Three less out of enthusiasm than to gain support for its own policy of economic security.

From the British standpoint dismantlement appeared quite reasonable, if only because no other approach to economic disarmament seemed very realistic: The United Kingdom was not prepared to make the long-term commitments needed for the success of a policy like that of contrôle organique. As previously, payment problems governed U.K. policy making. Although in the first two postwar years it had made sense to encourage the revival of German production in order to cut down on the costs of the occupation, the British feared after mid-1947 that further output increases would expand exports and cut into foreign markets. They therefore stuck to their original figure of allowable steel production, thus shifting from the least to the most restrictive of the Western Allies as regards German output. As quid pro quo for the revised Level of Industry Agreement of August 1947 raising permissible maximum steel production to 10.7 million annual tons, Britain received assurances that dismantlement would begin in earnest. On 17 October 1947 the two military governors published a list scheduling 682 factories for dismantlement, of which 496 were in the state of North Rhine–Westphalia and 294 in the Ruhr. Wrecking crews set about work at once.[84]

Although the Americans and British frequently invoked the Potsdam agreements as the juridical basis of the dismantlement policy, the reference was misleading. The breakup of Ruhr steel factories was not the product of Four Power agreement, nor did it result in the shipment of goods to the USSR. Potsdam had assigned the Soviets 15 percent of capital equipment from the Western zones against compensation, plus another 10 percent in uncompensated "surplus" equipment. The 1946 Level of Industry Agreement declared 50 to 55 percent of rated 1938 steel capacity surplus and therefore available to the Soviets as reparations goods. Yet with the exception of one important shipment of 143 war plants to member nations of the Inter-Allied Reparations Agency (IARA), a common Allied reparations policy was never implemented.[85] In practice each

[84] USNA RG466 D(49)338, "Brief on Dismantling, Prepared for Mr. McCloy by US HICOG Staff for Use at Foreign Ministers' Conference"; Cairncross, Price of War, p. 172.
[85] USNA RG260 11/147-2/24-27, "Three Years of Reparations: Progress of Reparations from Germany in the Form of Capital Industrial Equipment," November 1948; Cairncross, Price of War, p. 191.

zone went its own way. For the USSR and France, "strip and ship" was the rule. The United States broke up Category 1 war plants but otherwise remained inactive. The British requisitioned plant but did not dismantle and remove it before 16 October 1947. The Soviets would get none of this material.[86] The amount received by each of the occupying powers from the Western zones was in fact inversely proportional to professed interest: The United States received $124 million in reparations from IARA, the United Kingdom $106 million, France $87 million, the USSR $26 million.[87]

It is difficult to describe the pace and amplitude of the dismantlement campaign accurately in a few words. Basic definitions, dates, and cost estimates vary enormously from account to account.[88] The scale of the program was nonetheless vast, and it went on for many months without any real indication that it would ever be broken off. The original list contained a total of 682 plants in the U.S. and U.K. zones scheduled for dismantlement. The French added another 236 to it. Roughly two-thirds of these factories were "war plants." The British Zone also contained 298 "surplus plants," meaning steel production units. By June of 1948 about 500 of the original 918 plants had been demolished, and the French had unilaterally removed another 42 from their list. Of those remaining apparently 381 belonged to the steel industry. By September 1949 a total of 873 tons of the originally scheduled 1,423 tons had been dismantled in the British Zone.[89] Though the full extent of the steel industry dismantlement cannot be determined, it included 38 percent of Bochumer Verein, Bochum; 38 percent of the Reichswerke, Watenstedt; 29 percent of August Thyssen, Hamborn; 57 percent of Dortmunder Brückenbau, Dortmund; 18 percent of Ruhrstahl, Hattingen; 44 percent of Klöckner, Düsseldorf; and 39 percent of August Thyssen, Duisburg. The pace of work apparently depended on the availability of wrecking crews. Once begun, it held steady at about forty-five thousand tons per month. All efforts to slow the pace were unsuc-

[86] Cairncross, *Price of War*, chap. 9, "Soviet Reparations from East Germany," pp. 194–218; Abelshauser, "Wirtschaft und Besatzungspolitik," in Scharf and Schröder, *Die Deutschlandpolitik Frankreichs*, pp. 111–139; SP, "Folgen der Stillegungen und Demontagen in der Eisenschaffenden Industrie," 6 November 1946.

[87] Cairncross, *Price of War*, p. 191.   [88] Ibid.

[89] USNA RG260 17/257-2/9, "The Significance for the German and European Economy of the Factories Remaining on the Dismantlement List," September 1949, p. 7.

cessful until September 1949. It proved extraordinarily difficult to dismantle dismantlement policy.[90]

The ECA provided the main opposition to the demolitions, but it encountered resistance to policy revision initially from the State Department and later from the Allies. Though the groundwork had been carefully laid for the adoption of a new line, not until April 1949 did the United States manage to get Britain and France to reconsider a change of position. The Washington conference held that month produced a drastically shortened dismantlement list but one that still contained the names of most Ruhr Konzerne. The campaign continued. The threat of massive resistance in September 1949 finally forced Britain to give in. This led two months later to the Petersberg Agreements, after which the campaign wound down.

Harriman and Hoffman needed nearly a year to persuade the Truman administration to reconsider the dismantlement policy. The cabinet first took up the issue in January 1948. After pondering the matter for a month President Truman decided to heed General Marshall's advise to drop the ECA's request rather than jeopardize Allied unity.[91] In a second initiative of March 1948 Harriman won an initial victory, the support of the secretaries of the army, commerce, and Interior to the dispatch of a technical mission (Collison Mission) to investigate the effects of the dismantlements on European recovery.[92] After hearing its report Marshall shifted ground. At a cabinet meeting of 9 August 1948 the secretary of state persuaded the president to seek congressional authorization for the establishment of an Industrial Advisory Committee (IAC) to review the progress of the dismantlement program[93]

The publication of the IAC's report in January 1949 signaled a change of course in American policy. It recommended retaining 167 plants slated for dismantlement, among them 37 steel plants, and raising the permitted level of output from 10.7 million to 13 million

[90] USNA RG466 D(49)338, "Brief on Dismantling, Prepared for Mr. McCloy by US HICOG Staff for Use at Foreign Ministers' Conference."
[91] FRUS 1948/II 740.00119 EW/2-1248, Memorandum Prepared by the Department of State, 12 February 1948, pp. 719–720.
[92] FRUS 1948/II 740.00119 EW/3-2248, Memorandum of Conversation, by Frank G. Wisner, Deputy to the Assistant Secretary of State for Occupied Areas, 22 March 1948, pp. 73–74; and 740.00119 EW/3-2548, Assistant Secretary for Occupied Areas (Saltzman) to Secretary of Commerce (Harriman), 25 March 1948.
[93] FRUS 1948/II 760.00119 EW/8-1648, Memorandum of conversation by the Undersecretary of State (Lovett), 16 August 1948, pp. 792–793.

tons annually.[94] The recommendations of the IAC, also known as
the Humphrey Committee, were actually identical to those made six
months earlier by the so-called Wolf Mission investigating the Ruhr
mills on behalf of the ERP. The Wolf Mission might more accurately
have been called the U.S. Steel Mission. Its head was president of
the U.S. Steel Export Company Ian F. L. Elliott. In the 1930s the
American responsible for maintaining liaison with the International
Steel Export Cartel was the mission's specialist for marketing and
distribution. Most of the other participants also had ties to big steel.
In early June Wolf's party met with a German team headed by
Hans-Günther Sohl, who after release from prison had taken over
responsibility for handling Vereinigte Stahlwerke's decartelization
problems.[95]

This was the first get-together of German and American steelmen
since the war. According to Sohl Wolf thought dismantlement stupid
and costly to the U.S. taxpayer. Sohl also claims that the documentation his team provided found its way into the Wolf Mission's report. Though this may well be true, the U.S. steelmen had already
reached their own conclusions concerning *demontage*.[96] The technical findings of their report curiously do not indicate that the dismantlement program, as then planned, interfered seriously with the
ability of the mills to reach the targeted figure of 10.7 million set in
August 1947, and they even pointed out a couple of areas such as
forgings where the number of plant reductions could have been increased. To raise outputs the mission report recommended a number of technical improvements such as better scrap preparation,
increased use of tilting furnaces, and better coordination of rolling
schedules.

The Wolf men nonetheless held the dismantlement program in
contempt. They thought it absurd that tube rolling mills should be
scheduled for dismantlement at a time of rising world shortages of
pipeline, drilling pipe, casing, and other tubular products required
for oil exploration, and they also sounded a general warning against
taking apart efficient integrated production units such as ATH,
Hamborn; Hüttenwerk Niederrhein AG, Duisburg; the Bochumer
Verein, Hontrop; and the Deutsche Edelstahlwerke, Krefeld. The
Wolf Mission's report called for a moratorium on all dismantlement

[94] FRUS 1949/III 740.00119 EW/1-1749, Secretary of State to French Ambassador (Bonnet), 25 January 1949, pp. 547–548.
[95] USNA RG260 3/246-1, "Wolf Mission Report: Recommendations for Increasing German Steel Production in Bizonia" (n.d.), pp. 111–112.
[96] Sohl, *Notizen*, pp. 111–112.

as well as lifting import and export controls, returning to free markets in goods and services, clearing up the ownership problem, and restoring management prerogatives. The positive-thinking Sohl regarded the visit of the U.S. steel industrialists as the first sign of light at the end of the tunnel.[97] Yet it was sputtering and still very distant. The Humphrey Committee was not established until August 1948. Its work would take another half-year to complete, and its recommendations then met head-on opposition from Bevin.

On 12 January 1949 the Humphrey Committee formally recommended to the State Department that, though dismantlement should begin at another 174 plants, 167 should be retained intact for production. Two weeks later Acheson promised Bevin and Schuman a new agreement for the so-called restricted and prohibited industries, for which negotiations had begun in December between the military governors, if they would agree to accept the Humphrey Committee's recommended reductions in the number of mills slated for removal.[98] The British foreign minister professed that he would contemplate a cut in British ERP aid before considering any such reductions but left the door open to later reconsideration.[99]

The matter was dealt with in the Three Power Washington conference of April 1949. The two deals arrived at there superficially resembled an American triumph, even though the first of them, the so-called Prohibited and Restricted Industries (PRI) Agreement, actually lengthened the list of banned or limited industries, including those for many artificial materials plants in the chemical industry such as the so-called Fischer–Tropsch refineries for artificial petroleum. The second agreement contained what appeared to be the major concession to the U.S. position on dismantlement: It reduced the list of factories facing the headache ball from 167 steel mills to 8. However, the plants remaining on the list were the largest and most important installations in Rhineland-Westphalia, employing approximately the same number as the 59 of those exempted, 40,000. The installations left on the list included the steelmaking and forging sections of the Bochumer Verein; the steelmaking equipment, foundry, and forge at the Deutsche Edelstahlwerke, Bochum; electric arc furnaces and converters of the Deutsche Eisen-

---

[97] USNA RG260 3/246-1, "Wolf Mission Report," passim; Sohl, *Notizen*, pp. 112, 114.

[98] FRUS 1949/III 740.00119 EW/1-1749, Secretary of State to French Ambassador (Bonnet), 25 January 1949, pp. 547–549.

[99] FRUS 1949/III 740.00119 EW/1-2749, Secretary of State to Embassy in France, 27 January 1949, p. 650.

werke AG, Mülheim; the steelworks and wire plant of the Klöckner
Werke, Düsseldorf; the bar-drawing plants of Hoesch AG, Dortmund; and the steelworks and three plate mills at the August Thyssen Hütte, Duisburg. The Wolf Mission and the Humphrey
Committee had both recommended sparing ATH, which had a capacity of 2.25 million tons and provided employment for 12,000
persons. As compared to the original dismantling list, the schedules
in effect after the Washington agreements meant increased exemp-
tions of only 10 percent in capacities of blast furnaces, 1.5 percent
in those of open hearth plants, 3 percent for those of rolling mills,
12 percent for those of hammer mills, and none whatsoever in those
for presses and rolling mills. The Washington agreement, if fully implemented, would have been hardly less crippling to German recovery than its predecessors.[100]

The Three Power agreements had no immediate impact on the
rate of plant breakups, which continued as before, and the disappointment they caused triggered a wave of German protests. The
first such demonstrations had occurred after the announcement of
the dismantlement program back in October 1947. To contain the
discontent the military government had threatened cutting off the
importation of foodstuffs. Displays of public opposition continued
to accompany factory demolitions. These actions could count on foreign sympathy. On 16 December 1947 President William Green of
the American Federation of Labor (AFL) protested the dismantlement program to President Truman. Pressure from organized labor
was partly responsible for the inclusion in the Foreign Assistance
Act of 1948 of a provision exempting German factories needed for
recovery from *demontage*. This was the stipulation under which the
Humphrey Committee was able to operate.[101]

The mass protests that began in late April and May 1949 required
the intervention of armed British troop units. Fearing that they
would undermine the Adenauer government and give rise to the
twin evils of communism and neonazism, the United States renewed
pressure on the British to break with dismantlement. Meeting with
the leaders of both the CDU and the SPD on 13 September 1949,
U.S. High Commissioner McCloy had to defend the policy against
the bitter accusations of Adenauer that it was of no value to anyone,
stirred up hatred on the right, and fanned the resentment of a pub-

[100] USNA RG260 17/257-2/9, "The Significance for the German and European
Economy of the Factories Remaining on the Dismantlement List," 8 September
1949, passim.
[101] Ibid.

lic convinced that it was motivated solely by a desire to eliminate competition. McCloy had to respond at the same time to SPD chief Kurt Schumacher's charges that the policy prevented the German working man from becoming internationalist-minded, increased unemployment, and set back recovery.[102]

McCloy was in an uncomfortable position; he thought the dismantling process had "little value . . . if any," and "its abrasive character [was] so great . . . as to risk some of our main objectives."[103] More could be gained, in his view, by giving solid support to the new government. McCloy advocated dropping the program as soon as the Germans had agreed to participate in the International Authority for the Ruhr. At the meeting of the Big Three held in Washington on 15 September 1949 the United States nonetheless failed to persuade Britain and France to give it up. Bevin led the charge against the Americans, arguing that the Germans "would take whatever we give them and then ask for more" and adding that if dismantlement were stopped the Ruhr would retain a dangerous fifteen million tons of capacity. The British foreign minister insisted that to complete the present program would take a year and a half and that it continued until the Germans had indeed been reformed.

But the protests continued, and within a month Bevin had to give in. On 26 October 1949 he told Acheson that "the moral authority of . . . the Allies in Germany is being rapidly destroyed [because] of the present dismantling program, which is arousing bitter resentment and opposition . . . particularly in the British Zone . . . The continuation of dismantling is [also] causing great disquiet among the Labour Party here and is becoming more and more unpopular in Parliament."[104] He added that the morale of the wrecking crews had been so seriously undermined by public censure that "it is only a matter of a few weeks before the dismantling collapses for lack of labor." Britain would suspend the policy, Bevin declared, if the Adenauer government agreed to join the International Authority for the Ruhr, (IAR), recognized the legitimacy of the military control boards, and acknowledged the need for maintaining additional security controls.

[102] FRUS 1949/III 740.00119 EW/9-1349, U.S. Commissioner for Germany (McCloy) to Secretary of State, 13 September 1949, p. 595.
[103] FRUS 1949/III 862.60/9-1449, Acting U.S. Political Advisor for Germany (Riddleberger) to Secretary of State, 14 September 1949, p. 597.
[104] FRUS 1949/III 740.00119 Control (Germany) 10-2949, British Secretary of State for Foreign Affairs (Bevin) to Secretary of State, 28 October 1949, pp. 618–619.

The Petersberg Agreements of 22 November 1949 brought the Ruhr dismantlements to an end but have an importance that transcends the issue: They represent the first sovereign act of the new state. The accords left much undone: They did not win West Germany the right to conduct an independent foreign policy or to organize a self-defense force, and important reserve powers resided with the high commissioners. The control groups, military security boards, and the coal and steel trusteeships all remained intact. Yet the accords enabled the Federal Republic to establish foreign consulates, ended its status as a protectorate, and cleared the way for its membership in international organizations, the first of which would be the International Authority for the Ruhr. Adenauer had actually resisted joining this body earlier; membership conceded recognition of its right to control of German coal exports. An invitation of 1 April 1950 to participate in the European Council in Strassbourg, a body whose prestige at the time compensated for its lack of power, soon followed German entrance into the IAR.

The working men and women of the Federal Republic were less impressed with improvements in diplomatic status than with the termination of the despised and destructive dismantlement campaign. This occurred unexpectedly, when both industry and the public were preparing for the worst. Adenauer had made the issue his top priority, warning at each of his meetings with the U.S. high commissioner that to Germans the word *demontage* was coming to mean "Versailles," to symbolize a harsh and unjust peace. The chancellor's dramatic success in securing the Petersberg Agreements constituted the first real proof that his government could protect jobs and create opportunities for economic growth.[105]

The dismantlement campaign strengthened the forces of political conservatism, weakened those of reform, and raised special problems for the union movement. The industrial co-determination championed by the Deutsche Gewerkschaftsbund extended no further than the realm of Allied control. Thus, though distinctly unenthusiastic about the decartelization plans of the Steel Trusteeship, the unions were reluctant to oppose them forcefully. Instead, they waffled. On 29 April 1948 Sohl, Schwede, and Wenzel of Vereinigte Stahlwerke met with Potthoff of the unions, Landrat Meier, and Deist of the Steel Trusteeship. Sohl's file minutes indicate complete agreement of all parties on four points: postponement of the property issue; the strictly temporary nature of both the coal–steel split

---

[105] USNA RG466 D(49)371, Political Liaison Bonn to HICOG Frankfurt (report of Bundestag foreign debate), 16 November 1949.

and the decentralization of steel production; the inviolability of the Verbundwirtschaft; and the indivisibility of manufacturing and heavy industry. "The goal must be," asserted the note, "to create efficient, vertically organized work groups, in which the frequently mentioned optimal 1 million tons of raw steel capacity is not necessarily to be a binding standard . . . In cases like [ATH] a larger capacity must be kept intact."[106] This went a bit far for Potthoff; in a demurrer to the minutes he reminded Sohl that the unions had neither renounced discussion of the property question except in the presence of occupation authorities nor agreed that the coal industry be necessarily tied back into steel. He also chided Sohl for failing to note that Wenzel had earlier acknowledged the need for parity in the boards of supervisors of steel firms.

The dismantlements put the unions and the SPD in the position of having to defend the sanctity of the coal–steel tie, the Verbundwirtschaft, and the vertically integrated trusts both publicly and privately. In a position paper of 8 September 1949, for example, the SPD found many powerful reasons for objecting to the continuation of the hated Allied policy: It would force Germany into long-term dependence on foreign subsidies, eliminate jobs, undermine the effort to make democracy work, and create labor unrest in France, Britain, and the United States. The core of the SPD's case was, however, the contention so often made by management that the Ruhr industrial complex was the product of undefiable iron laws of history, geography, and economics.[107] In a characteristic factory-by-factory criticism of the Washington agreement's dismantlement schedule, the SPD paper therefore attacked Allied plans to take apart the open-hearth steel plant and plate rolling mill at Hüttenwerk Niederrhein AG, Duisburg, on the grounds that "[the two] constitute essential links in a production chain which cannot be ripped out without damaging the whole work." Admitting that the open-hearth work constructed in 1909 was outmoded, the paper asserted that, though it had no monetary value, the plant "forms an indispensable link between the blast furnace and the rolling mills"[108] and thus must be retained; without it the flow of ore to the finished product would have been interrupted and the normal flow of mol-

[106] BA Nachlass Lehr 20, "Besprechung in Anwesenheit von Colonel Dieter," 29 April 1948.
[107] USNA RG260 17/257-2/9, "The Significance for the German and European Economy of the Factories Remaining on the Dismantlement List," 8 September 1949.
[108] Ibid.

ten pig iron to the open-hearth furnace cut, as would also that of hot bars to the rolling equipment. The removal of the open-hearth plant, the argument proceeded, would require cooling and forwarding the entire pig iron output of the blast furnace to an outside steel plant; the rolling mills retained at the plant would in addition be obliged to purchase their product from outside sources. The costs of such dislocations and readjustments would, the commentary concluded, be prohibitive. It is even more difficult now to judge the merits of the SPD's defense of Hüttenwerk Niederrhein, or similar pleas made on behalf of the other firms remaining on the list, than it was then: Neither the party's experts nor historians writing forty years later are in a position to determine the exact costs and consequences of the dismantlement campaign. This was less important, however, than the fact that Allied *demontage* policy forced the unions and the SPD into a defense of the existing technical and organizational structure of Ruhr industry: Decartelization, decentralization, or far-reaching change of any sort was excluded as a policy option. Dismantlement effectively disarmed the union movement. The future of co-determination rested with the old firms and the Adenauer government.

4.4   A new internationalism?

The dismantlement campaign did not turn the directors of Ruhr industry into instant internationalists; the steel chauvinism of 1914–1918 had long since faded into memory. With defeat in World War II came recognition that the Ruhr's future depended upon European acceptance and accommodation. Proposals for a new international authority to regulate heavy industry were much in vogue among the more advanced political thinkers of the producer associations and cartels, and some of them included a real transference of sovereign power. The Marshall Plan stimulated further thinking along these lines. The threat of physical destruction nonetheless made a real difference. In lonely desperation the titans of mine and mill threw themselves into the arms of the French and, to save what remained of their factories, declared themselves willing to be taken over. This sacrificial attitude naturally changed as the danger receded. Those who had confronted institutional extinction had no remaining doubt about the need for restoring and strengthening ties with foreign colleagues. Yet German producers could begin to take heart. The Petersberg accords were only one important milestone. Additionally, the steel glut mounting in 1949 gave rise to international exercises in planning to regulate worldwide outputs, and the Ruhr was included in them. A dialogue concerning Franco-German economic cooperation was initiated between the ministries of the two countries in the same year. In early 1950 French producers made a first groping attempt to reestablish old cartel ties. These were at least adumbrations of a new internationalism.

The announcement of the Marshall Plan first prompted the steel association's business manager, Wilhelm Salewski, to commit his ideas concerning economic cooperation to pen. His essay, "West European Heavy Industry Union," speculated about how to restore the unity of the area. The First World War, he began, broke off the capital connections developed in the late nineteenth century, and the French occupation of the Ruhr demonstrated the absurdity of plans for annexation like those of the Pan-Germans. Cartel cooperation in coal and steel had taken hold between 1926 and 1939, he emphasized, and to restore this positive development the once-thriving coke–minette exchange should be promoted. State authorities must invite collaboration, Salewski said, as the old cartels had disappeared, yet they too had to be revived if economic cooperation were to be restored. Otherwise there would be gluts, the Saar would become a bone of contention, and price controls would wreck the mines. Ruhr industry, he concluded, would eventually find the necessary capital from either state or private sources, if only economic cooperation were renewed: It was the essential prerequisite for recovery.[109]

The threat of dismantlements ended speculations like Salewski's and stimulated an effort to arrive at practical solutions. In October 1947 Konrad Mommsen, the business manager of one of the leading syndicates for steel products, announced a proposal for a "European Ruhr Industry," envisioning the transfer of authority to administer the industrial assets of the region to a European political federation with a legislative chamber elected by popular suffrage and governed by an economic council providing proportional representation to the various nations, including Germany. Mommsen's plans included co-determination and provisions making management answerable to a European board of supervisors but left industry in control of the production process. Businesslike procedures,

---

[109] SP, "Westeuropäische Montan-Union?" 9 June 1947; also, "Die europäische Eisenwirtschaft im Zeichen des Marshall-Plans," Stahl und Eisen 68 1–2:1948, pp. 15–19.

which Mommsen carefully described in the proposal, would have to
be followed and the European parliament would determine the distribution of profits.[110] Published in a journal of political ideas, *Die
Wandlung*, this scheme aroused less interest than it should have:
Ruhr industry, or at least one of its spokesmen, was now openly
committed to submit to European control in order to end the occupation and the threat of dismantlement.

The idea of offering the French major participation in the Ruhr
trusts first arose at a December 1947 meeting of industrialists held
in Mülheim to discuss the policies of Dr. Johannes Semler, who
then headed the bi-zonal economic administration in Frankfurt. At-
tending were Hermann Reusch of Gutehoffnungshütte (GHH),
Hermann Wenzel of Vereinigte Stahlwerke (VS), Günter Henle and
Karl Jarres of Klöckner, and Robert Pferdmenges, a member of the
VS board and president of the private banking house bearing his
name. Pferdmenges approached the de Wendel interests through
their representative in Germany, Armand Bureau, who could not,
however, generate any enthusiasm among the French for the Ruhr
proposal. Accounts of the episode fail to indicate precisely what the
Germans were prepared to offer (although it was rumored to have
been an astonishing 50 percent participation in Ruhr steel), and it is
equally difficult to fathom whether Pferdmenges was acting on behalf of the Ruhr or of ideas long cherished by his close friend, Konrad Adenauer.[111]

Adenauer was happy to take credit for sponsoring the initiative. If
the French had directors on the boards of the leading Ruhr firms,
he told an American counsul, they would have no reason to fear
German rearmament. Adenauer also boasted that back in 1925 he
and the soft-coal tycoon Dr. Silverberg had joined Hugo Stinnes in
an effort later scuttled by Stresemann to interest the French in a
mutual exchange of shares. After questioning by the same consul,
Reusch, Henle, Wenzel, and Jarres, though admitting that contacts
with de Wendel were indeed made, each emphasized that Pferdmenges had received no mandate to offer the French a partnership.
They all claimed to prefer international capital participation to exclusive arrangements with foreign interests.[112]

[110] USNA RG260 7/23-1/21, "The European Ruhr Industry" (excerpted from *Die
Wandlung*, 10–12 October 1947).
[111] USNA RG260 7/25-2/9-15, Consul-General in Bremen (Hildebrand) to Secretary of State, "Further Information Regarding Alleged Offer to French Industry to Participate in Ruhr Iron and Steel Industry," 25 March 1948.
[112] Ibid.

Less mystery surrounds the so-called Vereinigte Stahlwerke Plan of 1949. It was a straightforward proposal to give supervisory authority over the firm to foreign interests in exchange for an end to the dismantlements. No less than 80 percent of the installations slated for elimination after the Washington agreement belonged to VS, thirty of whose works had been wholly or partially dismantled already.[113] In summer 1949 crews began tearing down the August Thyssen Hütte, the pace of demolition stepping up month by month. As of October one ATH converter per day was being eliminated. Unaware that Britain was then prepared to abandon the dismantlement policy, Sohl warned that the wrecking had to be terminated by the beginning of November if anything were to be left worth saving; the survival of the firm, he said, was a question of days. On 6 October he implored Lord Mayor Lehr of Düsseldorf to do everything within his power to persuade the high commissioners to treat ATH as an emergency case and give immediate consideration to the "Proposal for Re-capitalizing Vereinigte Stahlwerke AG with the Inclusion of Foreign Investment and the Participation of Public Authority."[114] The VS Plan called for increasing firm capital from 460 million to 700 million Deutschmarks, through investment from abroad, the foreign interests to receive 180 million Deutschmarks of share values (and public authorities the remaining 60 million). The French would thus have controlled the largest single packet of shares in the firm, with some 25.7 percent of total holdings, enough for a blocking minority on the board of supervisors. To soothe their worries about economic security the French were also to receive unspecified statutory guarantees of the veto power.[115]

Adenauer was prepared to go faster and farther than the industrialists realized or perhaps wanted. He hoped to make political capital of VS's plight. Arguing that the French could impede American encouragement of German revival, he told the cabinet on 18 October 1949 that "an understanding with France is ... the fulcrum of the German-American relationship ... and for this reason [alone] the main task facing the government of the Federal Republic is to establish a cooperative relationship with France."[116] To facilitate this

[113] BA Nachlass Lehr 18, "Vorschlag zu einer Neuordnung der Vereinigte Stahlwerke A. G. unter Heranziehung ausländischen Kapitals und einer Beteiligung der öffentlichen Hand" (n.d.).
[114] BA Nachlass Lehr 18, Sohl to Lehr, 6 October 1949.
[115] BA Nachlass Lehr 18, "Vorschlag zu einer Neuordnung der Vereinigte Stahlwereke A. G."
[116] *Die Kabinettsprotokolle der Bundesregierung* (vol. 1, 1949), see Cabinet meetings, 20 September 1949, 18 October 1949, 25 October 1949.

he had already indicated to the U.S. high commissioner a readiness
to offer a 40 percent foreign participation in VS, which, he specified, could be directed by an "international consortium." The chancellor also had no objections to "internationalizing" management,
something never mentioned in the VS offer. Though McCloy encouraged Adenauer to present additional plans for ending the dismantlements, this proved not to be necessary; unbeknownst to him,
the Big Three had already decided to drop the policy.

Though the Ruhr steel industry appears to have been unaware
even as late as November 1949 that an end to the dismantlement
campaign was imminent, other evidence suggested that things
would get better. By mid-1948 the French had begun to seek a
closer economic relationship with the Ruhr. In September of that
year "major French industrialists" met at Essen with their German
counterparts to explore establishing "closer financial and economic
ties."[117] In early February 1949 President Richard Merton of the
Metallgesellschaft reported that he had received a surprise visit
from a diplomat attached to the French military government, who
had intimated that Foreign Minister Schuman and French High
Commissioner André François-Poncet were now ready to strike up
"lasting and completely forthright cooperation with Germany."[118]

This was more than idle speculation. In midsummer 1949 Monnet directed the planning commissariat to produce a series of comparative studies of the French and German economies for possible
use in "future cooperative projects." He also requested the ministries of the Federal Republic, which eagerly complied, to supply it
with raw data and statistics concerning population, land use, industrial outputs, levels of capacity utilization, energy consumption, cost
of living, and transportation. The studies were supposed to focus on
dual pricing, wage levels, transport, energy, and tariffs.[119] On 29
August 1949 France's foreign minister and high commissioner met
in Tübingen with a senior official of the German Marshall Plan ministry, Karl Albrecht, for discussions of economic cooperation. The
German suggested the possibility of a future arrangement like the
1921 Wiesbaden Accord, in which the French had offered technical
assistance to the mines and deliveries of bread grain in exchange for

[117] KA/Henle Wirtschaftspolitik, Deutsch-französische Zusammenarbeit (10.47-
31.12.55), Refferscheidt to Henle, 4 August 1948.
[118] BA Nachlass Dr. Hermann Pünder, "Deutsch-französische Beziehungen," 3
February 1949.
[119] BA B146/266, "Besprechung mit Herrn Aussenminister Schuman und Hoch-
kommissar Francois-Poncet," 29 August 1949.

guaranteed supplies of Ruhr coal, and raised the additional possibility of joint planning for the canalization of the Moselle. Schuman responded warmly to these proposals and recommended that discussions be continued between the French Commissariat du Plan and the responsible Bonn ministries.[120]

A dreaded steel glut, which began appearing in spring 1949, also helped end the international isolation of West German producers. In that year European steel tonnage of approximately forty-seven million passed the forty-six-million average obtaining between 1925 and 1929, and by 1953 it was expected to exceed interwar levels by several million. Yet world export markets had not recovered, nor were they expected to, and the cost reductions needed to stimulate increases in domestic demand were not in sight. The result was widespread fear that the surpluses beginning to develop would soon multiply. These predictions, which proved to be wildly off the mark, appeared in an extraordinarily influential publication of the UN's Economic Commission for Europe and put the Ruhr Problem on the very forefront of the European agenda.[121] The district had produced over twenty million tons before the war, had raised production to just over five million tons (almost none of it exported), and still retained capacities of about fifteen million tons. What would happen, contemporaries queried, if and when the West Germans decided to return to world steel markets?

The Westminster Economic Conference of the European Movement met to deal with these problems. Its steel committee recommended the institution of production control by means of an international agreement that could be enforced worldwide through publicly supervised cartels and producer associations. Though this proposal was not tantamount to advocacy of a new International Steel Cartel, it acknowledged the impossibility of regulating the European steel industry without producer cooperation: For the first time since the war it had become possible to discuss the old ISC in respectable company. The sentiments expressed by Pierre Dieterlinden in a paper prepared for the conference were widely shared. "Did not the international steel cartel," he asked rhetorically, "prefigure the birth of the European spirit? And is not the structure of the steel industry an expression of this spirit? Composed of few men

[120] Ibid.
[121] SP, Fachstelle Stahl und Eisen, "Stahlproduktion in Europa," 16 August 1949; also, USNA RG260 11/98-2/1-5 OEEC, Memorandum on the Iron and Steel Industry, 30 May 1949; United Nations, *European Steel Trends in the Setting of World Markets* (Geneva, 1949).

of similar background open to consideration of general issues and used to thinking long-term and in cooperation with public authority – no branch of industry is better suited than this one to synthesize industrial and international realities."[122]

Such nostalgic reflections naturally raised the question of the Ruhr. Unless Germany were to be treated as a case apart, the Westminster conference had to find some way to gear its steel production gradually back up to European speed. There was no lack of inventive attempts to evade the problem. Criticizing the Level of Industry Agreement as detached from reality and provocative, a delegate named Hugh Klare recommended dropping it altogether and instead broadening the International Authority for the Ruhr to include West Germany, giving it the power to set allowable outputs. Though prepared to admit that the impending glut might make it impossible to increase West German production levels above the authorized 10.7 million tons, Klare believed that the nation could find some consolation in having had a voice in keeping them there![123]

A more elegant and realistic solution to the steel dilemma was proposed by the international-minded French socialist, André Philip, at the Westminster conference. Why not persuade the Germans to put their industry under public control, he asked? There would then be no need to break up the old regulatory machinery, which, once the industry had been democratized, could be put to the service of an economically integrated Europe.[124] Philip presented his plans to a group of French and German parliamentarians in Bernkastel on 26 and 27 November 1949. Though not prepared to dispense with a special Allied office to supervise Ruhr industry, Philip proposed that a second more embracing control authority be created to administer steel production throughout Western Europe, including France. The new organization, as Philip described it, would have been responsible for overall investment policy as well as the allocation for coal, scrap, ore, and other materials needed for steelmaking. The principle of co-determination, he added, was to be binding: Unions would be represented at both the seat of the new

[122] AN 307 AP212 (Dautry Papers), "Mouvement européen: Preparation de la Conférence Economique Européene de Westminster," by Pierre Dieterlinden (n.d.).

[123] AN 307 AP273, "Préparation de la Conférence Economique Européene à Westminster: Les Industries du Charbon et de l'Acier en Europe" (n.d.), by Hugh Klare.

[124] MAE Y 1944–1949, "Note ... d'un exposé le 20 mai au Centre de Politique Etrangère," by André Philip.

authority and the firm level. The leading SPD representative in attendance, Carlo Schmid, welcomed Philip's remarks enthusiastically. Günter Henle, who in addition to being chief executive officer at Klöckner had become a CDU Bundestag representative, also acknowledged that labor should be included in a future European heavy industry settlement.[125]

The return to normal Franco-German commercial relations, though slow in developing, was another helpful development. In 1948 an association of French exporters to Germany was founded called ACIA (Association pour le Commerce et l'Industrie Française en Allemagne). It was forced to operate more or less underground and through private contacts. Plans to organize a German counterpart to ACIA had to be dropped for "political" reasons in October 1948, as did those to set up a joint council (Kuratorium). Things improved the following spring. In March French exporters set up AFREA (Association Française pour les relations économiques avec l'Allemagne) to succeed ACIA. The new group won official approval. Yet substantial progress was made only by business "study groups." A "Groupe d'Etudes Allemandes" initiated exchanges of letters and also generated a "twinned" Studienausschuss für die deutsch-französischen Wirtschaftsbeziehungen in Duesseldorf. The two organizations sponsored a fall conference at Bad Neuenahr attended by the head of the Patronat, Geroges Villiers, as well as leading German industrialists representing several different fields. Though a "good mood" was said to have prevailed, discussion remained on the plane of pious pronouncement.[126]

The first real evidence of a developing new business relationship resulted from the Franco-German tariff agreement of 11 February 1950. This had one serious drawback from the Ruhr's standpoint: Goods could pass freely from France through the French Zone onto German markets. To close the "gaping hole in the West" Ruhr pro-

[125] KA/Henle Wirtschaftspolitik, Deutsch-französische Zusammenarbeit (10.47-12.55), "Kurzer Vermerk über das deutsch-französiche Treffen in Bernkastel," 26–27 November 1949.

[126] See KA/Henle Wirtschaftspolitik, Deutsch-französische Zusammenarbeit (10.47-31.12.55), Refferscheidt to Henle, 4 August 1948; Hellwig note, 10 September 1948; Fritz Ludwig to Henle, 5 March 1949; "Gesichtspunkten für die Besprechung mit den Franzosen," 17 March 1949; and Kuhnke to Henle, 31 August 1949; also, USNA RG260 11/111-31/8-9, "Annex C: Copy of a letter dated June 16th, 1949, to the Secretary of the OEEC from the Chairman of the 'Conseil National du Patronat Francais' (with enclosures)"; BA B109, "Besprechung mit Miss Outhwaite aus London von "The Economist," 22 September 1949.

ducers both demanded rate concessions from the *Bundesbahn* and sought "private understandings" with French steel producers concerning South German markets. In early March 1950 a delegation from the Société Commerciale de Fonte headed by a M. Lequipé arrived in Bochum claiming to possess a special mandate to discuss a general export agreement. Wary of granting further concessions, the Germans refused to talk. Lequipé was not easily dismissed: The Société des Produits Sidérurgiques (the general industry cartel) had resolved and was determined, he said, to "pursue an export understanding with Germany." On 18 May 1950 representatives of the German steel industry again met with Lequipé, but this time discussion was limited to export arrangements for the Halberger Hütte, Brebach/Saar. A third round of talks began on 14 April and was still under way when the 9 May 1950 Schuman Plan announcement intervened.[127]

The shift toward interministerial cooperation, the marked increase of interest in the German market shown by French exporters, and the gradual return of the international cartel to a measure of respectability may indeed have been only straws in the wind, but they did show which way it was blowing. These phenomena were manifestations of a growing French weariness with the related preoccupations of war, occupation, Germans, and the Ruhr Problem. A return to normal peacetime conditions did not occur at once, however. The dismantlement of ATH, the flagship of the Ruhr steel industry, appeared unavoidable as late as October 1949: The Petersberg Agreements came as a last-minute reprieve.[128] The end of the occupation was still not in sight when Foreign Minister Schuman made his famous announcement.

Though the threat of dismantlement was over by 1950, other controls over the Ruhr remained in effect. The August 1947 Level of Industry Agreement, though its maximums had never been approached, was on the books and German coal exports were still being regulated. The military security boards held a veto power over new investment in mines and mills, and the control groups supervised production. The coal and steel industries were under sequester and would continue to be until the trustees had succeeded in drafting a reorganization plan under Law 75 acceptable to the

---

[127] SP "Sitzung über deutsch-französischen Handelsvertrag am 17. February 1950"; BA B109/97, "Übernahme von Roheisen aus Frankreich bzw. Saar," 8 March 1950; and "Übernahme von Roheisen aus Frankreich bzw. Saar," 22 October 1943.

[128] BA Nachlass Lehr 18, Lehr to Adenauer, 22 October 1949.

Allies, the German unions, and the government of the Federal Republic.

The steel trustee, Heinrich Dinkelbach, was right to have insisted in 1947 that the war had put the Ruhr out of business and that a hands-off policy as demanded by the old owners would have led nowhere. The restoration of industry required using the government to create a favorable business climate and involved a complicated and protracted process which could have been interrupted at many points by either political or economic changes in the Federal Republic: The role of Adenauer as director and executor was in this respect of immense importance. But a restoration required more than strong German leadership: The nations of the West had to create a framework of international economic and political institutions, making it possible to end the regime of subsidies by enabling industry to regain access to foreign capital and resume exportation on a large scale. The Ruhr restoration depended as much on external as on internal events.

In 1945 the Ruhr entered a kind of Dark Ages, a period of punishment and affliction from cruel and arbitrary powers acting on high. Like those who lived through those ancient times, the men of the Ruhr clung to their faith, hoping contrary to visible signs and pronouncements that a new and better era would eventually dawn. Yet the sufferers did preserve the sacred thread of organizational tradition linking them to an otherwise unrecoverable past: Thanks to confusion and scarcity they managed to keep essential cartel structures together and discharge quasi-official responsibilities on behalf of the zonal governments. With the erection of Fachstelle Eisen und Stahl in August 1949 the cloudy relationship between the steel producers' association, WVESI, and the occupation authority, Verwaltungsamt Stahl und Eisen (VSE), was cleared up. The Fachstelle was an agency of the federal government established to regulate the export trade and headed by the former steel executive, Max C. Müller. Müller's staff consisted of 143 officials seconded from WVESI who though on "service contracts" to the West German government enjoyed the right to return to previous employment and, in addition, receive their earlier salaries in the event that these were higher than the government compensation schedules in effect.[129]

After many months of inaction, the three military governors finally ordered German industry to form a peak association in May 1949. One of their purposes in so doing was to provide the recently

[129] SP, WFW to WVESI, 11 August 1949; SP, "Fachstelle Stahl und Eisen."

organized Federation of German Unions (Deutscher Gewerkschaftsbund) with an appropriate bargaining partner.[130] As originally set
up, the Federal Association of German Industry (BDI) was to be
responsible for labor matters. Its future was partly in Adenauer's
hands. His commitment to restoring private ownership in the Ruhr
was not unqualified because compromise with the unions was essential for social stability. On 9 January 1950 top-level discussions between the union federation (Deutscher Gewerkschaftsbund) and the
employer association (BDI) over co-determination began. For Ruhr
heavy industry the day of reckoning could no longer be postponed;
reform involving either socialization or some other form of public
control appeared unavoidable. Recombination involving a fundamental reworking of relationships with Western Europe was another
distinct possibility. Adenauer's abiding commitment to reconciliation
with France was well known, though its costs to the Ruhr in terms of
money, freedom, and power were not. As the new half-century
dawned the Dark Ages were only beginning to lift. Adenauer was
still at the mercy of the Allies. A Ruhr settlement would depend
more on them than on him.

[130] SP, "Kurzniederschrift über die erste Geschäftsführerbesprechung beim Bundesverband der deutschen Industrie," 31 January 1950; Gerard Braunthal, *The
Federation of German Industry in Politics* (Ithaca NY, 1965) pp. 24–25.

5

The end of the war against Germany: the coal–steel pool as treaty settlement

On 9 May 1950 Foreign Minister Robert Schuman interrupted the regularly scheduled broadcasts of French radio to make a historic announcement: In order to end the decades-long struggle over coal and steel, France was ready to become partners with its recent enemy, and other nations, in a new West European heavy industry community organized in such a way as to make war politically unthinkable and economically impossible. Little more than a month later negotiations for the coal–steel pool began. Joined by the three Benelux countries and Italy in addition to France and Germany, they would end eleven months later with the initialing of the Treaty of Paris by the foreign ministers of "The Six" on 18 April 1951. The document was the product of difficult and not always satisfactory compromises. Another fifteen months transpired before the new organization was installed in its Luxemburg headquarters, a delay due to the ratification process and the phase-out of occupation controls. By the time the new authority was able to commence operations, however, the senior civil servants involved in the coal–steel talks as well as representatives of national producers had arrived at understandings of their own that substantially modified the agreements arrived at during the official negotiations. The European Coal and Steel Community that came into being on 10 August 1952 was a profoundly different organization from the one first proposed on 9 May 1950.

The changes had both internal and external causes, the most important of which was the foreignness of Monnet's original notion. It had much in common with the thinking of American New Dealers like the progressive businessmen, lawyers, economists, and management specialists belonging to the Committee for Economic Development. John J. McCloy once called the Schuman Plan a Tennessee Valley Authority for Europe, a hardly accurate but nevertheless insightful description. Like TVA, the ECSC was conceived as a program for economic expansion anchored in a key sector of industry and governed by means of a strong central authority through a code of business conduct restricting the powers of private industry. This was a formula for neither socialism nor industrial self-government but what most American big businessmen stigmatized as state dictation and resented as a higher form of nagging. Monnet wanted the coal–steel pool to protect French security through international control of the Ruhr and with these aims in mind designed a powerful directorate to regulate European heavy industry, the High Authority. His blueprint for the pool was contained in the *Document de Travail*, a constitutional draft that would serve as the centerpiece of the Paris negotiations. The other delegations found the French proposals doctrinaire, radical, vague, inconsistent, unresponsive to their needs – *dirigiste*. The collapse of the negotiations would be narrowly averted thanks to special deals for individual nations, the exercise of American pressure, and compromises intended to offset or conceal the shift in power toward West Germany taking place concurrently.

This new prominence was due in part to political circumstances. The outbreak of the Korean War on 25 June 1950 stimulated a European armaments boom from which the Federal Republic profited enormously, the onsetting cold war accelerated Germany's moral rehabilitation, and the Schuman Plan was the catalyst to changes that strengthened the young nation's social and economic fiber. Monnet had no intention of founding "Germany Inc." and indeed, long after the coal–steel pool was operating, would continue to wage an intense decartelization campaign in order to weaken Ruhr industry. These efforts found favor only among the Americans. During the many months of negotiating in Paris common opposition to Monnet caused the representatives of the other parties, and French private interests as well, to turn to the Germans as defenders of traditional economic practices. Alliances born of convenience and opportunism developed into working relationships based on mutual interest and trust; for the German ministries, national isolation ended and European teamwork began. At the same time the dormant international cartels revived. Monnet had proposed the coal–steel pool in an attempt to control the Ruhr in the interests of Europe but succeeded in aligning it around a resurgent and only partly reformed ex-Reich.

5.1   From summit to swamp

Probably not even Monnet fully understood what prompted him to act in early 1950; he was not inclined to introspection and kept no diary or other record of his thoughts, instead relying heavily on able subordinates to express them on paper. Yet he had good reason to act when and how he actually did. Monnet had been troubled since 1948 by both the ineffectiveness of French policy and the lack of direction from Washington. Though the *Memoirs* point to his uneasiness about the onsetting cold-war mentality, at the time he appears to have worried more about the Soviet threat, being haunted after the Prague coup by fear that the Germans would eventually be driven into a Rapallo-like arrangement with the Russians.[1] Monnet was also distressed by evidence mounting in 1949 that the recession beginning in the United States would result in a restoration of international cartels, a development that would further disillusion the Americans, undermine his control over French industry, and (he was firmly convinced) endanger national security. The downturn also contained a silver lining: For the first time since the war coal had ceased to be scarce, and France could act without fear of economic blackmail.[2]

The specific timing of the 9 May 1950 announcement had less to do with cartelization than diplomacy. The steel crisis of 1949 was a warning signal rather than a sign of imminent danger. France's continuous wide-strip rolling mills were just beginning to come on line, and though the *maîtres de forges* had begun to clamor for protection a serious glut could have materialized only in the future. Monnet's immediate problem was one of control. In spring 1949 the Westminster conference had proposed establishing a public authority to supervise, yet also sanction, a fully cartelized industrial community. In August 1949 the newly constituted but still powerless Consultative Assembly of the Council of Europe adopted the recommendation, which was also endorsed by the Chambre Syndicale de la Sidérurgie Française.[3] Yet the chances of something developing out of this activity (or, for that matter, the efforts of Dr. Schwede of Vereinigte Stahlwerke, who in late 1949 was often sighted "taking tea" in Paris) were remote so long as Americans remained hostile to cartels.[4]

[1] See Monnet, *Memoirs*, " Europe in the Dark," pp. 271–287, and "A Bold, Constructive Act," pp. 288–298; JM AMG 2/3/8, "Exposé de M. Monnet devant le Conseil de la Haute-Commission Alliée Petersberg," 23 May 1950.
[2] See Pierre Gerbet, *La Genèse du Plan Schuman. Des origines à la déclaration du 9 Mai 1950* (Lausanne, 1962); also, C. H. Hahn, *Der Schuman Plan: Eine Untersuchung in besonderen Hinblick auf die deutsch-französische Stahlindustrie* (Munich, 1953), pp. 12–23.
[3] Gerbet, *Genèse du Plan Schuman*, pp. 8–9.
[4] See FRUS 1949/IV, Administrator for Economic Cooperation (Hoffman) to Chief of the ECA Mission in France (Bingham), 9 November 1949, pp. 443–444.

Lack of progress on the German question, however, threatened to undermine this traditional opposition. On his first official visit to Adenauer in January 1950, the French foreign minister demanded a fifty-year lease on the Ruhr. Bound by a narrow mandate, shy, and high-strung, Schuman was aloof, cold and rigid, and the meeting broke up in disarray. Adenauer compounded the difficulties by telling the American journalist Kingsbury Smith in early March that he envisioned a complete Franco-German political and economic union, with common citizenship, joint political institutions, and a single economic policy. The clumsy suggestion was ill received in France. With the Allies scheduled to meet on 10 May to reconsider the German question, Monnet decided to move.[5]

The Schuman Plan announcement was bold, simple, imaginative, and disarming – a public relations coup of heroic proportions. What France had to offer, according to the concise, finely crafted, and highly polished text spoken painstakingly into the microphone by the squeaky-voiced foreign minister, was no ordinary coal–steel deal but a proposal designed to end an ancient rivalry, prevent war, and build a better world. It was now possible, he said, to cast aside old differences and begin negotiations in a spirit of equality and without prior conditions. The nations of Europe were invited to join France in sacrificing a measure of national sovereignty to a new supranational authority created for the common good.

At the very moment of its announcement the Schuman Plan proposal became an established part of the context of events, a force for change, and a myth: The word "Europe" would never be spoken in quite the same way again. The power of the message impressed even skeptics at the time and has since made it difficult to disentangle the realities of the coal–steel negotiations from the aura enveloping them. Yet a couple of points are clear. The 9 May 1950 message was aimed at the public, particularly in the United States, and couched in language that barred easy exit from the negotiations and that established a framework for them. Second, the agenda was Monnet's, and it was not particularly generous to the Germans. Schuman spoke neither of restoring traditional economic relationships nor of a fraternal regrouping of parties, nor did he mention an end to the occupation. The promise of equality contained in his message – contrary to what the Germans wistfully later maintained – was something to be earned rather than granted, belonging to the future rather than the present.

[5] HST, Acheson Papers,Memorandums of conversations, "Ruhr Agreement, Cartel Control of German Steel," 27 March 1949.

Monnet meant to run the negotiations. He had learned as a result
of his experiences at the League of Nations that collective efforts to
strike compromises between diverse national interests had scant hope
of success and for this reason had been skeptical of the OEEC's
chances. He had little more than contempt for international parliamentary organizations like the Council of Europe, effective action in
his view requiring strong overall direction, a program, the backing
of powerful interests, and a reservoir of public support. His model
was the war production committee not the diplomatic conference.
Monnet would thus head the French delegation, chair its sessions,
and conduct negotiations within the framework of the Schuman
Plan Idea.

The essential elements of his policy appear in drafts of various
position papers prepared for the Paris conference, which leave no
doubt about the French intention to leave the Ruhr occupation machinery intact and underscore the importance of the New Deal
inspiration.[6] From these documents emerges the concept of a High
Authority whose mission is developmental and which acts by enforcing a code of industrial good conduct. The new agency was to seek
the advice of unions and other constructive elements but not the
"interests," as well as promote competition, devise and apply a common pricing policy, steer investment, and serve as joint buyer and
seller for the community. The future High Authority's operating
methods would be, according to a 9 May 1950 paper, the precise
opposite of those used by cartels: Whereas they promoted profit, it
encouraged productivity; whereas they operated by secret deals, it
favored open covenants; and whereas they were run by professional
managers in the service of private interests, it was the agent of the
public. The High Authority's mission was to be not protection of the
status quo but modernization.[7]

Adenauer responded to the initiative of the French foreign minister with alacrity, officially regretting during his first visit from Monnet (which took place on 23 May 1950) that the proposal had not
come earlier; he professed to having waited twenty-five years to hear
a Frenchman speak words like Schuman's.[8] The chancellor did in

[6] JM AMG 2/3/2 bis, "Note de P. Reuter, Problèmes posés par l'institution d'une
Haute Autorité internationale en regard du statut actuel de la Ruhr," 5 May
1950; JM AMG 2/3/3, "Note (Reuter) sur la proposition française dans ses rap-
ports avec le statut de l'Allemagne et celui de la Ruhr" (n.d.).

[7] JM AMG 2/4/4a, "Note concernant la Haute Autorité: Domaine, Objet" (n.d.);
JM AMG 17/8/62, "Note Anti-cartel," 9 May 1950.

[8] JM AMG 2/3/11, "Entrevue du 23 mai 1950 entre M. Jean Monnet et le
Chancelier Adenauer."

fact receive prior notice of the 9 May announcement and approved it without reservation,[9] a predictable reaction if only because the French were offering West Germans their first opportunity to participate as sovereign members of a diplomatic conference. The chancellor readily agreed to Monnet's request that occupation issues be excluded from the negotiations, raised no objections to the Frenchman's veto of his first two nominations for chief negotiator (Hermann J. Abs and a dual-national once associated with the Swedish match king, Ivar Kruger), and settled without complaint on the Frenchman's nomination to the post of an obscure Frankfurt law professor named Walter Hallstein. To him fell the unenviable task of serving as public spokesman for Adenauer's Europeanism while working as private advocate on behalf of Ruhr heavy industry.[10]

The chancellor was hardly a starry-eyed idealist. A shrewd calculation underlay his apparent acquiescence to the French. Confident of Germany's recuperative powers, Adenauer questioned neither the vigor of its producers nor their ability to stand up for their interests and was certain that time was on his side. The real danger as he saw it was that in trying to move too fast the Federal Republic could rekindle old fears and suspicions and prolong the occupation.[11] Adenauer was pleased at the impact of Schuman's announcement, not worried about being locked into negotiations, which he fully expected to end the occupation soon and satisfactorily. He would thus stand above the broils of the proceedings and demonstrate a conspicuous willingness to compromise while at the same time leaving it to subordinates to push, shove, evade, or otherwise maneuver in the national interest.

The men of the Ruhr looked upon the Schuman Plan as an opportunity to slip the noose placed around their collective neck in 1945 but did not count on a commutation of sentence or a simple reprieve and were well aware that the rope still draped loosely over their shoulders could at any time easily be tightened. They needed no reminder of the consequence of a botched escape attempt. Reacting skeptically to the Schuman Plan announcement, the business manager of WVESI feared that the principle of "equal protection"

[9] See Raymond Poidevin, *Robert Schuman: Homme d'état, 1886–1963* (Paris, 1986), pp. 256–257.

[10] JM AMG 2/3/4, Hans Schäffer to Monnet, 5 June 1950; JM AMG 2/3/19, "Plan Schuman: Entretien avec le Chancelier," 16 June 1950; JM AMG 2/3/20, "Plan Schuman: Entretien avec le Chancelier," by Leroy-Beaulieu, 16 June 1950.

[11] See Hans–Peter Schwarz, *Vom Reich zur Bundesrepublik: Deutschland im Widerstreit der aussenpolitischen Konzeptionen in den Jahren der Besatzungsherrschaft, 1945–1949,* 2nd ed. (Stuttgart, 1980), pp. 423–429.

to which it alluded could serve as a pretext for forcing the Ruhr to
consume more low-grade French ore and regarded the ban on "double pricing" which it advocated as an excuse for saddling the Ruhr
with higher energy costs and appropriating scarce scrap at giveaway
prices. Still worse, he warned, was the possibility that the "common
production" spoken of by Schuman would be invoked to suppress
"the greater dynamism which, even after World War II, our works
manifest by comparison to the French." He especially feared that
France, whose two continuous wide-strip rolling mills were nearly
finished, would prevent the Ruhr from ever building one. WVESI's
business manager did not recommend nonparticipation in the
coal–steel talks because, he reasoned, the cooperation called for by
Schuman would require the resurrection of international producer
agreements, if only to equalize costs and prices. He was confident
that the 9 May 1950 proposal would inevitably lead to a restoration
of the old Western European heavy industry community.[12]

American observers reached a similar conclusion and found it
troubling; from Secretary of State Acheson and President Truman
on down, in the offices and corridors of the State Department, in
newspaper reports and editorials, and in letters to the editor as well
as in comments overheard on the street, everyone seemed to agree
that the Schuman Plan proposal would result in a restoration of the
old cartels. Yet even those chronically suspicious of traditional European business methods like the trust-busting James S. Martin welcomed the initiative as a genuine and serious attempt by France to
deal with the old evil of Ruhr power. The ECA and State agreed
with him that the "cartel angle will need careful watching."[13]

Such concerns soon paled by comparison to the larger issues involved. A 21 May cable from Averell Harriman in Paris expressed
the all-but-unanimous American belief that the "proposal may well
prove most important step towards economic progress and peace of
Europe since original Marshall speech on ERP. It is first indication
of bold, imaginative, concrete initiative on part of European country
in attacking two basic problems . . . integration of European economy and conclusive alignment of Germany on side of West with
minimum political and military complications."[14] The proposal was

[12] BA 102/5134, WVESI to Schneider, RWM, "Montanunion Plan," 5 June 1950.
[13] FRUS 1950/III 396.1 Lot 15-1050, Acheson to Undersecretary of State Webb,
10 May 1950; Martin Papers (MP), Delbert Clark to James Martin, 13 May
1950.
[14] FRUS 1950/III 850.33/5-2050, U.S. Special Representative in Europe (Harriman) to Secretary of State, 1 May 1950.

held to be fully consistent with American efforts to liberalize trade and payments. Though, the cable proceeds, "successful implementation . . . depends on maintenance of initial momentum . . . gained by effective timing and dramatic announcement," the United States was not "to have an official association or even observers with the working committees engaged in the elaboration of . . . the plan . . . at this stage," as such an intervention could lead to accusations of an attempt to dominate the proceedings.[15]

There was no real need for an active American presence at the Paris conference. State and the ERP expected to work closely with Monnet behind the scenes and, if necessary, through him as well. A young Treasury official named William M. ("Tommy") Tomlinson would serve as liaison between Monnet and U.S. policy makers.[16] At his right hand was a special "Working Group on the Schuman Proposal" set up in June 1950 at the Paris embassy, whose staff of economists and management specialists would keep closely abreast of the negotiations. It became a hotbed of Monnet enthusiasts. Eager to make recommendations to promote the development of the new Europe, the men and women of the American working group soon began to advise him unofficially.[17] A proposal of July, for example, recommended fitting defense and security arrangements into the framework of economic institutions then being built around the coal–steel pool,[18] as later was the case with the French proposal for a European Defense Community. The thinking of Monnet and the young American integrationists was often so similar as to be indistinguishable.

The contrast between the reactions of the United States to the Schuman Plan and those of the French interests most directly affected by the 9 May 1950 proposal could hardly have been sharper. Monnet deliberately excluded the ministries from the Paris negotiations, which did not come before the cabinet even once in 1950. He also made little attempt to work through industry. The steel produc-

[15] Ibid.   [16] Monnet, *Memoirs*, pp. 269–270, 379.

[17] See USNA RG84 (500: Coal and Steel Pool), Working Group on the Schuman Plan Proposal, "Increasing Problems in European Steel and Coal Industries," 22 September 1950; Minutes of 18 October 1950 meeting re Lehr proposal: "Railway Tariff for Coal and Iron Ore," 20 October 1950; "Memorandum on the Institutions and the Permanent Economic and Social Dispositions of the Schuman Plan," 11 October 1950; "Restriction to Trade in Coal and Steel among Schuman Plan Countries," 10 July 1950; and "Emerging Problems in Connection with the Schuman Plan," 14 July 1950.

[18] USNA RG84 (500: Coal and Steel Pool), "Modification in the Authority's Powers Required by Defense Considerations," 26 July 1950.

ers showed Monnet scant gratitude. In spite of his efforts to assure
the steady supply of coke to France, not to mention the thirty-five
billion francs of a total of fifty-three billion for industry as a whole
that the French plan had provided to finance the two continuous
wide-strip rolling mills, the producers considered him a usurper.
This resentment cannot be identified with the usual liberal objections to governmental interference in the marketplace. "Permit me
to observe," Monnet wrote President Aubrun of the Chambre Syndicale de la Sidérurgie Française, "that there is no such thing as free
enterprise in the steel industry and that the alternative you offer to
administrative *dirigisme* is a *dirigisme* of the interests."[19]

Though the Chambre reacted favorably "in principle" to the Schuman Plan idea, it was prepared to cooperate with Monnet only on
unacceptable terms. The producers looked favorably upon rapprochement with the Federal Republic, supported all measures facilitating
the flow of cheap German combustible to France, and approved
wholeheartedly of "the resumption of relationships with foreign
steel interests," but also feared that the proposal would force them
to compete with Germany on unfavorable terms.[20] The producers'
reply to the 9 May 1950 announcement, emphasizing the high costs
of the war, the lack of any mechanism for relieving them, the in-
complete state of the modernization program, the disappearance of
liquidity, and the weakness of links to domestic manufacturers, insisted that the industrial associations be represented at the conference proceedings.[21]

This never happened: Monnet negotiated on behalf of France
without the active support of any organized interests in the private
sector and in the face of bitter opposition from virtually all of it.
Initially, one delegate of the French mills *was* included on the working committee but as an individual rather than as representative of
its interests. This was A. Aron, a director of Chatillon-Commentry
who previously had belonged to the modernization commission for
the steel industry and was considered sympathetic to the Monnet
Plan. Aron also served on one of the major specialized subbodies
formed for the Paris conference, the working group for production.

<hr>

[19] JM AMG 18/4/2, Monnet to Aubrun, 12 October 1950.
[20] JM AMG 18/2/1, "Note de la Chambre syndicale de la sidérurgie française exprimant ses premières réactions face au Plan Schuman," 12 July 1950; also USNA
RG84 (500: Coal and Steel Pool), Paris Embassy to Secretary of State, 17 July
1950; USNA RG84 (500: Coal and Steel Pool: Tomlinson to Ambassador), "Views
of Chambre syndicale de la sidérurgie on Schuman proposal," 12 July 1950.
[21] JM AMG 18/4/2, Aubrun to Monnet, 5 August 1950.

Steel was otherwise represented only on committees of secondary importance such as the one for wages and salaries. Aron soon became disenchanted with his minor role in the negotiations and denounced the pool as socialistic, inconsistent, antipatriotic, and unnecessary.[22]

There was no cooperation whatsoever at the operational level between French steel producers and Monnet's negotiators, whom the industry even refused to provide with statistics and production information.[23] The French plan's own investigations invariably depicted the industry as better able to compete than it was prepared to admit, thereby fueling the fires of discontent.[24] Beginning in July 1950 the steel syndicate whipped up opposition to the Schuman Plan within industry, launching a campaign through the Conseil National du Patronat Français (CNPF). The organization's first vice-president, Pierre Ricard, a steel industrialist, was the man in charge. By the end of the year, protests of Monnet's conduct of the negotiations were registered by nearly every chamber of commerce in France.[25]

The British decision not to participate in the coal–steel pool was a foregone conclusion and could have come as no surprise to Monnet after the failure of the Plowdon talks. There is no need to agonize over whether the author of the 9 May 1950 proposal wanted the United Kingdom to join the pool or not. Monnet had set the price of admission, and the turnstiles had begun to spin. Who could be blamed but the British if they chose to stay out? If they came in on his terms, so much the better! Monnet won either way. The United Kingdom refused even to discuss terms of entry; haggling over them was inconceivable. Just prior to the 9 May 1950 announcement, the cabinet had after months of ministerial discussion

[22] JM AMG 18/4/10, Monnet to Aubrun, 17 November 1950; FRUS 1951/IV, Special Advisor to the Deputy Chief of the ECA Mission in France (Goldenberg) to Department of State, 14 March 1951, pp. 98–99.

[23] See JM AMG 18/4/10 quater, "Documents communiqués à la Sidérurgie" (n.d.); JM AMG 19/4/3, "Schéma pour la lettre du President du Conseil," 4 January 1951; JM AMG 44/3/1, "Capacité de concurrence des industries françaises du charbon et de l'acier" (n.d.).

[24] JM 18/4/17, Monnet to Aubrun, 11 January 1951; also, JM AMG 18/4/11, Aubrun to Monnet, 22 November 1950; JM AMG 18/4/23 Aubrun to Hirsch, 30 March 1951.

[25] USNA RG84 (500: Coal and Steel Pool), Paris Embassy to Secretary of State, 17 July 1950; see JM AMG 45/2/8, "Liste des arguments exprimant des inquiétudes des chambres de commerce" (n.d.). See, inter alia, JM AMG 45/2/1, "Note de presse, Conseil général du nord," 30 April 1951; also JM AMG 45/2/2 (Moselle), 45/2/9 (Lyon), 45/2/10 (Sedan), and 45/2/3 (Meurthe-et-Moselle).

reaffirmed the unacceptability of a cession of national sovereignty to any supranational authority. Though Attlee and Bevin both could have overridden this opposition, neither one of them intended to make such an attempt. Angered that Monnet had not consulted him prior to the, 9 May announcement and refusing to be disabused of the notion that it was a plot hatched to embarrass Britain, Bevin had a visceral dislike for the French coal–steel proposal. Neither he, the cabinet, nor the civil service could conceal its resentment that France, held in contempt for allegedly lacking a spine, had taken an initiative better befitting the nation that still unhesitatingly referred to itself as the "leading power of Europe."[26]

The strength of this resistance made it difficult for anyone in the Labour government to detect much good in the proposal. Monnet's insistence that, by tying West Germany's fortunes to its own, France would be stabilizing Europe and strengthening the Atlantic Alliance counted for little; purported fears that *Westintegration* would harden the division of Germany, by contrast, occasioned the shedding of a great many crocodile tears. In the cabinet only Sir Stafford Cripps appears to have seen any advantage to cooperating with Monnet, but he mistakenly believed that the 9 May 1950 proposal would reduce American power in Europe. The extensive examinations of the Schuman Plan proposal conducted at cabinet request under the auspices of the Economic Policy Committee, which pointed conclusively to certain advantages of membership to Britain, in the end amounted to no more than elaborate exercises.[27]

For the most part Britain wisely kept hands off instead of playing the "saboteur from without." Afraid of antagonizing the United States, Bevin soon dripped a smoke-screen proposal for a loose European coal–steel federation that the Foreign Office quickly cooked up in June 1950 to counter the French initiative and made no attempt to revive similar plans. Whitehall rather maintained a "correct and discreet" attitude until the concluding phase of the Paris negotiations. A policy shift came once it finally dawned upon London that the High Authority's takeover of responsibilities from the Allied high commissioner would effectively deprive Britain of a voice in the

[26] Bullock *Bevin*, pp. 766–790; also, H. Yasamee, "Britain, Germany, and Western European Integration, 1948–1951" (unpublished ms. HY2 AAC, June 1988); Roger Bullen, "The British Government and the Schuman Plan, May 1950–March 1951," in Schwabe, *Die Anfänge des Schuman-Plans*, pp. 199–210.
[27] PRO CAB134/293, Committee on Proposed Franco-German Steel Authority, Reports of 15 May 1950, 17 May 1950, 19 May 1950, 24 May 1950, 16 June 1950, Meeting of 12 June 1950; Bullen, *Bevin*, p. 202.

Ruhr. Rather than lose these powers by default, the United Kingdom unexpectedly started demanding special concessions as preconditions of any phase-out of Ruhr controls at the very moment that the Schuman Plan negotiations appeared, after months of negotiation, finally to be nearing a conclusion. The culmination of the new hard-line policy was the United Kingdom's blockage of the final agreement transferring powers to the new Coal and Steel Community in an attempt to gain preferential access to German scrap metal.[28] This was Britain's only important intervention in the Schuman Plan negotiations. It otherwise merely stood back, half hoping that "Europe" would fail and half afraid that it would succeed.

Beginning on 20 June 1950 at the summit of European diplomatic accomplishment, the Paris negotiations followed a tedious, meandering downhill path which by fall had led to swamps of bureaucratic maneuvering in the national self-interest. The inglorious descent was due in part to the inadequacies of the *Document de Travail*. After chipping away for months at it the other delegations, none of which had produced constitutional drafts of their own for the new community, arrived at common positions on key outstanding questions which, though not explicitly German, closely resembled proposals put forth by the Federal Republic in the course of the negotiations. The High Authority (HA) was the centerpiece of the French draft.[29] Monnet conceived of it as a muscular directorate acting collectively on the majority principle with executive power concentrated in the *primus inter pares*, the president. The new body was to be subject to only minimal external restraints, a court to which governments might appeal and a parliament limited to interpellation. Three separate committees composed of producers, consumers, and labor were to advise the new directorate upon request but otherwise lacked statutory authority. Regional producer organizations, whose composition was not specified, were to act as transmission belts for the directives of the High Authority and provide the information necessary for it to operate. The High Authority was to act on the basis of powers assigned it by the member states as defined in Article 19 of the *Document*. These were to include every-

[28] Boris Ruge, "The United States and the Creation of the ECSC, 1950–1952" (master's thesis, University of North Carolina, 1987), pp. 73–74.

[29] JM AMG 3/3/9, *Document de Travail*, 27 June 1950, section 1, articles 1–20; see Hanns Jürgen Küsters, "Die Verhandlungen über das institutionelle System zur Gründung der Europäischen Gemeinschaft für Kohle und Stahl," in Schwabe, *Die Anfänge des Schuman-Plans*, pp. 73–102, and Richard T. Griffiths, "The Schuman Plan Negotiations: The Economic Changes," in ibid, pp. 35–72.

thing necessary to create a "special market" (*marché unique*), "pool production," or specifically eliminate "all privileges of entry or exit, tax equivalents [or] quantitative restrictions on the circulation of coal and steel within the area of the member states." The High Authority was to have special authorization to abolish "all subventions or aids to industry," "all means of differentiation between foreign and domestic markets in transportation rates as well as all coal and steel prices," and all other "restrictive practices."

Article 17 set down the principles meant to guide the High Authority's actions. They were contradictory. On the one hand the coal–steel directorate was supposed to eliminate the "falsification" of competitive conditions, on the other, to equalize wages and working conditions and enforce "identical delivery conditions for coal and steel at the point of departure from mine or mill." It was not explained how the two were to be reconciled in practice. The articles dealing with prices, wages, production, investment, and rules of competition were even less reassuring. Pricing policy as outlined in Article 25 had noble but inconsistent and unattainable objectives. These were the protection of consumers against discrimination and of producers against "disloyal" practices, the "assurance" of steady expansion of markets and outputs, and the creation of conditions "guaranteeing the spontaneous allocation of output at the highest level of productivity." The wage policy set forth in Article 26 was similarly unrealistic. It committed the High Authority to preventing wage cuts during slumps, eliminating "exploitative competition," guaranteeing "coal and steel workers the highest standard of living compatible with economic equilibrium," and introducing "wage equality" by imposing assessments on production. The separate provisions were irreconcilable. To believe that the High Authority could, by its own efforts, assure the steady expansion needed to raise wages across the board to the highest level obtaining in heavy industry while simultaneously ending subsidies and lowering prices required a bounding leap of faith. The enumeration in Articles 28 to 30 of broad High Authority powers fueled fears of future usurpations. They empowered the HA to impose manufacturing programs upon firms, steer investments, recommend changes in customs duties, banking rules and transport tariffs, and fine wrongdoing producers.

The basic German working paper prepared for the opening of the Paris negotiations on 20 June 1950 contained the gist of an alternative scheme to the one Monnet presented to the conference. The Federal Republic wanted a High Authority too weak to interfere with either the restoration of the Ruhr or its traditional methods of operation and role in West European heavy industry. The working paper therefore took up a proposal already made by the Dutch to establish a Council of Ministers with veto authority, recommended giving the High Court broad powers of review, and suggested that the General Assembly be allowed to organize supervisory committees. The German draft deliberately precluded the High Authority from the exercise of social responsibilities and emphasized that its primary function was to be the promotion of efficiency. Here too, the High Authority's powers were to be curtailed: The HA was to be forbidden to set prices, establish manufacturing programs, and, with one significant exception, engage in long-term production planning. The exception was granted in order to enable it to review "recent new projects built in the wrong places," in other words, outside of Germany. The HA's means of financing new investments were also to be strictly limited and its authority to enforce "fair competition" drastically curtailed.[30]

The German paper assigned those powers stripped from the High Authority to producer associations. In calling for the organization of regional groups the French *Document* inadvertently provided an opening for the restoration of cartels. Monnet was deliberately vague about the composition of the proposed bodies, specifying only that they be formed on a transnational basis – a stipulation meant to threaten the Ruhr with a powerful Franco-Belgo-Luxembourgeois bloc. The German paper cannily accepted the French proposal "in principle" while introducing changes in wording that nullified it in practice arguing that the groups should "be created on a geographical basis without respect to national frontiers." The new language shifted the emphasis from the politics of federalization à la Monnet to the economics of location favoring the Ruhr. The Germans also described with great gusto the legal character, organization, and responsibilities of the regional groups. They were to be, first of all, not purely private associations that could be attacked as constituting restrictive agreements but semiofficial bodies (*öffentlich-rechtliche Körperschaften*) that, though self-governing, could be deputized to discharge responsibilities on behalf of the coal–steel pool. The German paper of course also strenuously rejected a proposal the French were then believed ready to make for dividing the Ruhr into two or more subgroups, claiming that this would destroy a model of technical efficiency, prevent large-scale rationalization of plant, and impair the operation of numerous scientific and technical organizations.[31]

[30] BA B146/263, "Grundsätzliches zum Schuman Plan," August 1950, and "Unterlagen zum Schuman Plan," 8 August 1950.
[31] Ibid.

These views were by no means those of only the smokestack barons: There was a solidarity of fact among industry, the ministries, and the West German government which, by the time the Schuman Plan was concluded, also embraced the coal and steel unions and a good portion of the social democratic movement. This unity was never unconditional and at times threatened to come unstuck, yet by comparison to the French, Italians, and Belgians it was exceptional.[32] The German policy-making consensus was not always evident to the casual observer. The country was still under occupation, mere mention of cartels taboo, and open advocacy of their restoration patently impossible. Assertions of national interest had to be muted or veiled: West Germany was not yet strong enough to attack Monnet head on. A sounder approach, than the direct charge was to applaud in public while working to undermine the *Document* behind closed doors. As put by Günter Henle, "The more we succeed in organizing economic support structures, the greater will be the chances of the [Schuman Plan's] political realization . . . Hopes will be dashed if the coal and steel union is not properly constructed or is overly ambitious."[33]

The various German participants took different roles in the negotiations. Adenauer beamed approval and, while keeping an Olympian distance from the hard, and sometimes nasty, elbowing for advantage that frequently took place, remained firmly in command of overall operations. Distrusting both the instincts and philosophy of the minister of economics, he introduced Erhard into the negotiations only when it was convenient to have a professed free-marketeer to defend German interests against the incursions of *dirigisme*. The chancellor directed Hallstein to kowtow to the Schuman Plan idea, let the secondary powers do most of the arguing against Monnet's team, and never press any point that might jeopardize the proceedings. Beyond this he seldom intervened, leaving ample room for maneuver to the men of the Ruhr, upon whom victory would ultimately depend.[34]

---

[32] See John Gillingham, "Solving the Ruhr Problem: German Heavy Industry and the Schuman Plan," in Schwabe, *Die Anfänge des Schuman-Plans*, pp. 309–336.

[33] KA/Henle Europäische Gemeinschaften, Schuman Plan, Wirtschaftsvereinigungen (1.7.50–30.9.50), "Vorschläge zur Montanunion," 4 July 1950.

[34] See BA B109/350, Besprechung zwischen . . . Hallstein, . . . Graf, . . . Solveen, . . . Schneider, "24 February 1951, and B109/348, Rechtsabteilung to Dinkelbach, "Monnets Vorschläge zur Inkraftsetzung des Schuman-Plans . . .," 16 November 1950.

Supporting them was cumbersome but powerful institutional machinery that relentlessly ground out analyses, studies, statistical surveys, think pieces, position papers, and countless memorandums. The Ministry of Economics housed a Schuman Plan desk (Schuman Plan Referat), which received support from the Ministry of Transport and the Iron and Steel Office (Fachstelle Eisen und Stahl). Aiding it was the steelmakers' association, WVESI, whose statistical section of 250 persons kept data on all producing nations. The Deutsche Kohlenbergbau Leitung provided similar information concerning fuel. A "coordinating committee" (Koordinierungsausschuss), to which belonged numerous prominent figures from diverse sectors of the economy, linked public and private policy. It included the banker Hermann J. Abs, Klöckner managing director Henle, the industrial statistician Dr. Rolf Wagenführ, general director Dr. Wilhelm Roelen, and the SPD economist and party leader Dr. Ernst Nölting. Reporting to the coordinating committee were specialized subcommittees for wages, social policy, ferrous metal, and coal, whose membership lists were equally impressive. Constellations of academic advisory groups surrounded these influential bodies.[35]

On 8 August 1950 a diplomat attached to the German delegation in Paris wrote that "the French document de travail has become outmoded in every single point as a result of the recent negotiations and remains of mere historical value."[36] This was wishful thinking. The negotiating process would continue for months over arcane but vital issues, and much of the form, if little of the substance, of the original Schuman Plan idea would remain intact. The most important survival was the proposal to create a High Authority, even though a Council of Ministers representing the governments would be attached to it. Though empowered to issue directives to the HA at times of crisis and, in general, inform it of the national policies of the participating nations, the council was really a supervisory body. Executive power still resided with the High Authority. Its exercise would be determined in part by the technical committees set up to deal with the different aspects of economic policy.

Monnet expected the negotiations to be over by August. They began on 20 June with general discussion of the Schuman Plan idea by

[35] BA B102/4382, "Organisatorische Vorbereitungen für die deutsche Beteiligung am Schuman-Plan," 28 June 1951, and B102/5132, "Themen für Diskussion im Wissenschaftlichen Beirat zu Fragen des Schuman-Planes," 9 July 1950; see also BA B109/347.
[36] BA B102/5132, "Sekretariat für Fragen des Schuman-Planes: Ulrich 449/50," 8 August 1950.

the heads of delegations. Sleeves were first rolled up on 3 July, when various working groups got down to business. Monnet chaired the one dealing with institutions, and his deputies those for production, price, and investment policy, trade policy, and social questions. The working groups remained in session until 12 August, then adjourned for the summer, having accomplished very little. When the negotiators resumed work on the thirty-first of the month they found themselves on a tighter leash. Unable to reach agreement on many technical issues, the working groups were abolished on 10 October. From then until the end of the year the heads of delegations met in successive attempts to work the inconclusive and frequently confusing recommendations made by the working groups into acceptable constitutional language.[37] Since the French controlled the conference secretariat, the wording of the treaty tended to reflect the Monnet position. Yet this did not greatly matter: Few of the issues raised in the negotiations would be resolved in the treaty.

In the six months of their operation the working groups reduced the variegated terrain of the *Document de Travail* to a kind of featureless rubble. The reduction process began with wage policy. It had to contend with vast differences in pay rates and productivity prevailing among The Six. Only the Belgians, whose earnings were the highest in Western Europe, were enthusiastic about the upscale equalization proposed in the *Document*. To their disappointment the French head of the relevant working group, Pierre Uri, soon dropped the idea. By October the Germans had won general agreement to minimizing the High Authority's powers in the field. In the final draft the HA's interventions were limited to prohibiting wage reductions in coal and steel except when this was ordered by the national governments as a component of overall economic policy. Virtually nothing remained of the original visionary plans.[38]

All of the non-French delegations registered objections to the proposed investment controls: The Dutch and Italians opposed interference in their modernization plans, Belgians and the Luxemburgers disliked public intrusions into the private sphere, and the Germans resented being continuously subject to regulation and supervision. Etienne Hirsch, the Monnet deputy who chaired the working group for investment controls, conceded in July that the High Authority would exercise only indirect powers in the area, an

---

[37] JM AMG 10/6/4a,b, "Memorandum for George W. Ball, Schuman Plan Chronology," 28 December 1950.
[38] Griffiths, "Schuman Plan Negotiations," pp. 40–41.

all but meaningless formulation that left unclear whether the authority to intervene directly would belong to governments or to the regional groups, whose composition was still undetermined.[39]

In the case of production controls, the other delegations succeeded on 2 August 1950 in limiting the HA interventions to times of threatened surplus or scarcity, the Germans and the Dutch maintaining that even then its order should be nonbinding and based on the directives of the regional groups. The Belgians and Luxemburgers actually demanded that they be vested with responsibility for production planning. The relationship between the coal–steel executive and the producer associations thus hinged on the structure, composition, and legal authority of the regional groups. This subject was not discussed during the summer, presumably because Monnet did not feel strong enough to tackle the issue. Whether in the future they would become transmission belts as he wished or merely provide new institutional shells for the old cartels remained unclear.[40]

The working group that dealt with price policy deserved credit, if that is the appropriate word, for having conducted the most complicated and confused discussions in the entire negotiating process. In attempting to arrive at some form of common, communitywide pricing that would simultaneously provide producers an adequate return, keep costs to consumers low, and support wages at socially acceptable levels while eliminating market distortions, Monnet was trying to square the circle. His instructions call for the adoption of a uniform system of price maintenance which, though allowing limited variations on a product basis, forbade increases and permitted decreases only when justified by cost reductions. Beyond this they were vague. Because it was unclear whether the costs of the least efficient or the average producer should be governing, the actual workings of the proposal were impossible to predict. If set too high, they would have protected inefficient producers; if too low, they could have caused dislocation and unemployment.[41] The single price standard raised further problems, the first of them being that cost variations ruled out consideration of either market pricing or a system ex-mine, ex-mill. The discussions of the working group therefore soon turned to basing points. In this approach, which was used by the American steel industry, standard prices were set for

[39] Ibid., p. 43.   [40] Ibid., p. 46.

[41] Ibid., p. 47; JM AMG 18/0/4, "Principes des prix de la Haute Autorité en matière d'acier," 27 July 1950; JM AMG 71/4/45, "Propositions au Comité Restreint sur le rôle de la Haute Autorité en matière de prix."

delivery of product to an industry-designated shipping point. The seller would absorb the costs of shipment from the factory to it – which were in fact built into the selling price – and the buyer would cover the costs of freight from there to the ultimate destination. Apart from the level at which prices were set and the extent of outside competition, the workings of the system depended on the number of basing points and the size of the regions in which they operated.[42] The working group devoted much time and energy to working out a suitable basing-point system. The adoption of a single common basing point as sought by Monnet, the so-called *prix départ*, would have been nondiscriminatory in the sense that purchasers would pay a single list price, excluding freight, but had one basic flaw: Price levels would have had to be set very high in order not to bankrupt producers whose operations were distant from the common port or rail junction. The Italians, who were high-cost producers of steel, supported the French position in the hope of gaining protection, and the other delegations tried to devise a system of multiple basing points, so-called *prix parité*, while still maintaining a common range of product prices. This would have protected traditional sales areas but, depending on external factors, could have lowered overall prices and increased competition. Though the marketing problems that arose were too complicated to be resolved during the treaty negotiations, the working group agreed to adopt prix parité as a policy-making principle. At the same time, it refused to leave price determination to the High Authority: The HA's powers in this respect were limited in the treaty to periods of threatened over- or underproduction.[43]

Discussions of the structure and composition of the regional groups upon which so much depended never actually began. Monnet had initially hoped that after the High Authority's statutory powers had been fleshed out producer associations could be compelled to act as transmission belts for its policies. Yet on 4 October 1950 he abruptly and unexpectedly denounced the term "regional group" as a byword for "cartel" and dropped the idea of forming such new organizations. His hopes for eliminating the influence of producer associations rested thereafter almost entirely on the inclusion of strong anticartel provisions in the treaty. To enforce them he would turn to his only real ally in the negotiations, the United States.[44]

[42] Griffiths, "Schuman Plan Negotiations," p. 47.   [43] Ibid., pp. 48–53.
[44] KA/Henle EGKS, Schuman Plan, WVESI (1.10.50-31.12.50), Blankenagel to Henle, 9 October 1950; BA B146/268, Erhard to Hallstein (BWMIA3-8119/50), 4 November 1950.

The historical conflict between France and Germany was the main issue at stake in the Schuman Plan negotiations but by no means the only one. Though "Europe" could have been unmade only by France or Germany, its creation required the participation of Italy and the Benelux nations as well. To avert the breakdown of negotiations, Monnet had to cut special deals with each of them. Apart from tiny Luxemburg, the cooperation of the Dutch was probably the easiest to elicit; locked in fatal embrace with West Germany, the Netherlands could assert, but not exercise, economic independence from Bonn and Düsseldorf. Ultimately the Dutch had no choice but to go along with decisions made in those places.

Belgium, though less economically dependent upon West Germany than the Netherlands, was potentially more vulnerable to competition from the Ruhr. This peculiar binational state, with its oversized and inefficient coal and steel industries, was too weak to have remained outside of the Schuman Plan negotiations but could not have allowed them to develop along the lines set down by Monnet. At the same time, no one could have forced it to accept this outcome. On the one hand, the latent language conflict always made it possible for Belgium to threaten self-destruction. On the other, the absence of competition from Germany together with general worldwide shortages gave the divided land a unique prominence on international steel export markets; never before or since would its share be as large as during the postwar years. The Belgians were well positioned either to wreck the negotiations or to demand a high price for saving them. This took the form of a special deal for the nation's coal mines.[45]

Contrary to claims, the problems of the Belgian pits had less to do with either a meager resource endowment or the effects of the war than with a long history of mismanagement by the big bank holding companies that controlled both coal and steel. The efficient Kempen district of Flemish-speaking Limburg could have been put into operation earlier and on a larger scale than was the case. In the 1930s mine outputs had been heavily subsidized, but profits had flowed into other sectors of the economy as well as overseas. The war greatly aggravated problems due to undercapitalization and lack of new investment.[46] With a view to profiting from the high postwar demand for steel, the Belgian government stepped up mine output by giving massive rebates to domestic consumers of Belgian coal and providing wage subsidies. Almost overnight a nation noteworthy for

---

[45] See Alan S. Milward, "The Belgian Coal and Steel Industries and the Schuman Plan," in Schwabe, *Die Anfänge des Schuman-Plans*, pp. 437–453.

[46] See Gillingham, *Belgian Business*, passim.

exploiting labor found itself saddled with the highest mine wages in
Europe, Belgian labor costs per ton of mined coal being no less than
40 percent above those of France and 60 percent above those for
West Germany. In 1950 the average price of coal to the domestic
consumer was 690 Belgian francs (BF) per ton as compared to an
equivalent to 422 BF in the Federal Republic. The Schuman Plan
obviously required major adjustments.[47]

Since Belgium would have walked out of the Paris conference if
forced to compete, the French generally supported the demands
of its delegation for payment by "efficient producers," meaning
Germany, of an equalization tax. The so-called Hirsch–Vinck Plan,
which was devised outside the official negotiations, called upon the
Federal Republic to pay Belgium 3,000 million BF over a five-year
period in order to subsidize wages and new investment. As a *con-
trepartie* the Belgians were asked merely to make 3 percent annual
decreases in output, a stipulation included only because France's
own mines had scheduled such reductions. Controls over use of the
payments were wholly inadequate.[48] The Hirsch–Vinck deal was
incredibly inequitable. It required hungry Germans to foot the bill
for decades of inattention and profit milking by the big Belgian
banks and to subsidize the wages of the best-paid labor force in Europe. On behalf of a treaty concluded in the name of fair trading
practices, it directed the mines of the Ruhr to pay competitors to
sell in markets that under conditions of equality would have provided a rich yield. The Federal Republic forebore from seeking
compensation for itself on the Belgians' grounds of geological disadvantage, with its negotiators wisely dropping a demand that the
high production costs of the low-grade Salzgitter ores from fields
put into operation to supply the Hermann Göring Works be offset
by nations recently victimized by the Nazis. Swallowing hard, they
instead agreed to pay Belgium the required 2 percent tax on coal
turnover even though this meant keeping a competitor in business,
raising German steel costs, and draining profits from the Ruhr
mines.[49] Belgian coal and steel publicly vilified the Hirsch–Vinck
deal, crudely branding it as a sellout of the national patrimony to *les
Boches*. Rejection of the proposal by the cabinet was only narrowly
averted. The coal deal nevertheless served its purpose of making
Belgium dependent upon financial transfers from Germany by cre-

[47] Milward, "Belgian Coal and Steel Industries," p. 438.
[48] USNA RG84 (500: Coal and Steel Pool), Paris Embassy to Secretary of State/
2817, 17 November 1950.
[49] BA B146/263, "Unterlagen zum Schuman Plan," (vol. 3), 22 August 1950.

ating a constituency for the Schuman Plan in a nation whose industrial leaders might otherwise have succeeded in their openly professed objective of reestablishing the old international cartels of the prewar period.[50]

Italy was the odd man out among The Six, a part of neither Western Europe nor the Ruhr-dominated European heavy industry economy. Everyone in Paris agreed that this should be changed, lest Italy be left orphaned and isolated, and recognized as well that adjustments, and therefore costs, would be involved. The agreements enabling Italy to enter the coal–steel pool were a solid if undramatic diplomatic achievement, an important step in the industrial development of the country, and a precedent for the expansion and extension of European community institutions.[51]

Italy was not blessed with substantial domestic sources of coal and ore, and its steel industry was divided and for the most part inefficient. There were in fact three quite distinct steel industries. The first, consisting of numerous small-scale finishers of steel products scattered throughout the Alpine valleys of the northern part of the country, were actually quite competitive. Almost outspokenly enthusiastic about the prospect of a European common market, they played no role in the negotiations because their manufactures did not fall within the definition of steel as decided upon by the conference. The Falck interests constituted the second steel industry. They had no blast furnaces but relied heavily upon imported scrap and semifinished steel to supply their Siemens-Martin and electric hearth ovens. Though this group demanded some protection, its main concern was access to foreign supplies of scrap and semis. Finally, the three works erected in consequence of the Sinagaglia Plan adopted in the 1930s represented Italian attempts to create vertically integrated steel production complexes like those of the Ruhr. The main objectives of this group were to secure reliable supplies of coke and ore as well as price protection during the transition to competitive conditions.[52]

The rational economic choice of the Paris conveners would have been to shut down at least two-thirds of the Italian industry, and the more technocratically minded of the French delegates as well as cost-conscious spokesmen for Ruhr interests advocated doing nothing less. Fortunately in this case, the politicians overruled the voices

[50] Milward, "Belgian Coal and Steel Industries," pp. 451–452.
[51] See Ruggero Ranieri, "The Italian Steel Industry and the Schuman Plan Negotiations," in Schwabe, *Die Anfänge des Schuman-Plans*, pp. 345–356.
[52] Ibid., pp. 346–347.

of market reason, and the Italians succeeded in negotiating deals that adequately safeguarded the development of the national economy. The settlement reached at the Santa Margherita conference in February 1951 guaranteed Italy a five-year supply of ore from Algeria at prices in line with intracommunity deliveries. In addition, a special equalization arrangement, though only one-third the size of the one envisaged for the Belgian mines, protected the Italian market from a flood of steel imports. In order to finance modernization the Italians were also allowed to maintain their steel tariffs for another five years. The formation by the High Authority of a European scrap cartel in 1953 offered further guarantees to domestic producers.[53]

These arrangements were hardly consistent with the creation of the common market, and if in the end they promoted economic liberalization it is because of developments that those who negotiated the settlement could have known little about. Long-term market and supply trends would turn out to be unexpectedly favorable to small-scale tidewater producers, and Italian steel management would prove to be determined, flexible, and highly entrepreneurial. The Italian negotiators nonetheless deserve credit for having left Paris with agreements offering a reasonable assurance that their nation could join Europe without becoming an economic colony of it.

Bilateral deals with weak members could strengthen the coal steel community but not hold it together, and by definition they eroded its substance. The discussions in the working groups ground it down still further; the only consensus on the Schuman Plan to emerge from them was negative. Without a core, or hub, of some kind to replace the directorate envisaged by Monnet, the coal–steel negotiations would have at best reached a meaningless, face-saving outcome and at worst might have fallen apart altogether. Only West Germany was in a position to take over the central role in the coal – steel pool, but only the United States had the strength to apply the heat and pressure required to fuse that nation into the framework of Europe. Whether it had the finesse to do so without breaking the fragile structure remained to be seen.

5.2   The bombshell at the Waldorf

On 25 June 1950 North Korea invaded the Republic of South Korea across the Thirty-eighth Parallel, and the United States soon found

[53] Ibid., pp. 350–351.

itself enmeshed in a "police action" fought on a very warlike scale that it would twice come perilously close to losing. The Korean War would cost the lives of 54,246 American servicemen. Hostilities would continue until a cease-fire agreement was reached on 27 November 1952. An armistice was eventually concluded on 27 July 1953. The events in East Asia changed the nature of the American commitment to Europe, supplanting the Marshall Plan policy of developing the economic and political strength of the free-world nations with one aimed at building up their military power in the face of a perceived Soviet threat; economic cooperation, in other words, retreated before mutual security.[54] The war also brought about a dramatic improvement in the status of the Federal Republic. Originally excluded from the North Atlantic Treaty Organization (NATO), an alliance created in large part to protect Europe from the German threat, the young nation would soon find itself called upon to rearm and gear up its defense effort. The Federal Republic could therefore no longer be treated like a *quantité négligeable* and its delegates at the Schuman Plan negotiations be expected merely to smile and applaud.

The United States embarked upon rearming the Germans with great trepidation but felt compelled to do so. The decision was made in an atmosphere of panic. It was true that the British relied on the French to protect Europe, undeniable that half of France's 300,000-man army was tied up in Indochina, and evident to anyone that between them the Allies could muster up no more than twelve divisions to defend the Continent against a Red army believed to be about ten times as large. There was no indication, however, of a Russian buildup in Eastern Europe either before or after the North Korean invasion, nor would one occur in the near future. Prior to 25 June 1950 most American analysts had in fact judged the USSR incapable of launching a major war for several years.[55] There would be no more such talk in Washington after the outbreak of hostilities in Asia. The Soviets, it was feared, were poised and ready to strike and could be "deterred" – to employ the unavoidable jargon – from doing so only by the strength of Western Europe's defenses.

Though prior to the outbreak of the Korean affair a top-level planning directive known as NSC-68, which foresaw the need for stationing the U.S. army in Europe, had reached the president's desk, he had not yet signed it. An "air of unreality" pervaded NATO staff planning, according to Lawrence Kaplan, because the Medium-Term

---

[54] See Hogan, *Marshall Plan*, pp. 380–426.
[55] See HST, Oral History Interview with Theodore Achilles, 13 November 1972.

Defense Plan, which was the operational directive, assumed the existence of ninety divisions, only a fraction of which could be funded through existing appropriations.[56] The assumption made by many Europeans that somehow the Marshall Plan would make up the difference was unwarranted and unrealistic given the fiscal conservatism of Secretary of Defense Louis Johnson. An actual Soviet invasion would have been met by an attempted U.S. withdrawal to the Pyrenees.[57] Once hostilities in Korea had begun, this situation would change dramatically: The defense buildup became the preeminent concern of American policy, and the political and economic integration of Western Europe, which had been the primary objective, was reduced to a mere means to this end.

Before the outbreak of the Korean War the State Department had steadfastly opposed the formation of a military force in the Federal Republic unless and until it had become a part of a strong, united Europe. First the one and then the other – in that order: the sequence had to be maintained.[58] Emotional resistance to any form of German armament remained strong, and its influence on decision making was still significant. Fear of the Russians was not yet great enough to erase memories of the recent past. There was thus considerable inconsistency in American views. This was evidenced by U.S. High Commissioner John J. McCloy who, on the one hand, disagreed outright with the notion that the West Germans needed their own defense forces and said so repeatedly. In an authoritative statement of February 1950 he ruled out the creation of a "German army or air force"; the attachment of a "German foreign legion" to the Allied armies was equally unacceptable to him. On the other hand, McCloy was genuinely sympathetic to Adenauer's pleas for a national police force, though also insistent that it not be allowed to develop into a paramilitary formation. Yet the defenselessness of West Germany continued to trouble him, and he appears to have been skeptical that the official "tripwire theory" would work. It held that the Federal Republic had no need for its own defense forces because the Russians, aware that the death of even a few GIs at the hands of the Red army would trigger a massive U.S. response, would think it too risky to invade West Germany and thus be de-

---

[56] Lawrence S. Kaplan, *The United States and NATO: The Enduring Alliance* (Boston, 1988), pp. 39–40.

[57] Ibid.

[58] See HST, Acheson Papers, Memorandums of conversations, "The Position of Germany in the Defense of Western Europe," 31 July 1950.

terred from starting a war.[59] McCloy still could not shake his doubts about the Germans. On 28 June 1950, three days after the North Korean attack, he wrote a long, reflective personal letter to his mentor, Henry Stimson, decrying their "arrogance" and "traditionalism" as well as the tendency "to blame others when theirs was the greatest crime of all because they had the greatest capacity to know," and admitted in closing to being "reluctantly of the view that we should not now make any attempt to create or use a German army."[60]

A month later Secretary of State Acheson presented the main lines of new American policy to President Truman, who directed him to proceed. German rearmament was still subordinated to European integration, but the two objectives would hence be pursued simultaneously rather than sequentially. The linchpin of the American policy structure would be a European defense force in which the Germans would participate but which they would not be allowed to dominate. As Acheson told the president,

> The question [is] not whether Germany should be brought into the general defensive plan but rather how this could be done without disrupting anything else that we [are] doing and without putting Germany in a position to act as the balance of power in Europe . . . To create a German General Staff . . . and a German military supply center in the Ruhr would be the worst possible move, would not strengthen but rather weaken Western Europe and would repeat errors made a number of times in the past . . . [We are] thinking along the lines of the possible creation of a European army or a North Atlantic Army . . . made up in part of national contingents and in part by recruits from a number of countries who could act under a Central European or North Atlantic command. German economic power might be integrated with the production of the other Atlantic powers so that it would not be a separate and complete source of military equipment but would have to operate with others. In such an arrangement, Germans might be enlisted in a European army which would not be subject to the orders of Bonn but would follow the decisions reached in accordance with the North Atlantic Treaty procedure.[61]

[59] Thomas A. Schwartz, "From Occupation to Alliance: John J. McCloy and the Allied High Commission in the Federal Republic of Germany" (Ph.D. diss., Harvard University, 1985), pp. 286–287.
[60] Cited in ibid., p. 290.
[61] HST, Acheson Papers, Memorandums of conversations, " The Position of Germany in the Defense of Western Europe," 31 July 1950.

In late August Acheson informed the Pentagon that the State Department was ready to proceed with the organization of a European defense force composed of units from the United States, United Kingdom, France, and the other nations of the Continent, including Germany, commanded by an international general staff with full command authority. Participation in this multinational military alliance would convey West German membership in NATO and further implied an eventual centralized direction of military production. The U.S. plans did not, however, authorize the Federal Republic to manufacture critical military hardware or develop a capacity for independent operations. Overall U.S. control was meant to assure that these conditions were met.[62]

The new defense policy had an immediate impact on American approaches to the coal–steel negotiations. In July the Working Group of the Schuman Proposal, the informal think tank operating in the U.S. embassy in Paris, had produced a study examining the heavy industry negotiations in light of the anticipated economic mobilization and reached a remarkable conclusion: German rearmament could be turned from a threat into an opportunity. The author of the paper, Raymond Vernon (later prominent on the faculty of Harvard Business School), recommended expanding the High Authority as proposed by the *Document de Travail* into something akin to Speer's Zentrale Planung, which had run the Nazi war economy through steel allocation. Vernon advised that the HA "be the administrative agency for the achievement of common defense objectives" and that a consultative committee representing the member governments be established to help it. The High Authority should have the power to set production targets, allocate resources, fix maximum prices, ration raw materials used by the coal and steel industries, and limit exports.[63]

Vernon cited a number of justifications for this proposal. One was to contain West Germany's economy, whose growth would be stimulated by rearmament. Another, more subtle one was the need to protect the common market concept. Economic mobilization, the paper reasoned, would inevitably undermine the principle of competition. Rather than leave the adoption of antimarket measures to the national governments, the High Authority should do the job it-

[62] USNA RG466 D(50)2030/TS(50)94, Acheson to Douglas, Bruce, and McCloy/ CN-34097, 22 August 1950.
[63] USNA RG84 (500: Coal and Steel Pool), Working Group on the Schuman Proposal, "Modification in the Authority's Powers Required by Defense Consideration," 1 August 1950.

self on the theory that what it could put together at one time it could take apart at another. The paper's author also expressed the hope that because during emergencies power flows toward the center the High Authority could become an agent of liberalization throughout the economy.[64]

Though Monnet would later pick up on ideas very much like those of the American Working Group on the Schuman Proposal, speculations such as Vernon's were mere shots in the wind compared to what was now to happen. On 12 September 1950 Acheson informed Bevin and Schuman at the New York foreign ministers' conference that the United States had taken four important decisions: It had decided to commit troops to Europe, organize an integrated command structure for the Atlantic alliance, fit German military units into it, and remove economic restrictions limiting the nation's defense contribution. These decisions, he added, were parts of an indivisible package that the United States would not discuss piecemeal with its allies.

Acheson was completely unprepared for Schuman's horrified reaction to his remarks. Had France not recently begged for the stationing of American troops – not to mention the financial assistance promised by the new policy? Had it not also itself brought up the need for a West German defense contribution? Such thoughts paled by comparison to the political nightmare conjured up by the bombshell that Acheson had dropped at the Waldorf. Was this, the French public would protest, where the European initiative named after their foreign minister was to lead – to a new Wehrmacht stationed along the Rhine? The secretary of state's few words threatened to destroy the Paris negotiations at a single stroke. To save them Monnet would soon let loose a couple of shellbursts of his own. Rather than continue to oppose rearmament for Germany, he would attempt to assert direction over it, just as with the integration process. To prevent the initiative from again slipping out of his hands, he would also turn the screws on the Ruhr. Decartelization was back with a vengeance.[65]

The most critical phase of the coal–steel talks now began, would last until March 1951, and would turn not only on the official Paris negotiations but on a decartelization struggle between the government of the Federal Republic and the Office of the U.S. High Commissioner, acting in this case as agent not only of American interests

[64] Ibid; also, Schwartz "Occupation to Alliance," pp. 337–369.
[65] Gillingham, "Solving the Ruhr Problem," pp. 422–423; Schwartz, "Occupation to Alliance," pp. 361–371.

but of Jean Monnet. The official mood of Euro-affability prevailing
in Paris had no counterpart in Bonn: the discussions held there
were tough, anything but friendly, and the day was saved only at
the last minute thanks to compromises reached by the German
chancellor and the American high commissioner. The feuding on
the Petersberg delayed the conclusion of the treaty by three months
and left problems in their wake that resulted in many more months
of rancorous debate. Yet the outcome could have been far worse.
The breakdown of the Bonn talks would surely have meant the end
of the Schuman Plan and a delay of years in any progress toward
integration.

Monnet's angry memorandum of 28 September 1950, the purpose behind which was to pressure the working groups to return to
the original Schuman Plan idea, opened the battle. In the form of
"clarifications," the Frenchman gruffly asserted the HA's right to
exercise a number of powers allegedly contained in, or implied by,
the *Document de Travail*. Where the *Document* forbade firms from entering price and production accords, he claimed for the HA the
right to approve mergers and specialization agreements. By extending the scope of provisions in the draft treaty Monnet also demanded the power to establish and regulate so-called *organismes
communs*, a term used to cover a broad array of joint buying and
selling syndicates, freight conventions, and pressure groups. As for
producer associations in the strict sense of the word, the regional
groups, Monnet insisted that the High Authority stood under no
obligation to consult with these bodies but enjoyed a wide discre-
tionary authority and could specifically deal with individual firms.
In a final clarification he further demanded that the so-called advisory committee be broken down into two bodies, one for coal, the
other for steel, each providing one-third representation to labor and
an equal share to the coal- and steel-consuming industries. No one
of course knew whether Monnet could enforce positions taken in
the 28 September memorandum, but he had clearly demonstrated
the determination to do so.[66]

The Germans fired the next salvo. The French should have been
prepared for it. After listening with "visible satisfaction" to McCloy's

[66] JM AMG 3/1/6, "Observations sur le mémorandum du 28 Septembre 1950 ex-
posées par M. Jean Monnet au cours de la réunion restreinte des chefs de délégation le 4/10/50"; AN 81 AJ137, "Note pour M. Schuman," 28 September
1950; also JM AMG 8/3/19, "Stellungnahme zur Preisgebung im Memorandum über die Einrichtungen und die wirtschaftlichen und sozialen Dauer-
bestimmungen des Schuman Plans vom 28. September 1950."

report on the results of the New York foreign ministers' conference, Adenauer told the assembled high commissioners with uncharacteristic bluntness that the just-published Regulations I–III to Law 27 jeopardized the Schuman Plan negotiations and might necessitate the recall of the German delegation from Paris.[67] Law 27 itself had been published back on 20 May 1950, after many months of redrafting, as a tripartite successor to the U.S.–U.K. Law 75 and differed from the earlier version in only one important respect: By empowering the government of the Federal Republic to make ultimate disposition of coal and steel assets, it rendered indemnification of shareholders virtually inevitable. For this reason the French high commissioner refused to initial the proposed law, which took effect only because of a statutory technicality. Still, for six months no action was taken under Law 27, and it might never have been, had the coal-steel talks proceeded as Monnet wished.[68] The three regulations published in September to which Adenauer so vehemently objected destroyed the German hope that the equal treatment referred to in Schuman's proposal meant that decartelization would soon be forgotten: They ordered the immediate breakup of the six largest Konzerne, a cut in the ties between the coal and steel industries, and the organization of no less than fifty-four separate mining companies.

The chancellor's threat to recall the German delegation from Paris was not taken seriously. Neither was the resignation a few days later of Gutehoffnungshütte's managing director Hermann Reusch from an important Schuman Plan coordination committee.[69] The speech of Robert Lehr to the Munich Export Club on 1 October 1950 provided the French with the first unmistakable evidence of hardening Ruhr attitudes. They were perturbed by the publicity given the address of the ex–lord mayor of Düsseldorf and particularly angered by the thought that Lehr, who was the interior minister–designate in the Adenauer government, had received official approval for his actions.[70] Though couched in the respectable lan-

[67] AN 81 AJ138, "Copie d'un télégramme réservé [François-Poncet]," 24 September 1950.
[68] See Hubert G. Schmidt, "Reorganization of the West German Coal and Steel Industries under the Allied High Commission for Germany, 1949–1952" (preliminary draft), Historical Division, HCCOM, September 1952.
[69] JM AMG 20/7/5, Telegram of François-Poncet, 4 October 1950.
[70] JM AMG 18/o/6, "L'importance de l'industrie de l'acier pour l'union économique européene," 14 October 1950; AN 81 AJ138, "Fiche [Leroy-Beaulieu] pour M. le Ambassadeur" (n.d.); AN 81 AJ138, "Entretien [Leroy-Beaulieu] avec le Dr. Lehr, aujourd'hui Ministère de l'Intérieur et un des Dirigeants des Vereinigte Stahlwerke," 13 October 1950.

guage customary at gatherings of German business dignitaries, Lehr conveyed an unmistakable warning to the French that, like it or not, the occupation was over. West Germany, he said, could no longer accept the tonnage level of 11.1 million authorized for steel production, which should be raised at once to 16 million. Or did the Allies really want the Federal Republic to remain an economic dependent? Europe needed the Ruhr to grow, he reminded his audience, and so recovery should be allowed to occur naturally even if this meant curtailing France's investment plans. Lehr warned that Germany would not allow the French to use the Schuman Plan to perpetuate their temporary, war-gained industrial advantage and recommended that they specialize in lines of production complementary to the Ruhr. He hinted in closing his remarks that the best solution for everyone might be to revive something along the lines of the 1926 agreement.

In order to preempt action by the Combined Steel Group, which was known to have been preparing a decartelization schedule, the Federal Republic presented its own program for implementing Law 27 in early November. Confusion beset their preparation. Under the terms of the occupation statute, the Ministry of Economics was unable to maintain contact with the trusteeship organizations and had little time to distill their voluminous studies into a workable action plan. It also had to contend with demands from the "coal side" that the mines enjoy a greater degree of independence than previously. The government's proposal called for the formation of seventeen unit steel companies and possible consideration of another five cases and a linkage of mines to mills equal to 25 percent of total output but did not address the issue of coal reorganization.[71]

The operating principles behind the German plan, which Adenauer disclosed to McCloy on 3 November 1950, were to preserve the coal–steel tie to the extent necessary to make the mills self-sufficient in coke and as required for the marketing of low-grade fuel; to introduce organizational changes only where necessary; and to get over the decartelization process as rapidly as possible. An internal memorandum of the following day provided a more detailed explanation as to what could be expected. "Regional associations" for coal would still be allowed; new mining companies could be organized so as to remain efficient and competitive; and "technical

[71] Isabel Warner, "Allied-German Reconcentration Negotiations, 1950–1951" (unpublished ms., 1987); KA/Henle Umgestaltung des deutschen Kohlenbergbaus und der deutschen Eisen- und Stahlindustrie, Gesetz 27, Schriftwechsel vom 1.10.50 bis 31.13.50, "Vereinbarung über die Zusammenarbeit Kohle/ Eisen," 29 November 1950.

and economic" ties between coal and steel, as well as with the chemical, energy, and transportation sectors, would be taken into account in the restructuring effort. The plans also contained reassurances to the steel industry: Though nine "core companies" would remain in being, another seventeen, mainly non-Ruhr companies would be left alone. Finally, the six big producers facing reorganization were promised the cooperation of the federal government in implementing the plan in a satisfactory manner.[72]

While the Germans elaborated their proposal, Monnet lobbied vigorously for the adoption of the 28 September memorandum, stipulating that the High Authority be empowered to authorize or forbid all past, present, and future formal and informal arrangements between producers, or between them and their clients, as well as approve or disapprove all future mergers or fusions.[73]

Monnet's memorandum aroused angry comments on all sides. His explanation that the purpose behind the measure was the prevention of a postoccupation revival of international producer syndicates pacified only Heinrich Dinkelbach. In notes prepared for a 1 November 1950 speech the steel trustee emphasized that only through rigorous enforcement of antitrust law could the Schuman Plan organization be prevented from developing into a cartel. Monnet's "initiative," as he termed it, was nothing more than an attempt to apply the equitable trade principles of the Havana Charter. By following them as well, he implied, Ruhr industry could have a healthy influence on the coal–steel pool.[74]

Though Monnet did not take the advice of his friend René Mayer to declare that the IAR and the rest of the control machinery would remain in operation until the conclusion of the decartelization, he did call in the heavy artillery.[75] It was an open secret that the group at the American embassy in Paris run by Tommy Tomlinson and supervised by Ambassador David Bruce had been acting as Monnet's personal staff in matters relating to the anticartel and anticoncentration provisions of the draft treaty, Articles 60 and 61. The

---

[72] BA B102/60686, Adenauer to McCloy, 3 November 1950; BA B102/60686, Adenauer to McCloy, 4 November 1950.

[73] JM AMG 8/1/5, "Observations sur le mémorandum du 28 septembre relatives aux institutions et aux positions économiques et sociales permanent du Plan Schuman," 4 October 1950.

[74] BA B109/347, "Herrn Dinkelbach Schuman Plan: Disposition für einen Vortrag," 1 November 1950.

[75] AN 81 AJ138, René Mayer to Schuman, 23 November 1950; USNA RG84 (500: Coal and Steel Pool), Bruce to Secretary of State/2531, 7 November 1950.

ERP was also ready if need be to use its funds to influence the outcome of the settlement.[76] Yet this effort remained strictly behind the scenes. Things were different in Bonn. There the French would work in the shadows as McCloy and his general counsel Robert Bowie brought the Germans to heel.

Why did the Americans front for Monnet? Unaware of his hidden hand in the Bonn negotiations, the Germans were convinced that they had fallen victim to a recurrent outburst of American cartel-phobia. Yet not even Monnet's most devoted followers doubted that the coal–steel community would turn into something other than a cartel. Tommy Tomlinson's November 1950 "Observations on the Treaty" are full of warnings about what was emerging from the negotiations.[77] Tomlinson did not like the way the Belgian coal problem was being solved, wanted to limit the High Authority's powers of intervention to "exceptional circumstances," worried that the proposed directorate would strip national governments of essential regulatory powers, and did not think that it should be allowed to influence investment decisions. Tomlinson admitted to not understanding French proposals for pricing policy, objected to allowing the High Authority to set production quotas during times of emergency, and thought that giving it power to make distinctions between good and bad cartels would lead to abuse.

As for action under Law 27, there is no reason to doubt the commitments of either Bowie or his deputy, Sidney Willner, to decartelization; both were well schooled in antitrust law.[78] At the same time, their boss, McCloy, was hardly an avid enthusiast for such policies. A leading Wall Street lawyer and a Republican, upon assuming office as U.S. high commissioner in Germany, McCloy professed to have "no knowledge of the decartelization problem except what I have read in the paper and except as I have read in certain reports about it as well as the recent memorandum issued by the War Department. That is all I know."[79]

McCloy supported Monnet partly out of personal conviction. Like his friend, the high commissioner believed that international control

---

[76] USNA RG84 (500: Coal and Steel Pool), Working Group on the Schuman Plan Proposal, Telegram to Embassy, Paris (WGS D-2/23), 10 November 1950.

[77] JM AMG 9/3/15, "Observations on the Treaty," by Tomlinson, November 1950; see also FRUS 1950/III 850.33/10-350, Acting Secretary of State to Embassy in France, 3 October 1950, pp. 754-755, and 850.33/12-850, Secretary of State to Certain Diplomatic Offices, 8 December 1950.

[78] See USNA RG460 D(50)2019, Robert R. Bowie, "Freedom of Trade," *Information Bulletin*, October 1950.

[79] USNA D(49)4E "Press Conference of Mr. John J. McCloy," 1 July 1949.

of the Ruhr provided the key to European integration and that it
in turn was the solution to the German problem. He too had been
an early advocate of the constructive Europeanization of German
heavy industry, having as president of the World Bank (International Bank for Reconstruction and Development) recommended to
the State Department in June 1947 that his organization, backed by
the United States and supported by the coal-using nations of the
Continent, set up an agency to operate and develop the Ruhr mines.
Production, he estimated, could be doubled and supply improved in
both Germany and Europe. But more than merely material benefits
would flow from the proposal, as McCloy believed that the Ruhr
could become the base of an integrated European economy: "With
this in mind . . . I made the suggestion [because] building strip mills
or even road or transportation systems throughout Europe is just so
much unrelated effort until we have the basic economy which, in
both England and Europe, is coal."[80]

Yet the U.S. high commissioner, though by 1950 having completely given up on Britain, did not expect France to seize the
initiative in the integration process. Early in the year, he recommended that the United States take a hard line, in effect forcing the
French to relinquish occupation controls. In the same week that
Monnet was hatching the Schuman Plan, McCloy appealed for a
bold new initiative as regards Germany. American policy, he said,
was repeating the mistakes of the 1920s. The United States had
imposed Versailles-like restrictions on rearmament in order to prevent "another von Seekt conspiracy" and break German economic
hegemony but in the effort set production limits and prohibitions,
installed international control over the Ruhr, and adopted a deconcentration policy, which was in his view self-defeating. The greatest
danger facing Europe, as McCloy described it, was that a badly
treated Germany would throw in its lot with the East; to prevent
this, living standards would have to be raised, sovereignty restored,
and at least an initial step taken toward eventual rearmament. A
timetable must be set for the gradual phase-out of controls over
German industry, he added, including the Ruhr Authority, which
might evolve into "an international [agency] for control of all Western European coal and steel."[81] In another position paper penned
for use at the 10 May 1950 London conference of foreign ministers,

[80] USNA, Department of State 862.6562/7-747, McCloy (International Bank for
    Reconstruction and Development) to Lovett, 7 July 1947, and 862.6362/7-735,
    McCloy to Douglas (cable), 10 June 1947.
[81] USNA RG466 D(50)1235/TS(50)28, McCloy to Byroade, 25 April 1950.

McCloy prescribed a phase-out period of eighteen to twenty-four months and recommended that France be told politely but firmly to take it or leave it.[82]

The U.S. decision to advance German rearmament transformed McCloy from reluctant decartelizer to enforcer of Law 27, from a proponent of an early elimination of controls to prolonger of the occupation, and from advocate of a hard line toward France to instrument of its policy. The logic behind the high commissioner's actions was perverse but elementary. German rearmament required European integration on terms acceptable to the French, and these involved restraining the power of the Ruhr, at least until this seriously interfered with military security. Since it was difficult to determine the precise location of such a point, it would remain common for Allied authorities to encourage with one hand the very developments they were trying to suppress with the other. The United States did, however, impose sacrifices upon the Ruhr trusts for the good of Europe. Rearmament of the Continent surely accelerated the Federal Republic's return to sovereign status, increased its importance internationally, and created the conditions for unprecedented prosperity, yet what came as a blessing to the many was a curse for the few. In the short run at least, rearmament compounded the woes that the German coal and steel industries had earlier hoped would end with the adoption of the Schuman Plan.

Necessity rather than choice dictated that the next phase of integration have a military character. According to George Ball, who was with Monnet at his home outside of Paris on the day of the North Korean invasion, the author of the Schuman Plan realized immediately that the United States would demand a large role for Germany in the defense of the West.[83] Though he was initially reluctant to launch a new initiative that might compete with the Schuman Plan, a 5 August 1950 speech of Truman's calling for a greater European acceptance of German rearmament caused Monnet to throw the machinery of the French planning commissariat into gear.[84] His group of technocrats was soon busy working out policies for common budgeting, taxation, procurement, standardization, and rationalization of military production. Monnet also kept closely abreast of American planning, especially that being conducted at

[82] USNA RG466 D(50)1299, "Draft Memorandum: Program for Progress on the German Problem" by McCloy, May 1950.
[83] Ball, *Another Pattern*, pp. 90–91.
[84] JM AMI 0/1/3, "Rapport CED: Génèse, évolution, échec," by H. Burgelin, September 1951.

the Paris embassy. France's subsequent proposal for the European Defense Community (EDC) was in principle quite similar to the schemes worked out in the State Department in early summer. Their main features were overall U.S. command within the NATO framework, multinational staffs, integrated chain of command, nationally recruited units at the lowest militarily feasible level of operation, and, in addition, common overall procurement, a separate budget, and overall political control.[85]

The bombshell dropped by Acheson fell nearer than Monnet would have liked but did not catch him unprepared. "We must now make a fundamental choice," he wrote the French foreign minister on 16 September 1950: Do nothing, deal unilaterally with Germany thereby wrecking Europe, or "integrate Germany into Europe through an enlarged Schuman Plan, thereby preempting national decisions at the European level."[86] As a first step Monnet instituted the setup of a committee under Prime Minister René Pleven, a trusted deputy who had served as his personal assistant back in the Polish loan negotiations of the late 1920s. The formulation of what became known as the Pleven Plan involved "an internal conspiracy within the Cabinet. Schuman, Pleven, René Mayer . . . and [certain representatives] of the Quai . . . worked with Monnet's group [from the French plan] in order to present [Defense Minister] Moch and [Finance Minister Petsche] with a proposal that they could support."[87] First presented to the Americans on 15 October 1950, it differed from theirs chiefly in being dovetailed into the Schuman Plan. Though Monnet had previously opposed bringing defense issues into the coal–steel negotiations, the French now proposed that the defense community come into effect only after the new heavy industry authority had begun operations.[88] The two were also supposed, at some point, to be brought together in a common institution, thus moving Europe from economic, through military, to ultimate political union.

Most military men hated the EDC proposal: Very few agreed that the political advantages of integration outweighed the organizational complexities of multinationalism, and almost none appreciated the potentially immense economic benefits of common budget-

[85] USNA RG466 D(50)1903/TS(50)69, Byroade to McCloy, 4 August 1950; also, Schwartz, "Occupation to Alliance," pp. 307-335, passim.
[86] JM AMI 4/4/3, Monnet to Schuman, 16 September 1950.
[87] Schwartz, "Occupation to Alliance," p. 379.
[88] USNA RG84 (500: Coal and Steel Pool), Bruce to Secretary of State, 12 September 1950.

ing, centralized procurement, and standardized manufacture of
armaments. Civilians were hardly more enthusiastic. The Germano-
phobic French minister of defense Jules Moch, though responsible
for negotiations with the NATO Allies, could barely conceal his contempt for the common army proposal. Pleven only narrowly won
the assent of the Assembly to proceed with diplomatic discussions.
The Germans had even less faith in the military merits of the EDC.
Its purposes were nonetheless political, not military,[89] and many of
them were eventually achieved: The EDC provided Adenauer with a
convenient reentry vehicle into Europe, saved the Schuman Plan,
expanded European political horizons, and later diverted attention
from the practical difficulties of coal–steel integration.

As a result of preliminary discussions conducted in Paris, Washington, and London as well as agreements formally concluded at the
19 December 1950 meeting of NATO foreign ministers held in
Brussels, the Pleven Plan, which had been officially announced on
24 October 1950, began to bear fruit very soon. SACEUR (Supreme
Allied Commander, Europe) blossomed almost overnight. The training of a new German army as well as its integration into EDC and
NATO required the organization of a powerful and permanent
European headquarters. The huge political obstacles in the way of
creating such a thing were removed at a stroke when General Eisenhower, whose personal prestige was unequaled, allowed himself to
be persuaded by the new secretary of defense, General Marshall –
his former commander and the single individual most respected by
him – to serve as supreme commander of the proposed new organization. A development of immense importance had occurred:
NATO had been instantaneously turned from a planning into an
operational organization.[90]

The so-called Spofford Compromise further reduced French opposition. Initially the French demanded that the Germans be restricted to national formations at the level of the battalion rather
than the division, which the United States favored as militarily more
effective, and insisted as well that the Germans be allowed to join
the EDC only after its command structures had been assembled
and put into place. To blur the operational and organizational dis-

[89] See HST, PSF/General Files, Box 132, NATO (folder 1), Eisenhower to Marshall, 3 August 1951.
[90] See Lawrence S. Kaplan, *A Community of Interests: NATO and the Military Assistance Program, 1948–1951* (Washington DC: Government Printing Office, for
the Office of the Secretary of Defense, Historical Office, 1980), pp. 129–130;
Schwartz, "Occupation to Alliance," pp. 372–373.

tinctions between the two kinds of military unit, the U.S. special representative to the so-called Deputies Council of the Defense Committee, Charles M. Spofford, proposed building the new army around hybrid regimental combat teams of varied national and functional composition. Adding the proviso that German units not be allowed to exceed 20 percent of total manpower, the French agreed to drop their opposition to the simultaneous rearmament of the Germans and reorganization of the EDC. This was a second unspectacular but important breakthrough.[91]

Though Adenauer did not get formal Allied agreement to the request first made in July 1950 that the phase-out of the occupation be put on a contractual basis, he now had reason to be confident that the United States would eliminate it step by step, in the measure that Germany rearmed, or at least agreed to be rearmed, so long as progress was made on democratization.[92] Nor could Monnet have been disappointed at the results of the first round of EDC negotiations: He had secured an increased American commitment to French security policy.[93]

American distrust of the Germans went deep, and the planned rearmament of the former enemy was not a reassuring thought. On 15 December 1950 Undersecretary of State Lovett informed Acheson that Eisenhower had told him that the Germans should "get over the feeling that they were in command of a trading position," adding that he agreed that "we should ease up saying how very important they were."[94] Acheson concurred. He thought that the German situation should be put "on ice" and promised to tell Adenauer only "in general terms" about arrangements concluded at the Brussels meeting. Even while the United States orchestrated a restoration, it welcomed decartelization to remind the Germans who had won, and who had lost, the war.

There would be other reminders as well. German hopes were soon dashed that Acheson's announcement of an American commitment to raising Germany's defense contribution would bring an end to the hated occupation clearly into sight. At a 20 November 1950 meeting held at the French Plan to consider the "harmonization of the Schuman Plan and the controls presently exercised in Germany,"

[91] Kaplan, *Community of Interests*, p. 131.
[92] Ibid.; also, HST, Acheson Papers, Memorandums of conversations, Acheson to Truman, 5 January 1951.
[93] Kaplan, *Community of Interests*, p. 131.
[94] HST, Acheson Papers, Memorandums of conversations, Telephone conversation with Undersecretary Lovett, 15 December 1950.

Tomlinson and Bowie curtly informed Hallstein that he was laboring "under an illusion to think that the operation of the Schuman Plan would automatically entail an immediate end to [them]." Americans, one of them chimed in sarcastically, were not afraid to "take the plunge," but they preferred doing so wearing lifejackets.[95] Distancing himself from this kind of insulting remark, Monnet reassured the German delegate the following day that occupation controls would be lifted once the Schuman Plan had gone into effect. Hallstein would not have been mollified had he been aware of what Monnet had in mind. The bashing of the Ruhr had just begun.

5.3   The bashing of the Ruhr

To prevent a revived Germany from overwhelming Europe economically and politically Monnet and McCloy would try to stun Adenauer into agreeing to prolong the controls imposed on Ruhr heavy industry after the war. Decartelization would thus continue, or rather be made effective for the first time, while responsibility for imposing and maintaining it would pass from the governments of occupation to the new High Authority. The bashing had gone on for three months and brought the coal–steel negotiations to the brink of collapse by the time Adenauer finally gave in and accepted the Franco-American terms for both decartelization and the detrustification provisions of the treaty. Yet Monnet did not win: The settlement bought at such a high price could not be maintained; the treaty was full of holes; and in defending itself against the Franco-Americans the Ruhr gained both the foreign and the domestic support needed in order to burst the bonds placed upon it.

Though their victim was too punchy to have realized it, Monnet's was the brain directing the avalanche of blows that descended on the Ruhr. He first reopened the cartel discussions in Paris by confronting the Germans with a harsh *Diktat* that would have repealed the gains they had managed to eke out over the month and a half since the Waldorf explosion, then moved the scene of action to the Petersberg where discussions surrounding Law 27 were just getting

---

[95] AN 81 AJ148, "Compte rendu de la conversation ayant lieu au Commissariat au Plan le 20 November 1950 au sujet de l'harmonisation du Plan Schuman et des contrôles actuellement éxercés en Allemagne," 21 November 1950; USNA RG84 (500: Coal and Steel Pool), Bruce to Secretary of State/1531, 21 November 1950.

under way. By Christmas the fight strategy was set: The Americans would work the body, landing flurry after flurry while Monnet, feinting and occasionally jabbing, readied the knock-out punch.

The strategy was all-important. "What we are doing today," Monnet wrote René Mayer on 2 February 1951, "will give us our last chance to transform . . . the structure of the Ruhr and establish supranational managerial authority over it."[96] Monnet knew that under Law 27 the German government had the ultimate authority to determine the property settlement and could, by nationalizing basic industry, have escaped detrustification altogether. France's only hope, he wrote, was to force the Germans to enact a law consistent with international agreements for deconcentration and decartelization as required by the United States. Notwithstanding this pessimism, Monnet thought it possible to restructure the Ruhr so as to prevent "any single person or interest from occupying or acquiring a dominant position" in heavy industry.[97]

Having no intention of phasing out the occupation, Monnet sought a transfer of power from the Allies to the new coal and steel community. On this point he was inflexible, setting down his terms in a draft letter to Hallstein dated 5 December 1950, which remained on his desk for two months until the opportune moment had come to send it. France, said Monnet, would wind down the Ruhr Authority as soon as the pool proved able to take over responsibilities from it and to this end would urge the Allies to revise the relevant portions of the 28 April 1949 London accords. The letter added that his nation would also recognize that the steel control groups could be disbanded once the planned reorganizations had taken effect and the necessary power to enforce them been vested with the High Authority, adding that the French government would consider limits on production superfluous once "the Community is able to exercise its functions and treat the entirety of the production of the member nations as a single unit."[98] After this condition was met, finally, France would propose that the high commissioners voluntarily relinquish the exercise of the Ruhr-related reserve powers retained under paragraph 2g of the occupation statute, thus dissolving the military security boards, whose power to veto investments would pass to the High Authority along with occupation controls in the Ruhr.

[96] AN 81 AJ148, Monnet to Mayer, 2 February 1951.   [97] Ibid.
[98] JM AMG 10/2/13, "Project, 5.12.50 [Texte remis à titre personnel de M. Monnet à M. Hallstein] le 1/2/51."

The Paris cartel discussions resumed in the first week of December 1950, when the French submitted a new proposal for Articles 60 and 61 that was in fact all but identical to the one of October.[99] The Germans, who thought the matter had already been settled, were outraged. Their reaction was understandable. The cartel question had come up in connection with the issue of regional groups, and at that time the Germans had secured the support of the other negotiating parties for their position. They further maintained that all the non-French delegations had rejected Monnet's nearly identical proposal of October even as a basis for discussion. Erhard even claimed general agreement for basing Articles 60 and 61 on a proposed German law specifically sanctioning syndicates that improved efficiencies in organization and production. The West German Ministry of Economics hoped that the consensus on cartels would make it possible to "turn Paris against Petersberg."[100]

What really upset the Germans was that the anticartel and anticoncentration provisions were directed unilaterally against them, the Ruhr mines being the only ones affected because those of the remaining Six (excluding Belgium, which was treated separately) were publicly owned. Under Monnet's plan the selling syndicate originally set up by the British, the Deutschen Kohlen Verkauf (DKV) – in reality the wizened old coal cartel in heavy makeup – would have to disappear and be replaced by sixty-five new merchandisers. How, one official asked sarcastically, was a sector so loosely organized supposed to compete with the nationalized industries of England, Poland, France, and the Netherlands? He further objected that the stipulation requiring the separation of sales agencies from mines was meaningless in the case of the state-owned industries of the other member nations, since both production and distribution would remain in the same hands.[101]

Though the French proposal for Articles 60 and 61 did not directly affect the organization of the steel industry or the nature of the coal–steel tie, both of which came under Law 27, the power they assigned the High Authority to approve or disapprove fusions was clearly aimed at preventing reconcentration, and this too was discriminatory in the German view. Thus the circumspect Henle expressed "surprise" to Bowie on 13 December 1950 that the vertical

[99] JM AMG 17/8/82, "Dispositions proposées rélatives aux cartels et aux concentrations industrielles," December 1950.

[100] AA Schuman Plan/Bd. 737, Erhard to Adenauer ("Stellungnahme zum Schuman-Plan"), 11 December 1950.

[101] BA B146/265, "Sitzung des interministeriellen Ausschusses am 7.12.1950."

integration of Ruhr industry had aroused so many objections. He pointed out that "in the USA combinations of coal and steel . . . are not unusual [and in] the case of France [there was] a particularly close combination of ore and steel as well as a close relationship to the German mining industry." Why, he wondered, was "sauce for the goose not sauce for the gander?"[102]

By this time ready to walk out of the negotiations, Erhard recommended on 12 December 1950 that the treaty not be concluded until the key economic issues were resolved. The discriminatory provisions proposed for Articles 60 and 61 were in his judgment impossible to reconcile with the equal treatment promised to the negotiating parties. He insisted that the Ruhr Authority would thus have to be dissolved and the new community assume responsibilities for coal and steel allocation as well as for enforcing nondiscrimination. Erhard thought the limits on production ridiculous and obsolete. Though having been lifted to permit Germany to make a defense contribution, they were, he noted sarcastically, once again actually being outstripped. The veto power over investments held by the military security boards was in his view merely an obstacle to expansion that prevented the construction of the continuous wide-strip rolling mill needed for modernization. The draft treaty's proposed anticartel and anticoncentration articles merely added to these inequities, he asserted: The time for downplaying German dissatisfaction was therefore over, and special negotiations should begin for correcting the situation.[103]

The business leaders were also growing more restive because the rearmament-induced demand for steel was creating terrific money-making opportunities that threatened to elude them. Ruhr steel output, Sohl wrote Lehr at the end of November 1950, was up to a rate of 13.2 million tons and mills were operating at capacity. Though a small amount of new investment would be needed to raise outputs, the necessary plant could be constructed from the remains of the dismantled works and put quickly into operation. An impatient man, Sohl thought the time had come to call the bluff of the Allies and the "Schuman people" by giving them a chance to demonstrate the seriousness of their proclaimed intent to raise production and lower costs. He proposed that as a "test case" the high commissioners be requested to authorize the huge new mill on the

---

[102] USNA RG466 D(50)2742-A, Henle to Bowie, 13 December 1950.
[103] BA B146/265, Erhard to Adenauer ("Stellungnahme zum Schuman-Plan"), 11 December 1950; BA B146/265, "Stellungnahme des Bundesministers für Wirtschaft zum Abschluss des Schuman-Planes," 12 December 1950.

drawing board at August Thyssen Hütte. We could then, he told his friend, draw the appropriate conclusions about our further participation in the Schuman Plan.[104]

On 22 December 1950 Adenauer intimated to Monnet that to head off demands for co-determination he had decided to nationalize a portion of heavy industry. The chancellor assured him that he had in fact already secured the agreement of labor and capital to the setup of twenty-four steel companies, of which the government would hold a 35 percent share in a deal based upon a plan drafted jointly by the coal and steel trusteeships. For Adenauer such an arrangement would have been the last line of defense. Never an enthusiast for "mixed" ventures, the chancellor retained close personal and political ties to private industry. To nationalize, even partially, in order to prevent co-determination, was to put the cart before the horse. Monnet knew, however, that even halfway socialization would render detrustification irrelevant legally and make it impossible to impose politically. He understood all too clearly the message that Adenauer was sending: Press us too hard, and the match will be over.[105]

It is difficult to determine precisely when Monnet decided to place the fate of the Schuman Plan in the hands of the U.S. high commissioner. Fearing, as an American cable put it, that the Germans would "insist on linking Schuman Plan to German defense contribution," Monnet moved on 17 December 1950 to adjourn the meeting of the heads of delegations even though agreement on the cartel provisions had still not been reached.[106] On the eighteenth of the month Ambassador Bruce in Paris informed Bowie that French industry along with a portion of the cabinet was undermining Monnet's position and would continue to do so unless action were taken at once against the Ruhr under Law 27.[107] This warning sounds suspiciously like a pretext for bringing well-laid plans into operation.

The following day France and the United States – that is, Monnet and his American friends – quickly reached a well-reported "fundamental agreement" on decartelization. It had four points: Law 27 was understood as not only requiring deconcentration but forbid-

[104] BA Nachlass Lehr, 20 Vermerk (Sohl), "Deutsche Bedingungen für die Unterzeichnung des Schuman-Plans," 24 November 1950.
[105] JM AMG 10/3/7, "Aperçus sur le situation économique," 22 December 1950.
[106] USNA RG84 (500: Coal and Steel Pool), U.S. Embassy, Paris, to Secretary of State/1157, 18 December 1950.
[107] USNA RG84 (500: Coal and Steel Pool), American Embassy, Paris, to Secretary of State/1158, 18 December 1950.

ding reconcentration by financial manipulation; deconcentration was to proceed on the basis of the high commissioner's plans calling for the further dissolution of certain trusts; the centralized coal sales agency (DKV) was held to be incompatible with the Schuman Plan and therefore would have to be phased out; and coal–steel ties were to be allowed in only a limited number of cases. Though events would later prove that this deal masked rather than eliminated certain basic differences in French and American approaches to industrial reorganization, the U.S. high commissioner had nonetheless assumed responsibility for saving the Schuman Plan on behalf of Monnet.[108] "McCloy and Bowie," he wrote Schuman, "are going to take the initiative in discussions with the Germans. It is therefore important that our representatives receive formal instructions to confirm the line of action that I am going to lay down for them."[109]

The fate of the Schuman Plan was now firmly pinned to the outcome of the struggle over Law 27. On 22 December 1950 Monnet wrote to Schuman, "From the political point of view it is essential to prevent the Ruhr from again dominating German policy. From the economic point of view, our industries will fight at a disadvantage if the trusts are able to reestablish things as they were."[110] "Without the reorganization of the coal and steel industries of the Ruhr," he wrote in a follow-up letter to the foreign minister, "the Schuman Plan will not work because it will . . . consecrate the dominant position in Europe as a whole of those powers and resources concentrated in only one region of it."[111]

To defend themselves against the Franco-Americans, the Germans backpedaled and counterpunched, absorbed the easy blow, held in the clinch, pushed, shoved, and even head-butted, hoping to wear down and exhaust their stronger, more skillful, but less determined opponent by the midrounds of what they soon learned would be a long fight. Adenauer decided not to get involved in a war of words over Articles 60 and 61 of the draft treaty, even though this meant accepting the breakup of the DKV; he recognized that on the terrain of ideology the Germans would continue to lose until Americans had learned to love cartels. The chancellor rather than Monnet actually first suggested shifting the decartelization battle from Paris

[108] AN 81 AJ137, "Memorandum of Meeting Held 19 December (1950) on Coal and Steel Problems."
[109] AN 81 AJ137/JM AMG 10/3/11, Monnet to Schuman, 22 December 1950.
[110] Ibid. [111] JM AMG 20/5/5, Monnet to Schuman, 22 December 1950.

to Petersberg. Adenauer's goal was not, however, to devise a workable plan for taking apart the organizational structures of Ruhr coal and steel but to tie up the issue in technicalities so as to gain time. He thus also insisted shortly before Christmas upon breaking off discussions until the beginning of the new year, let the negotiations drag on through January, and finally brought them to collapse by the end of the month.[112]

The French were extremely suspicious of German maneuvering from behind the scenes. According to a 2 January 1951 report from their representative on the Control Group for steel, Armand Bureau, the decartelization proposal made by the federal government allowed for the restoration of Klöckner and Gutehoffnungshütte and the augmentation of Mannesmann as well as the main units of the Vereinigte Stahlwerke in the Duisburg region. The German proposal also permitted each of these firms to maintain tied-in mines. "If the juridical and fiscal systems are not changed," Bureau added, "the spirit of corporative organization and the [traditional] commercial practices [will remain] and the dismantlement of factories will have no effect. The reconstitution of the trusts will be but a matter of time and the deconcentration of the Ruhr prove to have been an illusion."[113] Bureau also warned that if the Federal Republic succeeded in restoring the dismantled factories of ATH, the Bochumer Verein, and Hattingen and joining them to the successor firms of the Rheinische Röhrenwerke in Witten and Stahlwerke Bochum the remaining unit companies would soon be forced to merge back into the traditionally dominant firms.

On 8 January 1951 a meeting to hammer out the details of the German negotiating strategy was held at the office of the chancellor. Hallstein emphasized that there was no need to arrive at a "perfect" settlement under Law 27 because the "equal-treatment" premise of the Schuman Plan would make it difficult for the High Authority to impose discriminatory policies in the future. Pointing out that the fate of the coal–steel pool depended on the negotiations for the decartelization law, the head of the German delegation bade the members of his team to conduct themselves in such a way as to "shift the blame for the non-fulfillment of the Schuman Plan to the other side." The negotiations should not, however, in any case be allowed to fail because, as a Herr Kattenstroth added,

[112] JM AMG 11/1/9, Monnet to Hallstein, 22 January 1951.
[113] AN 81 AJ127, Bureau to Monnet, 2 January 1951; also AN 81 AJ137, "Entretien entre le Groupe de Contrôle de l'Acier et les Experts du Gouvernement Fédéral . . . ," 16 January 1951.

"Mr. Bowie will still have Law 27 available to him."[114] Hallstein recommended that, rather than state publicly the conditions that
would be unsatisfactory to Germany, Mr. Bowie should be let known
unofficially that by pressing too hard he could jeopardize a compromise settlement.

As for the specifics of decartelization, the Germans were not worried by the "25 percent clause," which limited the amount of coke
that each steel firm could control, because they expected that it
could be watered down. Coal was a more difficult matter, but the
delegate responsible, Walter Bauer, hoped to stretch out the decartelization period to five years and strongly emphasized the undesir-
ability of a commitment to a fixed phase-out schedule. Any attempt
to rescue DKV by revising Articles 60 and 61 was judged to be futile
by everyone present. No mention whatsoever was made of steel reorganization during the discussion, which suggests that Bureau's
suspicions that the Germans took it lightly were well founded.[115]

The Ruhr found support for its position from the other parties
and interests involved in the Paris negotiations. On 18 December
1950 U.S. Ambassador Bruce cabled from the Paris embassy that
"Monnet is having great difficulty in obtaining French government support on [the] anti-cartel provisions of [the] Schuman Plan
Treaty due to increasingly strong opposition from both industrial
circles and certain members of the cabinet. While the fight on these
provisions is led by the steel cartel, it has the support of other
highly cartelized industries which fear that such provisions might
set a precedent [for them]." French attitudes were no secret to the
Germans.[116] On 28 November a staff attorney told his colleagues at
the Ministry of Economics that "the French steel industry has no
influence whatsoever on the negotiations and is not even kept informed by the office of the [French] Plan. One hears talk that [the
steel industries] mean to have Monnet sacked, but are afraid of
socialization like that occurring recently in England."[117] On 1 January 1951 a "persistent rumor" circulating among the delegations
reached the Paris embassy that Tomlinson and Bowie were jeopar-
dizing the Schuman Plan by being too tough on the Germans. Am-

[114] AA Schuman Plan Bd. 84/159, "Niederschrift über die Sitzung im Haus des
Bundeskanzlers am 8. January 1951."
[115] Ibid.
[116] USNA RG84 (500: Coal and Steel Pool), American Embassy, Paris, to Secretary
of State/1158, 18 December 1950.
[117] BA B146/265, "Schuman–Plan, Kartelle und Konzerne, Preise und Ausgleichs-
kassen, Regionalen Gruppen" (28 November 1950 meeting of BWM).

bassador Bruce had to cable Washington to request the issuance of
an official denial that Americans on the spot had exceeded their instructions and to warn that any attenuation of the statement's wording could have fatal consequences for the negotiations.[118]

French steel did its part in turning what had been a serious dispute into a major crisis, having already managed to arouse fears in
the cabinet and technical ministries that the lifting of the PLI (Permitted Level of Industries) Agreement would result in an explosive
expansion of German steel capacity, thereby curtailing the French
coke supply. The *maîtres de forges* also demanded their own version of
the special deals secured by the Belgian mines and the Italian steel
industry. This was troublesome to Monnet, who on 8 January 1951
tried to persuade the skeptical Americans that to win public support
for German rearmament France would need guaranteed access to
Ruhr coal. He acted less to prevent shortages than to control the
coal supply, a week later informing Schuman that if the DKV were
left intact the French steel syndicate would become the sole importer of German combustible and the Charbonnages de France would
be delivered to its tender mercies. To prevent this development, he
would soon try to make his own deal with the DKV.[119]

Though by mid-January 1951 the opposition had begun to wear
down Monnet, it seemed at the time to have come too late to change
the outcome of the negotiations. On the seventeenth of the month
McCloy reported that Bowie's talks with Erhard were developing
satisfactorily: Both the economics minister and Hallstein had agreed
that DKV would have to accept a transitional period lasting until 1
October 1952 – far longer than necessary in the U.S. view but far
shorter than what the Germans had had in mind. The United
States, he said, had for its part agreed to persuade the French to be
"flexible" in applying Law 27. The U.S. high commissioner thought
that unless the Germans tried to obtain concessions "through the
back door" agreement was in sight.[120] McCloy did not know it, but
they had already caught Monnet by surprise with a hard and punishing head butt. On 15 January 1957 Hallstein told him bluntly
that the Federal Republic could not accept the treaty as written and
would sign only under protest that the clauses relating to deconcen-

tration and decartelization had been imposed rather than freely negotiated.[121]

The German delegation chief can only have acted on orders from above. The move was too drastic to have been decided upon at the ministerial level, and such a bolt out of the blue appears never to have even been discussed as a possible tactic by the relevant working parties. Adenauer's effectiveness in the negotiations would have been at an end had he admitted responsibility for the initiative. The vagueness he often displayed about the coal–steel talks was, however, too studied to have been completely genuine. On 17 January 1951 the French High Commission's director of economic affairs reported that Adenauer seemed barely in touch with the relevant events in Paris and Bonn. Yet the notes from their discussion indicate that the chancellor managed to lure the Frenchman into a complicated denial of a Franco-American division over the dissolution of the DKV, while simultaneously leaving him with the impression that he was ready to negotiate separately on the matter with France.[122] The only real hint that Adenauer was indeed the man behind Hallstein's threat was provided by an interview that the chancellor's spokesman, Pferdmenges, gave to the French high commissioner, in which the Cologne banker vaguely related the hardball tactic to Adenauer's worries about the Russians. Regardless of how the idea originated, by threatening to sign the Paris treaty under protest the Germans forced the negotiations into the final stage.[123]

Monnet understood immediately that Hallstein's few words could bring negotiations crashing down. "If [the Germans]," he told Bowie over the telephone, "do not want to agree to a settlement which is reasonable – which you think is reasonable – and [they] are willing to accept, that fits in with the Schuman Plan, then I think it [best] to know that and jut blow off the whole thing."[124] But Monnet was not yet ready to throw in the towel and told Bowie to get tougher, while in the meantime protecting himself by reaching a special deal with the DKV that would undercut the position of his domestic enemies during the critical phase of the negotiations.

[121] AN 81 AJ137, Extracts of a telephone conversation between Monnet and Bowie, 15 January 1951.

[122] AN 81 AJ137, Note (Leroy-Beaulieu), 17 January 1951.

[123] AN 81 AJ137, "Note pour M. L'Ambassadeur de France; Haut Commissaire de la République en Allemagne; Entretien avec M. Pferdmenges" (n.d.).

[124] AN 81 AJ137, Extracts of a telephone conversation between Monnet and Bowie, 15 January 1951.

The decision to enter discussions with the DKV must have been acutely painful for Monnet, but, like many others who had tried long and hard to undermine the power of the Ruhr coal cartel, he expected to make only temporary concessions. His German interlocutors must have been reminded of the essential truth of the French adage to the effect that the more things change the more they remain the same. Monnet's initiative, for which he claimed U.S. authorization, brought him face to face with the man who had directed the coal syndicate's exports in the 1930s, Ernst Russel. The French planning commissioner was impressed by the gentleman's "remarkably competent and clear exposition" but told him confidently that once the treaty had been ratified organizations like his would belong to the past. For the present, however, Monnet admitted that he was ready to deal in order to reach "an equitable international allocation . . . assuring that in the period of transition [to the Coal and Steel Community] the different steel industries would be treated equally until the creation of the common market and guaranteeing, notably, a favorable reference point for French producers . . . This [agreement] will also establish a base of production for a month or a quarter for German steel." Negotiations for the coal deal proceeded over the following weeks.[125]

On 29 January 1951 Monnet sent Hallstein an ultimatum in the form of a personal letter. The Germans, he said, were now in control of the conference and therefore responsible for using their influence on the other delegations to bring it to a successful conclusion. The Frenchman declared that France had been generous in giving the Federal Republic enough time to work out a mutually satisfactory settlement, admitted that the absence of one was due in part to his lack of support at home, and hastened to add that the United States was more determined than ever that the issue be settled on his terms. Complaining that instead of agreeing to a plan to implement the dissolution of the DKV West Germany had threatened to sign the treaty under protest, Monnet warned Hallstein that if the Federal Republic did not immediately drop its opposition to his plans he would advise the governments not to sign the treaty; the Schuman Plan and the very idea of supranationalism in Europe would then be dead.[126]

---

[125] JM AMG 20/5/13, Monnet to Hallstein, 10 January 1951; also, USNA RG84 (500: Coal and Steel Pool); also, U.S. Embassy, Paris, to Secretary of State/342, 8 January 1951.
[126] JM AMG 11/1/9, Monnet to Hallstein, 29 January 1951.

In February it nearly did die. Monnet's ultimatum had no effect. Instead, opposition that had previously been muffled became strident, and rumbles from behind the scene soon mounted to a dull roar. The Frenchman had been overoptimistic in thinking that the German delegation could assert control over the Paris proceedings: It risked being repudiated by Ruhr industry. In neither Paris nor Bonn was anyone in control of the coal–steel negotiations. It fell to McCloy to conclude them.

Ruhr dissatisfaction with the placatory approach followed thus far was mounting. The normally even-tempered Henle, one of Adenauer's two main links to the Ruhr, wrote Pferdmenges on 20 February 1951 that his patience was at an end. He found it "nightmarish" that an issue as important for Europe as the Schuman Plan could get bound up with unfair demands on Germany and galling that decartelization clouded Ruhr heavy industry's otherwise favorable prospects. Henle also warned Pferdmenges of the danger of men like Walter Bauer, who in his view had conceded too readily the dismemberment of DKV, and of the statistical tricks that Bowie and his team would play in an effort to wreck the Ruhr trusts. In their work Henle detected the "last whiffs . . . of the Morgenthau poison."[127] The Klöckner chief nonetheless found some consolation in having averted the worst: Five years of suffering were nothing compared to what might have happened had the Americans not been staved off. For Henle times had definitely changed. When Mr. Bowie reproaches us for having growing appetites, he told Pferdmenges, "I am going to tell him that as Adenauer says we no longer use the 1945 calendar."[128]

The mistrust was mutual. Bowie's negotiations with the Ministry of Economics over Law 27 soon got hopelessly bogged down in disagreement about such technical issues as calculating coal consumption, determining requirements for the transitional period, and measuring industrial concentration. For this he blamed men like Henle. Bowie reported to Monnet on 22 February that "the compromises set in motion by Mr. Erhard and the Ministry of Economics have been blocked by the industrialists. When it is a matter of applying general principles to specific cases, the German experts demonstrate ill-will and use [estimates] of industrial capacity that do

[127] KA/Henle Umgestaltung des deutschen Kohlenbergbaus und der deutschen Eisen- und Stahlindustrie, 1.10.50–31.12.50, Henle to Pferdmenges, 20 February 1951.
[128] Ibid.

not come even close to corresponding to reality . . . The discussions are a dead letter."[129]

Negotiations for the special French coal deal, which the Germans had initially regarded with enthusiasm, also got nowhere. On 15 February 1951 Hirsch met with the German representative von Dewall at the rue de Martignac. There was disagreement about reference periods as well as confusion about the role of the International Authority for the Ruhr (IAR) and the relationship between it and the future High Authority. In negotiations two days later the Germans refused to discuss any special arrangement pending the settlement of broader issues. They wanted an explicit promise from the French government to abolish the Ruhr Authority after the initialing of the Paris treaty as well as additional assurances, which Hirsch refused to give, that the control groups would disappear at the same time. Von Dewall further proposed that rather than break up the DKV the High Authority should create a special office to supervise it, a suggestion Monnet rejected out of hand. Soon thereafter Hallstein ordered an end to bilateral coal discussions.[130]

The Americans also blamed the Germans for causing the breakdown of the Paris negotiations, overlooking the responsibility of the other delegations. Starting in mid-February the Germans tried to reopen the issue of Articles 60 and 61, using, according to an American account, the "unwitting Dutch as their tool." The Belgians supported the German effort quite openly, as did the steel industries of each of The Six.[131] McCloy told Monnet over the telephone on 19 February 1951 that at his most recent meeting with Ruhr industrialists he could not be firm because they had confidently assured him that their French colleagues were happily making their own coal deals.[132] U.S. Ambassador Bruce thus reported on 21 February that Monnet was trying desperately to avoid further discussion of antitrust matters in Paris and "struggling to close the conference before additional ground is lost." "The only real bargaining counter

---

[129] AN 81 AJ137, "Entretien avec M. McCloy au sujet du Plan Schuman," 22 February 1951; see also USNA RG466 D(51)269A, Adenauer to McCloy, 28 February 1951.

[130] BA B109/97, "Kurzprotokoll über Besprechung mit Herrn Hirsch im Planungsamt," 15 February 1951; also USNA RG84 (500: Coal and Steel Pool), U.S. Embassy, Paris, to Secretary of State/149, 2 February 1951, and U.S. Embassy, Paris to Secretary of State/476, 8 March 1951.

[131] USNA RG84 (500: Coal and Steel Pool), U.S. Embassy, Paris, to Secretary of State/1121, 21 February 1951.

[132] AN 81 AJ137, "Conversation téléphonique entre M. Monnet et M. McCloy," 19 February 1951.

the French now have," he concluded, "is the threat of Monnet's resignation from the conference ... Ruhr industrialists are definitely of the opinion that if they succeed in postponing decisions they will inevitably gain complete elimination of all restrictions and achieve preferred positions that they seek. Events are already demonstrating this."[133]

To prevent the collapse of the Schuman Plan, McCloy dictated the terms of settlement under Law 27. In one respect it was easy to do so: The Germans had conceded the main points early on – the existence of antitrust provisions in the draft treaty, the dissolution of the coal cartel, the limitation of the coal–steel tie, and the deconcentration of steel production units. The sticking points were of a technical nature, such as defining what constituted 75 percent of a particular firm's coal supply. Behind these semantic disputes, however, loomed the larger issue of how long, and under what guise, the occupation would last. All parties understood that because of the new American security policy its days were numbered and that the Germans could only gain by prolonging the coal–steel talks. Thus for two months, Bruce protested, a "general understanding with the Germans has supposedly been just on the point of achievement ... but [they] have managed to keep questions opened to their own advantage by swinging back and forth between conversations in Germany on deconcentration and negotiations here [in Paris] on the Schuman Plan."[134]

McCloy had to end this situation in order to protect his remaining authority. The failure of the Schuman Plan, he said on 22 February, would set back the clock by two years, make it impossible to revise agreements limiting production, hopelessly compromise the European Defense Community, and, in general, undermine *Westintegration*. He even went so far as to speculate on the possible merits on an "Austrian solution" for Germany – Four Power rule and neutralization.[135] McCloy had begun to pressure for a settlement back on 12 February 1951, by tossing Adenauer a bone in the form of an agreement to allow three previously unauthorized combinations (the merger of Dortmund Union with Hörde and the amalgamation of

[133] USNA RG84 (500: Coal and Steel Pool), U.S. Embassy, Paris, to Secretary of State/1121, 21 February 1951; also in FRUS 1951/IV 862.19 Ruhr 2-2151, Ambassador in France (Bruce) to Secretary of State, 21 February 1951, pp. 93–94.
[134] Ibid.
[135] AN 81 AJ137, "Entretien avec M. McCloy au sujet du plan Schuman" (François-Poncet), 22 February 1951.

the Klöckner works, as well as of Mülheim Meiderich, with the Rheinische Röhrenwerke) while at the same time stating unequivocally that "any effort to press for further concessions will not only fail to obtain them but will also jeopardize the proposed compromises."[136] When this ultimatum went unanswered, Monnet and McCloy concluded that Ruhr industry was the source of the problem. They would be brought in for the next round.

On 3 March 1951, less than twenty-four hours after receiving a letter from Adenauer that to McCloy's mind represented no advance from his earlier position, the U.S. high commissioner summoned the chancellor to his offices for a dressing-down. The American reminded the chancellor that the Federal Republic had delayed the conclusion of the treaty for two months and claimed that France and the United States were left with no alternative but to impose their own decartelization scheme. On the following day he presented it to Adenauer and on 5 March, with the concurrence of the chancellor, confronted the leaders of heavy industry with it. Their frenzied protests were in vain. Now only the coal and steel producers, supported by cartellists from the rest of The Six, stood in the way of the treaty, according to a Monnet memorandum of 8 March. Five days later Adenauer accepted the settlement.[137] The treaty was ready for initialing. On 20 March 1951 the Paris conference ended, Bruce reported, in the same "team spirit" in which it had begun and to the accompaniment of the Belgian delegation singing in chorus a hymn to Europe of their own composition.[138] On 18 April 1951 the foreign ministers of The Six committed their pens to parchment. The conference was over.

The treaty could not take effect, however, until detrustification got under way. Recognizing, as Monnet put it, that the "monolothic structure of the Ruhr" had given it "artificial advantages," the treaty was meant to redress the imbalances of nature and organization between coal-rich Germany and the remaining six fuel-importing nations. Its terms were simply those to which the Franco-Americans had forced the Germans to subscribe earlier. The DKV was to be

[136] USNA RG466 D(51)19D/B, McCloy to Adenauer, 12 February 1951.
[137] JM AMG 13/27/10, "Memorandum sur la déconcentration de la Ruhr et la conclusion des négotiations sur le Plan Schuman," 8 March 1951; BA Nachlass Lehr, "Besprechung mit Mr. McCloy am 5.3.51"; AN 81 AJ142 Adenauer to President of the Allied High Commission, 14 March 1951; also, FRUS 1951/IV, U.S. High Commissioner in Germany (McCloy) to Secretary of State," 3 March 1951, pp. 97–98.
[138] USNA RG466 D(51)353, U.S. Embassy, Paris, to Department of State, 20 March 1951.

taken apart by 1 October 1952 and during the period of transition to be subject to Allied authority. Twenty-seven steel companies were to be created, none of them larger than those in the rest of The Six or with capacities above 2.2 million annual tons. Only eleven steel enterprises would be allowed to retain mines, none of which could provide more than 75 percent of a mill's consumption requirements, as calculated on the basis of the American formula. This would reduce steel's control of the mines, according to French estimates, from the prewar 56 percent to 15 percent.[139]

Though Ruhr producers rejected this settlement outright and demanded its revision, Adenauer offered them no guarantees of support. To him the Treaty of Paris was a way station on the road to sovereign status, a slightly leaky but still welcome shelter in which to rest and regain energy before moving on. The treaty did not end the occupation. The International Authority for the Ruhr was to remain in being until the High Authority was in a position to take over its functions, a somewhat farcical stipulation since it had never been able to exercise any of them effectively. When in the first quarter of 1950 it was finally able to set export levels, coal surpluses made this unnecessary. IAR's other statutory powers remained unused. Nor did the treaty abolish the PLI (Permitted Level of Industries Agreement), which limited outputs to 11.1 million tons, though the French consented to their being exceeded temporarily for security reasons. Both the military security boards, which could veto new investment, and the control groups, which directed detrustification, also were to remain in operation until receiving adequate assurance of the High Authority's ability to guarantee the terms of the settlement. Finally, decartelization was to continue as before.[140]

The "Treaty Instituting the European Coal and Steel Community" signed in April 1951 by the foreign ministers of The Six consisted of an even 100 articles to which was attached a lengthy appendix describing the transitional arrangements for Belgian coal and Italian steel as well as others that were to remain in effect for a five-year period following the opening of the common market. To be binding for fifty years, the treaty described the basic structures of the new organization, defined its rules, and was intended to serve as

[139] JM AMG 13/27/10, "Memorandum sur la déconcentration de la Ruhr et la conclusion des négociations sur le Plan Schuman," 8 March 1951.
[140] See BA B102/60686, "Durchführung des Gesetzes Nr. 27 . . . ," 27 March 1951; USNA RG84 (500: Coal and Steel Pool), U.S. Embassy, Paris, to Secretary of State/1347, 26 February 1951; USNA RG466 D(51)216 HICOM to U.S. Embassy, Paris, 26 February 1951.

a kind of constitution. Governing in this way – through a written document containing an implicit economic philosophy applied supranationally – was something untried and fraught with difficulty: The new pool did not operate within a well-established legal framework; the document itself reflected rather than resolved the conflicts that had arisen between Monnet and the producers during the negotiations; and it also provided only sketchy guidelines for dealing with the practical problems of integration that would arise in the future.

The outlines of Monnet's original design were left intact in the treaty, with the High Authority remaining the exclusive source of executive power. Directing it was a college of nine members, one of whom was to serve as president. Voting procedures were not specified except that the majority principle was to be observed. The High Authority was made responsible for internal organization, though this subject was with one important exception not otherwise mentioned. The exception was for an Advisory Committee to be composed of between thirty and fifty-one members representing by thirds the interests of producers, labor, and consumers and dealers. Appointment was to be the responsibility of the national producer associations. As indicated by its name, the committee's powers were to be nonbinding.

The other institutions of the community were defined as extensions of HA power rather than checks on it. The Council of Ministers, an idea the Dutch originally introduced into the negotiations, was to be a purely advisory body and to meet only when convoked by the High Authority or by one of the member governments. The only responsibility specifically assigned the council in the treaty was to harmonize policy made by the executive with those of the member governments. The Common Assembly was merely to serve as a forum for discussion of proposals made by the High Authority. The powers of the High Court were also closely circumscribed: to try legal actions taken against the High Authority, initiate prosecution of encroachments suffered by it, and hear appeals from the national governments.

The economic section of the document was strongly influenced by antitrust ideas, with the first four articles establishing modernization as the purpose of the community: It was to create common markets, reduce prices, increase efficiencies, and eliminate all forms of restriction and discrimination. The High Authority could exercise numerous disciplinary and regulatory powers in pursuit of these principles: set price minimums and maximums for sales within the common market (and under certain conditions for exports as well), require the publication of prices, and penalize those who refused to do so. It could further impose punishment for falsification of data, rebating, and price gouging and, finally, subsidize mines forced to sell at a loss by the steel industry and take other necessary measures against monopsonistic buyers. The treaty also assigned the executive a broad array of production controls: to impose quotas for limiting excess production, allocate raw materials at times of scarcity, and fine and indemnify as necessary. The deconcentration and decartelization articles (65 and 66) were the lengthiest and most detailed in the entire document, enumerating a long list of varied circumstances under which the High Authority could intervene. Among the other provisions of the treaty, the extensive articles for labor and social policy – concerning wage discrimination, retraining, and readaption – would be relatively unimportant, as would those dealing with commercial policy. So too would the financial provisions enabling the High Authority to tax and borrow. The articles concerning transport would have an unsuspected significance in promoting integration.

The treaty's omissions were also significant. Regional groups were unmentioned, and thus the nature of the relationship between producer associations and the High Authority was left undetermined and open to dispute. Though the creation of the Advisory Committee was specifically called for in the treaty, its powers and procedures were not discussed. Nor does the document describe the internal organization of the High Authority, apart from indicating its responsibility for the matter. Finally, the treaty resolved none of the disagreements concerning economic policy that had vexed and imperiled the negotiations. Though intended to serve as a constitution, the Treaty of Paris in fact became little more than a cease-fire agreement. The initialing of it by the foreign ministers on 18 April 1951 would be a mere interlude in a continuing struggle over control of the coal–steel pool.

5.4 Confusion, conflict, and corporatism

While Monnet attempted to weaken the Ruhr, hostility to him drove French industry into its hands, the American attempt to raise the Federal Republic's defense contribution strengthened the position of the coal–steel producers, and the Schuman Plan set in motion a process of institutional reformation that restored the old Konzerne to prominence. The industry of the Federal Republic emerged from the negotiations having acquired many of the distinctive neocorporatist features that commentators often credit for having sustained high rates of growth, stimulated innovativeness, and established an impressive record of labor peace. One by-product of the coal–steel talks was the legal enactment of co-determination in heavy industry, which institutionalized a new tradition of labor–management cooperation. Another feature for which the Schuman Plan provided an essential part of the background has never found an appropriate name but grew out of what previously had been called "industrial self-determination" and actually meant business management of the economy. Whereas during Weimar the expression represented more a hope than an actuality and under the Third Reich its scope was restricted by the power of the state, in the Federal Republic it developed into the secret motor of economic policy – the thing derided by detractors as the *Verbändediktatur*. This "dictatorship of associations" was enshrined in the extraordinary investment aid law of 1952. The Schuman Plan also increased the prominence of Ruhr industry in Europe. The steel producers of France, Germany, and Bel-Lux slid back into old habits faster than they would have without it, and business diplomacy like that of the 1930s revived as well. After the conclusion of the official negotiations in Paris, moreover, ministerial bureaucrats took charge of the integration process, thereby changing the character of the coal–steel pool and contributing to the unification of Europe around a West German core.

The investment aid law of 1952 was no ordinary subsidy but a profound declaration of customer loyalty on the part of the German manufacturing community to Ruhr heavy industry. The Adenauer government endorsed and supplemented the arrangement. The package contained two large and lovely presents, direct payments of 1.16 billion Reichsmarks raised from contributions made by 130,510 different firms, and tax credits, provided by the government, worth another 3.21 billion Reichsmarks, all of which were destined for coal and steel and the affiliated electrical utilities of the Ruhr. The write-offs provided by the measure initially amounted to two-thirds of those taken for industry as a whole and ran for four years, diminishing in value over that period to about a third of the total. The investment aid law provided a private equivalent to the publicly sponsored heavy industry modernization program of the French Plan.[41]

[41] JM AMG 16/1/1, "Traité instituant la Communauté Européene du Charbon et de l'Acier."

The initiative behind it did not come from the Federal Government, which so long as Erhard ran the Ministry of Economics gave priority to the consumer industries. The United States, however, made both a positive and a negative contribution to its genesis. Rearmament-generated coal shortages, which would be endemic for the next few years, first appeared in October 1950. To prevent them from disrupting the defense buildup, the U.S. high commissioner not only insisted on maintaining price controls but on 31 October 1950 reintroduced fuel rationing. A senior official of the old syndicate, Dr. Martin Sogemaier, was called upon to serve as temporary commissar.[142] In the same month that McCloy forced Adenauer to agree to detrustification, he also pressured the chancellor into urging Erhard to accept an "action program" involving "significant modification of the free market system" as well as the introduction of prioritization and industrial rationing. In March 1951 McCloy further directed WVESI to present plans for expanding capacity to 16.5 million tons. The likelihood that the reregulation waves being beamed from the left side of the American brain would effectively nullify the deregulation impulses emanating from its right did not prevent the pursuance of these contradictory policies.[143]

The continuation of controls as demanded by the United States together with the thinness of German capital markets excluded financing basic industry expansion and modernization either internally or over the exchanges. At the same time, the support given by the United States to decartelization made it politically impossible either for heavy industry to raise capital on its own or for the government to provide it directly. Ruhr producers were not in fact involved in putting the investment aid package together; nor did they ever comment officially on it. The idea was conceived and carried to fulfillment by Fritz Berg, the president of the Bundesverband der Deutschen Industrie (BDI).[144]

The date of Berg's initial proposal, 2 April 1951, coincided with the conclusion of the coal–steel talks: German manufacturers were no more willing to sacrifice their basic industries or entrust their fortunes to a supranational authority than were the French. The aid package (approved on 18 April by representatives of no less than 450 different firms) was more than an extraordinary vote of confi-

[142] Heiner R. Adamsen, *Investitionshilfe für die Ruhr: Wiederaufbau, Verbände, und Soziale Marktwirtschaft, 1948–1952* (Wuppertal, 1981), p. 23.
[143] Ibid., pp. 143, 90.
[144] Ibid., p. 128–129; BA146/265, Fritz Berg to Chancellor Adenauer, 2 April 1951.

dence by the business community in the traditional industrial leadership; it was a statement "of the solidarity and resoluteness of
industry *vis-à-vis* the state and the unions" as well as of its assertion of
the right to a central role in the formulation of economic policy for
the nation.[145] The BDI was soon playing it, with Berg serving as
Adenauer's economics minister in everything but name. There was,
the leading historian of the investment aid law concludes, "nothing
unusual or unique about the measure; it is rather evidence that a
new 'style' [of doing business] had been discovered, one that made
superfluous purely fiscal approaches, or whatever else one chooses
to call such doctrinaire (*unorganische*) methods."[146] The *Ruhrhilfe*
package was a cornerstone of the new corporatism.

So too was co-determination, and it likewise strengthened the
Konzerne. When it was introduced by Dinkelbach's trusteeship organization into the steel industry after the war, the employers had successfully resisted its extension both to other industries and beyond
the plant. U.S. High Commissioner McCloy's decision of September
1949 to lift U.S. Military Governor Clay's earlier suspension of the
co-determination laws enacted by Hesse and Baden-Württemberg
reopened the issue. Discussions resumed in January 1950 in Hattenheim. Encouraged by the receptivity of the president of the employers' association, Dr. Raymond, the union federation again set out
optimistically to arrive at a new understanding. The employers soon
agreed to discuss co-determination in national organizations, "concede the right of co-determination at plant level," consider proposals for the establishment of a national labor tribunal, and establish
uniform policies on social insurance and for dismissal procedures.
The Hattenheim negotiations broke down on 30 March 1950 after
repeated employer reneging on promises to grant co-determination
at the plant level. Again having to leave the table empty-handed, the
union federation decided to let the Ministry of Labor take charge of
the next round of negotiations.[147]

Unlike the Social Democratic party, which officially endorsed Schumacher's unremitting hostility to the Schuman Plan, the Deutsche

---

[145] Adamsen, *Investitionshilfe*, p. 162.

[146] USNA RG466 D(50)11021, "History and Status of Co-determination in Labor,"
11 April 1950, and McCloy, 17 January 1950; USNA RG466 D(50)1023,
"Views of the Trade Unions Regarding the Reorganization of Basic Industries
in Germany," 28 March 1950.

[147] USNA RG84 (500: Coal and Steel Pool), HICOG dispatch 1063/1 June 1950,
"Trade Union Views of Schuman Proposals," and HICOG dispatch 1177/12
June 1950, "German Trade Unionists on Schuman Plan."

Gewerkschaftsbund (DGB) at first greeted it warmly.[148] The union federation had backed the Marshall Plan, was traditionally internationalist in outlook, and was intent upon strengthening ties to the AFL and CIO. In addition, the *Document de Travail* adopted the codetermination principle, assigning labor a one-third representation on the proposed advisory committees. Hans vom Hoff of the miners union was the unofficial chief of the labor element prominent in the German delegation to the Paris negotiations.

The publication of Regulations I, II, and III to Law 27 in September 1950 ended the unions' honeymoon with the French: Neither the federation nor its coal and steel affiliates were prepared to accept any attenuation of the bonds holding together the heavy industry complex of the Ruhr.[149] Their position on the issue was indistinguishable from management's. Nor did the union movement seriously contemplate socialization except as a means of evading occupation controls. The senior economic adviser to the French High Commission was appalled by "the reserve and the timidity . . . of the unions . . . [W]hile professing to be proponents of socializing the mines, they have agreed to accept the decision of Chancellor Adenauer, who certainly will oppose it." The Germans, he "candidly concluded, [are an] obedient people who obey the more readily when orders are passed down from high."[150] On 27 October 1950 Adenauer promised to secure co-determination for the unions in return for their support of the Schuman Plan.

To do this required political sacrifices. Industry adamantly opposed all concessions in the matter. On 11 October 1950 Sohl warned Lehr, who had just been appointed minister of the interior, that according to informants in the steel trusteeship the organization had come under the influence of socialists and unionists and was attempting to extend the co-determination principle throughout basic industry. The time had come, the industrialist from Thyssen told his friend the ex-mayor, for insisting to Erhard and Adenauer that the government block all undesirable appointments to the boards of the successor steel firms, as this would help nonradical board members stop further progress towards co-determination. To his note

[148] KA/Henle Umgestaltung des deutschen Kohlenbergbaus und der deutschen Eisen- und Stahlindustrie, 1.1.50–31.12.50, "Aktenvermerk über eine Besprechung mit dem Bundesvorstand des DGB am 20.11.50."
[149] AN 81 AJ138, "Extrait de la fiche [Leroy-Beaulieu] à M. l'Ambassadeur de France," 22 October 1950.
[150] Horst Thum, *Mitbestimmung in der Montanindustrie: Der Mythos vom Sieg der Gewerkschaften* (Stuttgart, 1982), pp. 49, 65–66.

Sohl thoughtfully appended a draft containing his ideas to the letter that he had written for Interior Minister Lehr to sign, seal, and deliver, as if by automatic transfer, to Minister of Economics Erhard.[51]

Adenauer was as good as his word. On 17 January 1951 he asked the industrialists to accept the fifty-fifty formula promised the unions, conceding that it should apply only to coal and steel. Negotiations between the BDI and the DGB, which began on the twenty-fifth, nonetheless broke down over the following week. Unable to secure voluntary compliance from management, the chancellor officially endorsed co-determination in heavy industry and submitted a draft law to the *Bundestag*. It passed on 10 April 1951.[52]

The chancellor's capitulation to the Allies on the antitrust issue made it all the more difficult for the coal and steel producers to accept co-determination. Yet on balance they came out on top in the power struggle with the labor movement. Though the settlement required the steel industry to accept 50 percent employee representation on the boards of supervisors and management and the coal operators to extend the same principle to the mines, the arrangement was limited to heavy industry; a co-determination law applying to other sectors of production would not be adopted for another year, and it would apply on the basis of one-third rather than one-half representation. Co-determination in industrial associations continued to be discussed but had ceased to be politically realistic. Moreover, the Schuman Plan would divide the labor movement for years to come. Within the DGB the pro–Schuman Plan "realists" and technocrats around Hans vom Hoff, whose meliorist foreign policy views were similar to the chancellor's, soon won a crucial vote of confidence in a confrontation with the independent-minded Dr. Willi Agartz and other movement dissidents. Between the pragmatists of the union federation and the ideologues of the SPD little basis for compromise existed. Party chief Kurt Schumacher regarded the Schuman Plan as hardly better than a Franco-American plot to cheat the German worker; by comparison to what he predicted might result from Allied designs, the old trusts were models of corporate enlightenment. "His outraged nationalism," one bemused French official observed, "has put him in the astonishing position for a socialist of opposing deconcentration and defending the theses of German big business. But he holds international, and es-

---

[51] BA Nachlass Lehr 18, Sohl to Lehr, 11 October 1950, and "Entwurf, Schreiben Dr. Lehr an Professor Erhard: Mitbestimmungsrecht im Rahmen der Neuordnung."

[52] Thum, *Mitbestimmung*, pp. 111–112.

pecially French, capital in outspoken contempt and accuses it of
having instigated the [Schuman] Plan."[153] On 25 May 1951 the SPD
chief gathered 1,100 top trade unionists and SPD officials for a
showdown over the coal–steel pool, and in a speech lasting an hour
and forty minutes cited thirty-six separate reasons for opposing it.
The heads of all but one of the unions represented abstained rather
than endorse his position.[154]

The rise of corporatism had an external complement. Between
May 1950 and April 1951 the isolation of Ruhr heavy industry
ended. Monnet could do nothing to propitiate the steelmasters of
Lorraine, and, like them, the *Patronat*, the chambers of commerce,
and most of French industry opposed him. Monnet's unpopularity
enabled the big businessmen of his land to act in ways that otherwise would have been either embarrassing or politically dangerous.
Thus what have somewhat euphemistically been called "private networks of cooperation and conflict resolution" soon began to be re-
spun: The cartels were indeed back.

In January 1951 Monnet made a final earnest attempt to reconcile the French steel industry to the Schuman Plan. In the course of
the month Etienne Hirsch conducted a succession of hearings at
which the directors of all the main steel companies were given the
opportunity to present their views regarding the impact of the Schuman Plan on firm operations. The minutes of these tense meetings
provide indisputable evidence that the frequent and heated protests
made by official leaders such as Aubrun, Villiers, and Ricard were
representative of steel producer sentiment as a whole. Aubrun was
the first to testify before the Hirsch committee, on 2 January, stating that it would be meaningless to provide the French delegation
with estimates of industry competitiveness because the effect of the
common market on prices was still completely unknown owing to
the confusion surrounding pricing policy. In response to Hirsch's
rejoinder that a transitional period was to be provided in which difficulties could be worked out, Aubrun maintained that standard
pricing would either wreck the *charbonnages de France*, whose costs
were still substantially above those in the Ruhr, or raise the overall
cost of coal to steel producers, thus forcing increases in their prices

---

[153] MAE Y, Europe, Généralités 67, Telegram, François-Poncet (1999/2004), 2
April 1951.
[154] USNA RG466 D(51)34, Liaison, Bonn, to Dept. 626 (CN-48776), 16 March
1951; USNA RG466 D(51)688 W. Brown to McCloy, 29 May 1949; MAE Y,
Europe 1949–1955, Généralités 67, Telegram, Hausaire (2451/2456/21), April
1951.

as well. The Chambre Syndicale's chief snidely added that if Monnet were really serious about lowering prices and promoting efficiency he would eliminate the so-called *rupture de charge* imposed by the French railways. This was a phantom expense supposedly incurred by transshipment at the French border, which had the effect of raising Ruhr coal prices, thus providing protection to the French mines.[155]

The meetings went from bad to worse. On 3 January 1951 the director of SIDELOR made objections like Aubrun's to the proposed price policy, scoffed at the notion that delivery and sales quotas could be eliminated without wreaking disaster on the industry, protested that the price control policy pursued since the war had nearly wrecked his company, and claimed that its investment requirements would be all but impossible to meet over the next several years. Two days later a director of Knuttange boasted to the committee that French steel could make a better deal with the Ruhr collieries than could the High Authority. On 10 January President J. Raty of Aciers de Longwy limited his comment on the Schuman Plan to the caustic observation that no sensible person would willingly take a leap into the unknown. Humbert de Wendel and his associates were well briefed for their meeting with Hirsch on 13 January: They demanded specific information about the transitional measures promised as protection from the opening of the common market, objected that though their position depended on exports nothing was being done to offset disadvantages due to distance from seaports as well as high duties on imports of production goods, and insisted upon new forms of compensation.[156]

For the next two weeks Hirsch was forced to endure continuous demands for special treatment piled up over foundations of complaint on principle. Among the gripers were steel producers of the North and Normandy, representatives of the ore mines, and laminators from the Loire and Centre.[157] At the committee's final meeting on 1 February, M. Fould, the director-general of the Hauts-Fourneaux, Forges et Aciéries de Pompey, frankly expressed his skepticism about Monnet's entire undertaking. "The enterprises de-

[155] JM AMG 18/3/2, "Procès-verbal de la réunion tenue de 2 janvier 1951."

[156] JM AMG 18/3/3, "Comité d'Etudes Sidérurgie française, Plan Schuman: Réunion de 3 janvier 1951"; AMG 18/3/17, "Réunion de 9 janvier 1951"; AMG 18/3/8, "Réunion tenue le 10 janvier 1951"; AMG 18/3/9, "Réunion, 13 janvier 1951."

[157] JM AMG 18/3/9, "Réunion 19 janvier 1951"; AMG 18/3/12, "Réunion, 27 janvier 1951"; AMG 18/3/13, "Réunion, 29 janvier 1951"; AMG 18/3/14, "Réunion 19 janvier 1951."

cartelized by the Allies in the Ruhr," he predicted, "will tend to re-group and re-concentrate in the fields of coal, steel, and manu-facturing. This concentration is what gives power to the magnates of the Ruhr. It has done so in the past and will continue to do so in the future. The hypothetical good will of the government of the Federal Republic offers only feeble protection. Economic forces alone will establish the future equilibrium in Europe."[158]

The U.S. embassy in Paris reported on 14 March 1951 that the hostility of French industry was mounting to new heights. In a recent interview A. Aron had blamed the Americans for the disaster of the Schuman Plan, claiming it could never have gained French approval without Monnet's frequent threats that rejection would result in the withdrawal of dollars. Aron rejected the notion that free competition in basic industry was possible and asserted that Americans had a mistaken idea of the international steel cartel. It had nothing to do, he said, "with the type of cartel that prevailed in the petroleum, rubber and tin industries." "If there were now an entente between European steel producers," he asserted, "the shameful situation of the extremely high prices at which steel in Europe is sold for export would not have taken place." The disgruntled ironmaster even denied Monnet credit for having improved French access to Ruhr coal: "If given the freedom to negotiate, [our] industrialists would have [had] no real difficulties in obtaining coke from Germany, since Germany is in need of many French products and a natural exchange would have taken place between the two countries." Aron stated in conclusion that the Germans had already made many approaches to renewing a steel entente but that further progress would depend upon "an understanding on coke and other matters."[159]

A producer consensus developed during the negotiations. In his brief history of the *Montanunion* Hans Dichgans, a German delegate to the advisory committee, cited September 1950 as the turning point in the treaty-framing process; it was then, in his view, that industry ceased to have any direct influence on the proceedings, and its delegates thus began to meet privately. Among those venerated by Dichgans as "steel-Europeans of the first hour" were several figures formerly prominent in interwar cartels. (Only the German delegation actually lacked a representative from such a background.)

[158] JM AMG 18/3/15, "Réunion 1 février 1951."
[159] FRUS 1951/IV 450.33/3, Special Advisor to the Deputy Chief of the ECA Mission in France (Goldenberg) to Department of State, 14 March 1951, pp. 88–89.

They included A. Aron on the French delegation; Arnold Ingen-
Housz, the managing director of Hoogovens as well as of the Dutch
steel cartel; Pierre van der Rest, the chief of its Belgian counterpart;
and Tony Rollmann and Eric Conrot from ARBED who represented
Luxemburg. Conrot had served as deputy to Alois Mayer in the
International Steel Export Cartel since the 1930s and was also a
member of the Luxemburg delegations to the European Coal Commission, the Paris Marshall Plan conference, and the steel committee of the OEEC with the Ruhr.[160]

On 10 November 1950 the chief economic adviser to the French
High Commission reported that M. Villiers, the director of the
French employers' association, and his deputy Ricard had paid a
visit to BDI chief Fritz Berg in order to discuss the formation of a
"permanent Franco-German committee to guarantee cooperation"
between the industries of the two countries. Berg suggested the
wisdom of establishing common production programs, doing joint
productivity research, and sharing markets. Villiers responded that,
though in principle French industry supported the Schuman Plan,
it "is never too late to amend [the treaty]" and concluded that "the
essential thing is to appoint sound men to the headquarters of the
High Authority."[161]

"In early December 1950," reported the later much-celebrated
American journalist Theodore H. White, "meetings of the major industrial cartels were quietly held with representatives of German
business. At the same occasion the directors of German and French
steel also arrived at a common position regarding the level of the
external tariff."[162] The upshot of these meetings was a secret memorandum drafted by the cartels and delivered to each government
denouncing the Schuman Plan and threatening vague but ominous
political reprisals. On 27 January 1951 National Association of the
Schuman Plan Nations was formed, an organization like those British and French business diplomats had hoped would prevent World
War II. The prologue to its charter contained a blast against the
"super-dirigism" of the proposed High Authority, which was reputed to be something "unknown in our democracies and drawn
from the most totalitarian regimes." The federation thereupon demanded the suppression of the HA's future powers to borrow, tax,
lend, and underwrite investment programs and insisted upon the

[160] Hans Dichgans, *Montanunion: Menschen und Institutionen* (Düsseldorf, 1980),
     p. 69.
[161] JM AMG 45/1/30, Telegram (François-Poncet), 10 November 1951.
[162] AN 81 AJ138, "Europe's Cartels Strike Back" (ms., n.d.).

devolution of remaining ones from the purportedly omnipotent and dictatorial planned directorate to the various national associations of producers on the grounds that only they could manage the transition to common market conditions. The resolution also urged that the national producer associations be empowered to establish joint production programs, set up common supply systems, and define prices, conditions, and terms of sale, further recommending that they be allowed to determine whether or not to inform the HA of the investment plans of their industries. The producers also sought to eliminate Articles 60 and 61 on decartelization and deconcentration and, finally, called for convoking a conference of states at the end of the transitional period in order to decide whether the plan should go into effect or simply be dropped.[163] They expected in the meantime that Monnet would allow them "to take an active hand in organizing the High Authority and the common market."[164]

Though such a conference was never held, the so-called Interim Committee achieved the desired result of preventing the chairman of the Paris conference from running the Coal and Steel Community in the manner of the French plan, as a directorate of gifted disciples under his personal command. Monnet never intended to organize "a large-scale and lasting establishment, because I . . . was determined to keep the High Authority's staff as small as possible . . . It was best to begin with the few people who had helped draft the Treaty. This nucleus could then be enlarged by co-optation when we saw what jobs had to be done. But the High Authority was to remain a nucleus and confine itself to reorganizing and stimulating the work of others."[165]

Not the man who inspired the Schuman Plan but the senior civil servants who tried to reshape it both during and after the Paris negotiations deserve credit for having created the Eurocracy that would soon flourish in Luxemburg. Knowing from experience that without their help the giddy peaks of European accomplishment could never be scaled, the Sherpas intended to guide the ascent along approaches with which they were familiar and at a pace of their own choosing. The industrialists also wanted a hand in orga-

---

[163] JM AMG 12/3/3, "Les Fédérations industrielles de l'Europe de l'Ouest et le Plan Schuman," 8 February 1951; JM AMG 11/2/5/BA146/265, "Observations des fédérations industrielles nationales des pays intéressés par le Plan Schuman sur les clauses économique de Projet de Traité en préparation," 17 January 1951; KA/Henle EGKS, Schuman Plan, WVESI 1.1.51-31.12.51, "Die Industrieverbände Westeuropas zum Schuman Plan" (n.d.).
[164] Ibid.   [165] Monnet, *Memoirs*, pp. 272–273.

nizing the new supranational entity. A paper issued at the founding of the Eurofederation of producers noted that "the High Authority will not be able to fulfill its tasks without the cooperation of industrial association," adding they "expected to secure positions in which to cooperate actively in the execution of the common market . . . under [its] supervision."[166] Max. C. Mueller, a German delegate to the Paris conference, put the matter more explicitly at a WVESI meeting of 1 March 1951. "The treaty," he complained, "contains countless empowerments (Kann-Vorschriften) . . . The organization of the High Authority's administrative machinery is therefore crucial. The man who initiated the Schuman Plan thinks that the sole qualification for the job should be that one is a 'good European.' It is all the more important that Germany set up an organization to deal with these problems because the French dispose of a smoothly functioning Monnet-apparatus."[167] Mueller suggested sending a special commission to Paris in order to make contact with delegations sharing common concerns.

Though other figures involved in the negotiations must have been arriving simultaneously at ideas similar to Mueller's, the exact origins of the Interim Committee remain a mystery. On 17 May 1951 its first meeting took place. Present were the heads of all the negotiating delegations except that of France, reportedly because Monnet was ill. The committee treaded softly. It declared itself responsible for working out public relations problems and language policy, finding a location for the High Authority, and dealing with myriads of legal, personnel, bookkeeping, and statistical issues. As for the real concern of the committee, the internal organization and personnel for the new administrative structure, "The German delegation, with the agreement of the other delegations, withdrew this issue from the agenda."[168]

The Interim Committee did not simply vanish, as Monnet had apparently hope it would. Fearing that the other delegations had already developed proposals for the organization of the High Authority, Adenauer told State Secretary Rust of the Ministry of Economics in September 1951 to begin at once with such efforts and

[166] BA 146/265, "Observations des fédérations industrielles nationales des pays intéressés par le Plan Schuman sur les clauses économiques du Projet de Traité en préparation," 1 January 1951.
[167] WVESI, Sitzungsberichte, "Sitzung Engerer Vorstand mit Gästen," 1 March 1951.
[168] BA B102/8603b, "Kurzprotokoll über die erste Sitzung des interministeriellen Ausschusses," 17–19 May 1951.

enter discussions with the French plan as soon as possible. No one
acted upon the second instruction. Instead, the responsible interministerial committee continued to circumvent Monnet and dealt directly with the other delegations as well as with industry.[169] The
request for cooperation had actually come from WVESI. On 26 September 1951 its representative, Dr. Blankenagel, asked the Ministry
of Economics for official authorization to enter negotiations with
the other steel producer associations of the community with a view
to working out a common agreement which could then be presented
at the Interim Committee meeting planned for 9 October 1951 at
the Hague.[170]

The Interim Committee continued to meet over the following six
months, gradually developing a life of its own. Those who sat on the
body intended its work to be permanent. According to a report of
February 1952, "A while ago the head of the Dutch delegation,
Spierenburg, visited State Secretary Hallstein and . . . there was
complete agreement that the Interim Committee should handle all
questions preliminary to the commencement of operations by the
High Authority."[171] The German delegation viewed its work as complementary to that of industry, on 26 September 1951 agreeing that
Dr. Blankenagel should be encouraged to continue with efforts to
strengthen the producer consensus regarding the organization of
the future High Authority. Success in this endeavor, the delegation
concluded, would strengthen the German position at the meeting
scheduled for 9 October at the Hague. All the delegations present
wanted to "minimize bureaucratic interference." They also disliked
the "collegial principle," which posed few restraints on the actions of
individual members, but thought a direct attack unwise. The Germans favored splitting the administration into subgroups for coal
and steel, thus allowing industry to be represented at the second
echelon and subordinating functional offices to it. The Dutch, who
could not see the point of the coal–steel subgrouping, preferred a
simple staff plan. The other delegations were less concerned with
the design of the administrative machinery than about who would
run it. Though agreement could not be reached on organizational

[169] BA B102/12608, "Vermerk über eine interministerielle Besprechung zur Frage
der Organisation der Hohen Behörde," 26 September 1951; KA/Henle EGKS
Schuman Plan, Wirtschaftsvereinigungen, 1.1.52–30.9.52, "Besprechung zwischen Bundeswirtschaftsministerium, Auswärtiges Amt, Kohle und Eisen am
5.6.1952."
[170] BA B102/12608, "Vermerk über eine interministerielle Besprechung zur Frage
der Organisation der Hohen Behörde," 26 September 1951.
[171] BA B102/8603a, "Besprechung in Den Haag am 26. und 27.2.52."

schemata, the ambitions of the Interim Committee were not easily satisfied. The German delegation demanded and secured acknowledgment of its competence from the national government, as apparently did the others from theirs.[172] By June 1952 its growing agenda included the naming of the eight members of the High Authority, the designation of the seven judges and two counsels proposed for the High Court, the allocation of seats in the Advisory Committee, and the establishment of the date for commencing High Authority operations as well as for the first meeting of the Council of Ministers.[173]

Monnet had no intention of accepting the work of the Interim Committee as a fait accompli and sought to bypass it, or rather overrule it at the meeting of foreign ministers scheduled for July during which the treaty, having been approved by the governments of each of The Six, was to be ratified. The issue never came before the gathering because an even more pressing problem remained unresolved: The Schuman Plan organization, due to begin operating in less than a month, needed a home! Only after hours of intense and angry discussion did Monnet, who had long cherished a hope to create a new federal district, a European analogue to Washington DC, relent and finally agree to designate Luxemburg as the seat of the new Coal and Steel Community. After this exhausting struggle there remained neither enough time nor enough energy to reach agreement about how the new enterprise should be administered.[174]

The head of the German Schuman Plan desk, Dr. Ulrich Sahm, described what happened next in a 1984 interview. In preparation for the commencement of operations Sahm had prepared "wonderful papers, full of grandiose organizational plans, personnel guidelines, compensation scales and so on" for the German vice-president-designate, Franz Etzel. Monnet would have none of this and "swept the whole thing off the table, saying that he was going to begin with a small team and work from there . . . on an informal basis – in other words . . . he wanted to keep his hands on everything and not get tied down by German perfectionism." Arriving at the ceremonial opening of the new organization Sahm and Etzel discovered that "there were French names tacked to every door. The

[172] Ibid.
[173] BA B162/8603a, "Monnet Präsident der Hohen Behörde?" from Le Monde, 17 June 1952, and B102/8603a, "Schuman Plan: Sitzung der Sachverständigen," 23, 24, 25 June 1952.
[174] Monnet, Memoirs, pp. 360–361, 372–373.

entire High Authority was in French hands ... they had planned
the whole thing out down to the last secretary and concierge." Etzel
immediately told every German in sight to attach his calling card to
the first door available, "no matter whether he planned to work
there or not." It was thus, Sahm deposed, that he found himself unexpectedly in the service of the High Authority.[175]

The birth of Europe resulted from a natural process but was not a
painless act: it involved more than what one would like to remember
and commemorate, the convocation of men of high ideals reaching
principled agreement to undertake the worthy venture called integration. France, West Germany, Italy, the Netherlands, Belgium,
and Luxemburg did not mesh effortlessly and automatically like
the gears of a precision machine; they had to be heavily lubricated,
sometimes reshaped, and occasionally even forced together in order
to get Europe to run. Tough issues requiring independent action
from Bonn and Washington also had to be resolved before the heavy
industry pool could begin to operate. Integration seemed to be
something that everyone wanted but no one knew how to get.

Only Monnet had the power and vision to break through the log-
jam of foreign and domestic problems that had piled up from the
war. The proposal of 9 May 1950 was a *coup de théâtre* of unparal-
leled boldness. Without his extraordinary American connection,
however, word would never had become deed. Wall Street and
Washington made Monnet an independent force in France, enabling
him to act on behalf of an immobilized and hostile industrial community. Without his uncanny ability to make the United States do
his bidding the coal–steel negotiations would have fallen through.
Instead, Europe made the transition from the aftermath of a catastrophic war to a peace that would be long and durable.

As a result of the Schuman Plan negotiations, a supranational authority had been created, a potential nucleus for a European federal
system. It would serve in lieu of a peace treaty concluding hostilities
between Germany and Western Europe. This was no grand settlement in the manner of Westphalia or Versailles. The agreement to
create a heavy industry pool changed no borders, created no new
military alliances, and reduced only a few commercial and financial
barriers. It did not even end the occupation of the Federal Republic; this process began with the conclusion of the contractual accords
of 26 and 27 May 1952. By resolving the coal and steel conflicts that

[175] JM "Interview mit Dr. Ulrich Sahm," 27 February 1984.

had stood between France and Germany since World War II, it did, however, remove the main obstacle to an economic partnership between the two nations.

The new Coal and Steel Community was still untested. The original design had been drastically altered: The centralizing reformist directorate was, by stages, turning into a corporate economic community. Whether and how it would operate were unknown. Success would depend equally upon the ability of an as yet only partially reconstructed Federal Republic to play positively the leadership role it had assumed in the course of the negotiations and on the spirit with which the other community nations accepted it. Defeat, occupation, and division seemed to have laid Germany's great power ambitions to rest, enforced a new sense of national solidarity, and created an almost fanatical determination to rebuild and prosper. It would require favorable circumstances, good luck, and a special kind of statecraft to harness these elements of a new nationhood to the welfare of Europe.

6

The success of a failure: the European Coal and Steel Community in action, 1952–1955

The European Coal and Steel Community (ECSC) made good only one promise, the most important one: It advanced the process of integration. This success grew out of failure. Little remained of Monnet's ambitious plans when he decided to step down as president of the High Authority in November 1954. Decartelization had given way to reconcentration. The High Authority, designed as suggested by its exalted and forbidding title to convey orders from above for execution from below, was beset by paralysis. The common markets created to discipline heavy industry had become its common property. Yet the disappointments of the ECSC did not result in a reversion to the nationalist-protectionist order of the 1930s. The "relaunching of Europe" that began after the French Senate's rejection of the European Defense Community treaty in August 1954 aimed at something far more ambitious than the common administration of a single sector of West European industry; the statesmen of The Six now sought "complete integration" (*Vollintegration*) – the eventual creation of a single communitywide environment for economic activity. This was to extend to the harmonization and coordination of tax structures, freight rates, labor practices, product standards, and administrative procedures of all sorts. No one, however, expected to "build Europe" overnight. A council representing the interests of the national governments rather than a powerful executive making policy on their behalf was to provide the institutional mechanism of integration; it was meant to operate not by schedules and bylaws but opportunistically, advancing the process whenever conditions were ripe. The Treaty of Rome contained no plan or program but merely enshrined a commitment.

The "relaunching of Europe" is something of a misnomer: This was actually the first such operation the Europeans conducted on their own. The Americans were on hand for the initial one. By 1955 The Six no longer needed the United States to act as silent partner, underwrite their venture, cohabit with them, provide reminders of the need for cooperation, or in any other way coerce or entreat them to work together for their own benefit. This is one measure of what had been accomplished since the announcement of the Schuman Plan five years earlier – proof that integration of some sort had occurred.

It is hard to find specific evidence for it. The constant squabbling at Luxemburg was hardly an example of the "new European spirit." Nor did the ECSC establish a novel form of economic and political cooperation: It neither reformed prevailing business practices, produced a new relationship between public authority and private power, nor shifted the locus of economic policy, even as regards heavy industry, from national state to supranational agency. The economic impact of the community was slight. Few of its policies had demonstrable effect; the overall increase in the coal and steel trade during the early years of its operation stemmed from the general growth of the European economy and in particular the boom in West Germany. Yet integration was no phantom; the term merely requires a qualifying adjective to take on historical meaning. It was *Westintegration* as sought by Adenauer. By 1955 the occupation was over, and the nation was committed to rearming. Its industry had become the engine of European growth. No one could seriously have doubted that the Federal Republic would be strong enough to dominate The Six in the future. The fact that France, as well as the others, accepted this when Europe was relaunched is proof that rapprochement had occurred.

It was by no means due entirely to the existence of the heavy industry pool. The Ruhr Problem was indeed rapidly becoming an anachonism. The 1950s witnessed neither crippling coal shortages nor chronic gluts of steel. Nor would huge new mills and modern mines be the pacesetters of economic growth; the automobile and oil were much more important. And strategic problems far greater than anything either Gallic or Teutonic, or having specifically to do with either coal or steel, loomed on the horizon. One threat came from the East, the other from overhead, and together they made old rivalries seem almost quaint.

The coal–steel pool completed the construction of the heavy industry community sought by producers and their governments since 1926, and efforts to surmount its problems led to the next stage of integration. The thinking that went into the Treaty of Rome grew out of disputes at the ECSC, which revealed the practical impossibilities of "partial integration" and started a line of inquiry indicating the route that would have to be followed to the next stage of the process.

6.1 From absurdity to farce:
the implementation of Law 27

The decartelization of the Ruhr, as Monnet never grew tired of repeating, was to be the indispensable precondition of the Schuman Plan. And so it was, after a fashion. The agreement of 14 March 1951 had called upon the Germans to divide the steel industry into twenty-four producing companies; limit each of them to the control of no more than 75 percent of its total coal consumption requirements; and eliminate the Deutschen Kohlen Verkauf (DKV), the distribution organization created by the Allies that for all practical purposes carried on the work of the old Ruhr coal syndicate. Though the decartelization process was expected to take no more than a year, it required twice as much time to complete officially and even then was not really over. Try though they might, Allied High Commission (HICOM) decartelizers could not keep abreast of the Ruhr's recartelizers. The failure of the detrustification campaign should have been evident even while it was under way.

The Allied attempt to take apart the DKV was the sore point in the decartelization process. Neither Adenauer nor the smokestack barons particularly worried about the breakup of the *steel* companies. Though Law 27 had not specified the precise form of compensation for seized assets, on 23 October 1951 HICOM made a ruling very favorable to shareholder interests: Rather than being saddled with long-term, low-interest government bonds or legally expropriated by some other confiscatory scheme, they would be assigned stock in the new companies issued pro rata at fair value. Restrictions were placed on sales of the new equities in order to protect their worth, but they were otherwise unencumbered.[1] Those fortunate enough to hold them could relax: The reconstitution of old holdings could now begin. This did not necessarily imply reconcentration on the scale of the 1930s; Vereinigte Stahlwerke was a dinosaur, and adjustment to conditions prevailing on the world steel markets would be necessary. The trusts would have to adapt.

Coal reorganization seemed to threaten the economic viability of the mines, the competitiveness of the steel industry, the value of the currency, and the balance of payments. German officials, industrialists, and union leaders were unanimous that any attempt to market coal in other than traditional ways would result in chaos; to many of

---

[1] Schmidt, "Reorganization," p. 192; AN 81 AJ142, "Aide-Mémoire sur l'historique de la Déconcentration de la Ruhr," 2 July 1952.

them compromise was not only economically unfeasible but down-right immoral. Rather than comply with Adenauer's March 1951 commitments, management, and labor too, made strenuous efforts to evade them. Only on the eve of the commencement of HA oper-ations did the coal managers agree to respect what they had signed earlier, and not until just before the very opening of the common market in fuel did they replace DKV with the new organization called Gemeinschaftsorganisation Ruhrkohle, or GEORG. GEORG would bear a strong paternal resemblance to DKV, and another ep-isode of the decartelization drama would thus soon begin.

It had long since degenerated into farce. In the months of the most vigorous trust-busting activity, from the initialing of the treaty to the commencement of operations in Luxemburg, the steel indus-try refined its rebating methods, tightened the operation of its car-tels, and made joint preparations with its foreign counterparts for the opening of the common market. The Americans preferred to overlook such developments but when no longer able to avert their gaze reacted with horror. French officials, while insisting officially that everything depended on the success of decartelization, doubted personally that it could be enforced. At a time when tactful compro-mises would have best served their interests, German stubborn-ness complicated and delayed a settlement. The publication of harsh new HICOM regulations and protests of them, the exchange of fi-ery memorandums from Petersberg to Bonn and Bonn to Peters-berg, and the emergency convocation of angry, wearying, stroke-inducing conferences, meetings, and discussions went on as never before. Those on the Allied side who thought the decartelization campaign ridiculous and destructive were overruled.[2] Few Germans were prepared to say openly that it was futile and should not be taken seriously. The frantic whirl surrounding decartelization seemed to have been based on a misunderstanding of something everybody had forgotten a long time ago.

Erhard's outright rejection of the 2 May 1951 Allied draft for the statutes of the new unit companies caused the first of several steel reorganization headaches. The proposal prohibited each of the main devices traditionally used in Germany to coordinate corporate pol-icy: interlocking directorates, banker control of proxies, and con-centration of shareholdings.[3] The Allies gave in after a month of German protests. Three of the offensive articles disappeared, and

[2] FRUS 1951/III, Kennan to Secretary of State, 5 September 1951.
[3] Schmidt, "Reorganization," p. 128.

another eleven were modified, leaving "only the bare minimum of provisions essential to allow the new units to develop as independent companies."[4] Legally, in other words, the new steel companies would be different from other German firms only in respect to the recently enacted co-determination.

The actual transfer of assets from the old owners to firms operating under the severance decree of 1946 now began. Between July and November 1951 twenty-two of the twenty-four planned companies were set up. One, Hochofen Lübeck, was dropped from the list, but the fate of the final firm on it, the August Thyssen Hütte, would remain open. This was a matter of considerable importance. ATH, which had been partially dismantled, was the logical site for a continuous wide-strip rolling mill and therefore the key piece in the Ruhr's growth strategy. Permission to go ahead, which was finally granted on 21 July 1952, rendered all further discussion of reorganization superfluous.[5]

Writing a year after the official conclusion (except at ATH) of the steel reorganization campaign, Günter Henle described the results as falling into three categories. Six firms he found "economically satisfactory": Dortmunder-Hörder Hütten-Union AG, Nordwestdeutschen Hütten- und Bergwerksverein AG, Rheinisch-Westfälischer Eisen- und Stahlwerke AG, Mannesmann AG, Hösch-Werke AG, and Stahlwerke Südwestfalen AG. A second group was "bearable under Allied pressure." It included Deutsche Edelstahlwerke Rasselstein-Andernach AG, Niederrheinische Hütte, and, once reconstruction had been completed, August Thyssen Hütte AG and Guszstahlwerke Bochumer Verein. A third group, however, he regarded as completely unacceptable. Hüttenwerke Ruhrort-Meiderich and Rheinische Röhrenwerke AG should have been combined, in his view, and the same was true in the cases of Stahlwerke Bochum AG, Ruhrstahl AG, and Guszstahlwerke Witten AG. Henle further objected to the splitting off of Hüttenwerke Oberhausen from its machine tool affiliate in neighboring Sterkrade.

---

4 Cited in ibid, p. 131.
5 Ibid., pp. 132–148, 142; Sohl, *Notizen*, pp. 127–128, 139–140, 145–146; BA B136/2376, "Ausbau der Stahlkapazität," *Frankfurter Allgemeine Zeitung*, 29 July 1952; BA Nachlass Lehr 17, "Anträge Thyssenhütte/Breitbandstrasse," 13 October 1950; idem, "Anträge betreffend Teilremontage der . . . ATH," 2 October 1950; BA NL Lehr 20, Sohl to Bowie, 6 January 1951, 4 January 1950; "Vermerk. Betr.: Breitbandstrasse," 19 January 1951; Lehr to Sohl, 4 January 1950; Sohl to Lehr, 3 November 1950; Sohl to Lehr, 24 November 1950; and Sohl to Pferdmenges, 30 December 1950.

These economic miscarriages did not worry him excessively because "the Schuman Plan and the contractual accords have cleared the way for the necessary corrections at the appropriate time."[6]

The DKV was not the only bone of contention between the decartelizers and the German government; ongoing disputes over the proper size and number of new mining companies, the allowable extent of coal–steel ties, and the permitted relationships between mills, finishers, and manufacturers made many heads turn gray.[7] The struggle over the existence of the coal cartel was the single issue that put the Schuman Plan most seriously in jeopardy. It began when the Federal Republic tried to back out of Adenauer's commitment of 14 March 1951 to dissolve the Ruhr's marketing machinery. The stalling worked. On 25 July 1951 the federal government questioned the competence of the special HICOM committee organized to devise an allocation system to replace the one used by the DKV. This resulted in the compromise inclusion of two Germans on a new eight-man committee responsible for producing the reorganization plan due by October 1951. It recommended the creation of a new, mainly supervisory federal coal office and six separate merchanting companies, each controlling nine regional and one export affiliate. In a dissenting brief the German members insisted upon retaining a central selling agency, and their government rejected the report altogether on 23 October. Two days later HICOM ordered Bonn to accept it. In the following months nothing happened. The Germans did not even budge after the Allies conceded on 13 March 1952 that a joint coal services organization could be maintained.[8]

After more than a year of futile efforts to strike a bargain with the Germans, the Allies decided on 27 May 1952 to force the issue and published a new dissolution decree for DKV under Law 27, Regulation 17. Adenauer had not been consulted and objected strenuously that HICOM's rashness would wreak economic havoc. Urging that the edict be revoked, he argued for transferring responsibility for coal reorganization to the ECSC for future settlement and for setting back the date for final dissolution to 31 March 1953, when the common market in coal was scheduled to commence op-

---

[6] KA/Henle Umgestaltung des deutschen Kohlenbergbaues und der deutschen Eisen- und Stahlindustrie, Gesetz 27, 1.1.52–31.10.54, "Fortgang der Entflechtung nach Gesetz Nr. 27 mit Unterzeichnung des Deutschlandvertrages," 22 January 1953.
[7] Schmidt, "Reorganization," pp. 74–84.   [8] Ibid, pp. 85–94.

erations. With the opening of the Luxemburg headquarters, which had been set for August 1952, then fast approaching, HICOM agreed to both conditions.[9]

The decartelization campaign could demonstrate some superficially impressive results. In place of nine prewar trusts there were twenty-four new companies controlling only 20 percent of overall coal output as opposed to 60 percent previously. Earlier there were ten big independent mines; now there were twenty-five. Enforcement had not been overlooked. The February 1952 London Conference of Foreign Ministers decided that the control groups should retain their decartelization authority until the coal–steel community was sufficiently strong to exercise it.[10] The "new structure of the Ruhr" was nevertheless a house built on sand, for the French lacked confidence in their own policy. Parisot, their representative on the Coal Control Group, spelled out his ojections to it in a 26 July 1951 letter to Monnet. He thought it unfair to deny the Germans monopoly sales agencies under the Schuman Plan when in France there was "a general accord between the Charbonnages de France, the Saar, and private importers limiting competition, demarking sales areas, and establishing special sales agents for [designated] clients." Parisot also thought that French interests were best served by maintaining a single privately controlled sales organization for Ruhr coal, since "Experience shows that a mining syndicate will always be able to export more at times of shortage than an official agency of government." The operators, he proceeded, "have tried in the past and will try in the future to maintain their foreign customers, whereas the government will always have to respond to pressures from the multitude of consumers, which in Germany means the steel producers."[11] Monnet fiercely upbraided Parisot (who incidentally was also very skeptical of *steel* reorganization) for "faithfully reproducing the arguments used constantly by the Germans" and for "manifesting total incomprehension of the Schuman Plan"[12] but could not force him out of office, possibly because his views were more representative than exceptional.

---

[9] Ibid, pp. 149–158.

[10] AN 81 AJ141, "Note [van Helmont] sur la déconcentration de la Ruhr," 15 March 1952; Schmidt, "Reorganization," pp. 201–202; AN 81 AJ141, "Meetings of Three Foreign Ministers in London," 17–18 February 1952; AN 81 AJ141, Monnet to Schuman, 15 February 1952.

[11] AN 81 AJ142, "Note Parisot: Réponse à la lettre du 19 juin 1951 de M. Hirsch."

[12] AN 81 AJ142, Monnet to Parisot, 26 July 1951.

Armand Bureau, Parisot's counterpart on the Steel Control Group from 1951 to 1953, was an anti-Ruhr zealot, who directed a continuous stream of scathing but well-informed memorandums to Monnet denouncing German efforts to evade control, warning of "cartelism," and appealing to the need for constant vigilance. Yet he too increasingly accepted the inevitability of a Ruhr restoration. A member of the High Authority's Advisory Committee after 1954, Bureau understood all too well what was happening. The main strategy in the Ruhr's restoration effort, according to his analysis, was the attempt to replace Stahlunion, Vereinigte Stahlwerke's merchanting company, with a new holding company strong enough to dominate the industry. A merchanting company, as his numerous reports explained, normally did more than merely maintain inventory of standard profiles; it also received special orders, which were then grouped together and placed with one or more firms in accordance with both the availability of spare capacity and agreements concerning market share. Its influence was particularly strong at times of weak demand when factories were idling.[13] In May 1952 Bureau alerted Monnet that plans were nearing completion to set up a new company under Dr. Walther Schwede, the former German representative on the international steel cartel and "a man well-known and favorably regarded in France for his attitude during the occupation." According to the Frenchman, the new company was intended to become the largest German steel exporter, one of the two largest sellers of scrap, and would give the Ruhr overwhelming power vis-à-vis its colleagues in the Schuman Plan nations. Bureau appealed to Willner, who had taken over Bowie's decartelization responsibilities, to prevent its formation, and language to this effect was included in a July 1952 agreement.[14]

The French lost interest in the merchanting company issue the following September after the explosive demand for steel in the Federal Republic caused the sales organization then still being dissolved, the "commercial services of the Vereinigte Stahlwerke under the direction of Dr. Schwede," to demonstrate what the French perceived as an "astonishing willingness" to cede them shares on South

[13] AN 81 AJ142, Bureau to Uri (personnal), 21 January 1953; AN 81 AJ141, "Systèmes des rabais et puissance des anciens Konzerns de la Ruhr," 7 February 1952; JM AMG 51/4/7, "Note (Bureau) pour M. Monnet," 6 June 1952; JM AMG 51/4/9, Bureau to Monnet, 25 June 1952; JM AMG 23/5/1, "Note pour M. de Vert, Conseiller Economique et Financier," 6 August 1952.
[14] AN 81 AJ, Note (Bureau), "Sociétés commerciales des Vereinigte Stahlwerke," 8 May 1952.

German markets. Not only did the boom reduce the relative economic importance of central order placement and eliminate any
reason to exercise organizational clout; it assuaged French fears of
an export offensive while making the Germans appear unusually
generous.[15]

Bureau clearly perceived that the current shortages of steel were
enabling cartels to resume operations under the guise of acting as
agents of the state. On 30 August 1951 he reported to Monnet that
"the old syndicates of the *Stahlwerksverband* are back under new
names at [WVESI] and have resumed their functions to the extent
that they are able to camouflage them from the Americans."[16] A
confusing division of functions between private and public responsibilities made cartel operations extremely difficult to detect. Many
an officially proscribed practice like rebating was enforced by the
German government as part of the price control system adopted
at the insistence of HICOM. Rebating was normally done at the
wholesale level by one of the three established operating companies.
Were these in fact cartels? In the judgment of the control groups,
"Each of [the three] associations is 'voluntary' in the sense that
membership in it is not a legal requirement for engaging in the iron
and steel trade. Since, however, these associations are in practice,
through governmental sanction, the governing bodies of the trade,
membership in an association is the only method for a merchant to
have any voice in trade matters."[17] The three companies were also
linked together by something called Arbeitsgemeinschaft Deutscher
Eisenhändler, an official body where policy could be coordinated
without arousing undue suspicion.

In May 1953 the powerful syndicate for rolled steel products cel-
ebrated its fifth anniversary. As described by Chairman Gerhard
Bruns in an article published in the financial daily *Handelsblatt*, the
organization's history was a "mirror of the postwar era." With the
old cartels banned after 1945, a new "rolled steel group" (*Walzstahlgruppe*) was formed within WVESI at Allied request to assure efficient order placement and allocation of scarce product. According to
Bruns, the new group soon was doing market analysis, "eliminating

[15] AN 81 AJ142, "Retransmission de télégramme [Armand Bérard]," 3 October
1952.
[16] AN 81 AJ139, "A. Bureau, Activités de la Chambre syndicale de la Sidérurgie
allemande," 30 August 1951.
[17] AN 81 AJ139, Combined Steel Group, "Rebate System in the Steel Industry,"
22 June 1955.

bottlenecks in cooperation with customers and dealers," protecting general industry interests, and advising member firms on how to set up mutually beneficial rolling programs. After the shortages brought on by the outbreak of the Korean War, Bruns proceeded, the Rolling Steel Syndicate, as it was by then officially named, was deputized by the government to steer production and enforce price controls, thus being transformed from a "self-administrative organization of industry" into "an agency supporting the state" (*Hilfsorgane des Staates*). The boom of 1951–1952 had an extraordinary impact on German markets, the steel industrialist noted, and halved the export of goods produced by his association; the syndicate thereafter devoted its efforts to building strong relationships with domestic customers. The lifting of price controls on rolled steel product in fall 1952, he hinted finally, had opened the way to new international agreements concerning German markets. Bruns concluded with a veritable celebration of the private preparations made for the impending opening of the common market. "What years ago seemed impossible now has actually come to pass in many big markets in a matter of a few weeks. Hundreds of experts from every [member] nation have come together in working parties that in open discussions have solved innumerable problems. This has created a basis for mutual understanding lacking previously and more impressive than even what grew out of a similar movement in 1925." Bruns predicted that in the near future new organizational forms would be discovered for international industry cooperation, as recent experience had involved a process of reshaping (*Umformung*) and repackaging.[18] Bruns was not boasting: The opening of the common market found industry well prepared, and new forms of business cooperation would in fact develop in connection with it.

Top-level interindustry discussions had actually begun within days of the initialing of the treaty and expanded in scope in similar fashion to those of the Interim Committee. On 17 April 1951 M. J. Pinot, the President of the Chambre Syndicale, delivered a friendly little after-dinner speech before Ruhr dignitaries hosting him at the Rheinarchiv. The time had come, he ruminated, for the coal and steel men of the two neighboring nations to get better acquainted with one another. After informally briefing his listeners on the structure of the French steel industry, Pinot made them privy to the fact that its cartel, the Comptoir Français des Produits Sidérurgique (CFPS), was less powerful than generally believed. Since most steel

[18] Gerhard Bruns, "Fünf Jahre Gruppe Walzstahl: Spiegelbild der Nachkriegsentwicklung in der deutschen Stahlindustrie," *Handelsblatt*, 15 May 1953.

firms had their own commercial operations, Pinot noted, CFPS did not play a large role in order placement as had the old Stahlwerksverband. The most important present function of the French syndicate, he added, stemmed from the workings of government price control policy: CFPS had to spread the losses from sales on the domestic market by redistributing profits from exports.

Pinot described the lack of a coal base as the most serious problem facing the steel industry. Swayed perhaps by the power of his own rhetoric, he even singled it out as the main source of French national decline during the previous 150 years! To overcome this historic vulnerability as well as the lack of new investment in steel after 1930, Monnet had set up the French plan, but to the orator's mind it had not fundamentally changed the situation of the nation's mills. Suggesting that the Schuman Plan could mark an end in the bids for autarchy so harmful in the past, Pinot emphasized the timeliness of a new agreement: The Ruhr should guarantee access to coal at times of high demand in return for a promise from French steel to cede a generous market share when business was bad. Pinot further proposed coordinating investments jointly and reworking freight tariffs to facilitate a geographically more rational division of markets, also expressing the hope that additional rationalization programs could be worked out through cooperation with the High Authority. Rounding off his talk, the speaker promised his interlocutors that they would find considerable goodwill in France, no less in fact than he had found in Germany during his frequent sojourns of the past two years.[19]

On 8 and 12 November 1951 the directors of the Bundesverband der Deutschen Industrie (BDI) and the Conseil National du Patronat Français (CNPF) met in Düsseldorf to form an entente. Present among others were Fritz Berg and Walther Schwede on the German side and Pierre Ricard, Albert-Roger Métral (president of the main manufacturers' association), and CNPF President Georges Villiers, who had originally suggested the meeting, on the French. Their wide-ranging talks touched on labor, public relations, patent and brand-name exchanges, as well as the improvement of productivity. As developed in Villiers's opening address, the main theme of the talks was "the European idea," which the CNPF chief declared himself ready to serve. In thanking the Frenchman for having initiated the meeting, President Berg underscored his agreement with "every point" made by him: German management felt equally strongly

[19] SP, "Conference faite au diner de 'Rheinarchiv' par M. J. Pinot," 17 April 1951.

about the need to create Europe, banish *Dirigismus*, prevent labor union abuses, and so forth. Though both the German and French federations held bilateral meetings with other national employer groups, it would be common for the two to arrive at a common position prior to entering discussions with other parties; within the community of The Six theirs would indeed be a privileged relationship.[20]

Representatives of WVESI and CFPS and of steel producers in the other four member nations met twelve times between 1 January 1952 and the opening of the common market to develop policy toward the community in conjunction with the Interim Committee. Walther Schwede first proposed forming a permanent organization that could meet regularly for the purpose, as the minutes of the 7 January 1952 meeting awkwardly phrase it, "of creating a basis that would make it easier for us to clear up anticipated problems with the High Authority in advance."[21] Everyone present applauded this idea as well as the suggestion that the next meeting be held in Düsseldorf.

The Germans took great care not to be too obtrusive. To quote from the text of the meeting, "Considering that Herr Chome is a Luxemburger and has special qualifications ... to serve as intermediary and because every effort must be made to avoid creating the impression that we want to play a leading role in everything, we should consider ... setting up the proposed future [steel producers'] association in Luxemburg."[22] The same tactful approach had been followed in 1926. Preferring to keep syndicate activity in the background, the Germans suggested that rather than make either nominations to the High Authority or appointments to the Advisory Committee directly the cartels should encourage the setup of national Schuman Plan committees that could serve as buffers between them and Luxemburg.

The industrialists' meetings did not always achieve complete unanimity or even agreement on any single point. The first to present a

[20] Institut für Weltwirtschaft, Kiel (WW), "Deutscher und französischer Industrie-Ausschuss," *Industriekurier*, 13 November 1951; also, *Handelsblatt*, 12 November 1951; SP, "Französisch/deutsche Industrieführergespräche am 8. und 9. Nov. 1951"; AN 81 AJ139, "Retransmission de télégramme [Arnal]," 14 December 1951.
[21] KA/Henle EGKS, Schuman Plan, Wirtschaftsvereinigungen, 1.1.52–30.9.52, "Besprechung mit Belgien und Luxemburg über Schumanplanfragen," 7 January 1952.
[22] Ibid.

worked-out organizational scheme for the High Authority, the Germans were disappointed that when France finally presented plans they were both very general and overcentralized. It would take considerable effort to arrive at a common proposal. Relations with the mines also presented a problem initially. The French collieries were socialized, politically close to Monnet, and unwelcome at confidential gatherings of businessmen: Fuel problems would have to be dealt with later.[23] The committee of steel industrialists nonetheless worked smoothly because everyone present thought that the syndicates should be as influential as possible in the operations of the community. The minutes of the 3 May 1952 meeting report agreement on three "basic principles," each of them embodying standard industry approaches: The collegial rather than bureaucratic (Ressort) structure should be adopted, men from industry be appointed whenever possible, and no "general economic section" be formed that could impose policy on specialized branches. Agreement was also soon reached with the colliery representatives, who had been conducting parallel meetings.[24]

Important new working relationships were established at the steel industry discussions. The Germans were particularly impressed with the recently appointed first vice-president of the CNPF, Pierre Ricard, and encouraged him to play a leading role in "political questions."[25] Long before the High Authority was in operation industrialists later prominent in the affairs of the community had become well acquainted. The names of those present at these meeting would come up again and again: Aubrun, Aron, Ricard, Charvet, and Ferry of France; van der Rest of Belgium; Conrot and Chome of Luxemburg; Ingen-Housz and van den Berg of the Netherlands; Fugmann, Sendler, Blankenagel, and Salewski of the Federal Republic; and Fumante and Vignuzzi of Italy.

The discussions are interesting less because of their results than because of the light they cast on the thinking of the participants.

[23] KA/Henle EGKS, Schuman Plan, Wirtschaftsvereinigungen, 1.1.52–30.9.52, "Pariser Besprechung der 6 Eisenschaffenden Industrien am 29.3.1952," and "Besprechung mit der Vertretern Belgiens am 18.4.1952."

[24] KA/Henle EGKS, Schuman Plan, Wirtschaftsvereinigungen, 1.1.52–30.9.52, "Besprechung über Schuman Plan-Fragen," 3 May 1952; "Sitzung des Ausschusses für Organisationsfragen," 14 May 1952; and "Besprechung des Organisationskomitees der Kohle- und Eisenverbände," 11 July 1952.

[25] KA/Henle EGKS, Schuman Plan, Wirtschaftsvereinigungen, 1.1.52–30.9.52, "Pariser Besprechung der 6 Eisenschaffenden Industrien am 29.3.1952."

The main concerns were to place as many business figures as possible in positions of influence in the High Authority or, failing this, clip its wings. The industrialists still hoped in June 1952 that it might be possible to avoid a Monnet presidency, with Ricard reporting uncertainty as to whether an industrialist or a politician would be appointed to the job.[26] He hoped that the former cognac vendor could be persuaded to serve as civilian head of the European Defense Community and either René Mayer or René Lacoste be named in his place. The steelmen also pressed hard to reserve a seat on the High Authority for someone from their ranks, agreement having been reached among them on 18 April 1952 that its ninth member should be "an industrialist already recognized before the war as a respected person."[27] This also was not to be. Though working closely with the Interim Committee, the producers were similarly unsuccessful in securing support outside of heavy industry for their many proposals for dividing the High Authority into coal and steel branches. A plan to immobilize the High Authority by requiring countersignatures to all directives of heads of division got nowhere, as was true of a similar one to create a "control apparatus" with veto power over all executive decisions.[28]

Things could have been much worse for the Ruhr in August 1952. Decartelization had failed and recartelization was already well under way. German leadership within European coal and steel was accepted when exercised quietly. Cooperation similar to that before the war had resumed painlessly, even naturally. It had not yet, however, begun to affect business conditions, nor was it enshrined in any really important official institutions. Though working with the Interim Committee, the steel producers had not gained official sanction for their proposals concerning the internal organization of the High Authority. The struggle over how the coal–steel industry should operate as well as who, or what, should be in charge would now be waged in Luxemburg.

[26] KA/Henle EGKS, Schuman Plan, Wirtschaftsvereinigungen, 1.1.52–30.9.52, "Besprechung über Schuman Plan-Fragen 3.5.1952."

[27] KA/Henle EGKS, Schuman Plan, Wirtschaftsvereinigungen, 1.1.52–30.9.52, "Besprechung mit den Vertretern Belgiens am 18.4.1952."

[28] KA/Henle EGKS, Schuman Plan, Wirtschaftsvereinigungen, 1.1.52–30.9.52, "Organisationsfragen auf dem Gebiete des Schuman-Plans," 28 April 1952; "Besprechungen des Organisationskomitees der Kohle- und Eisenverbände," 11 July 1952; "Besprechung am 26. Juli 1952 im BWM"; "Memorandum über die Gliederung der zuerst anfallenden Aufgaben der Hohen Behörde," 21 August 1952; and "Sitzung im Präsidentenkomitee der 6 Eisenverbände," 26 August 1952.

6.2   Birth pangs in Luxemburg

On 10 August 1952 the nine members of the High Authority came together for the first time. Three of them, Monnet, Dirk Spierenburg of the Netherlands, and Albert Wehrer of Luxemburg, had become acquainted during the Schuman Plan negotiations. Monnet had also met the first vice-president previously. He was Franz Etzel, a close associate of Adenauer. Albert Coppé, a well-connected Flemish economist, was second vice-president. The other members were Léon Daum, a distinguished elderly steel industrialist who had worked with Monnet in the French Plan; Heinz Potthoff, a representative of the German unions; Paul Finet, a conservative Walloon labor leader with good internationalist credentials; and Enzo Giacchero, an Italian Social Christian politician known for his strong advocacy of European federalism.

These men shared a belief in "Europe" and a readiness, not to say avid enthusiasm, for working with Monnet to build it, yet none of them could have known what to expect, or even quite what to do, once installed at their desks. The treaty was obviously an imperfect document. Advance arrangements were chaotic. Monnet had set no agenda, or at least had revealed none. The administrative machinery of the community was not in place. Those appointed to the High Authority had no realistic idea as to how it would operate. Though members were supposed to be "bound to no mandate," the composition of the body indicated that it was to be broadly representative; at the same time, they had been selected in part because they shared Monnet's belief in Europe and could be expected to cooperate with him.

Monnet launched the European Coal and Steel Community with a discourse so imperious that a former associate later likened it to a Speech from the Throne. Though the ECSC's mandate was limited to coal and steel, Monnet declared that it was a sovereign power responsible not to the states that had created it but only to the European Assembly elected by the national parliaments, and beyond this was subject only to the European Court of Justice. Monnet left no doubt about his intention to use this executive authority to create the kind of economy he thought Europe needed. The treaty was, he asserted, "Europe's first antitrust law," one that conveyed a "mandate to dissolve cartels, ban restrictive practices, and prevent any concentration of economic power." Monnet warned that the High Authority was "in direct relationship with all firms" and obtained its financial resources not from the contributions of the states but from mandated levies. Its decisions were to be "immediately binding throughout the territory." Monnet had escalated the war against "the interests." To wage it successfully he needed a controllable apparatus in Luxemburg. The attempt to fashion one consumed the first six months of the community's existence.[29]

External challenges also had to be faced. With the Common Assembly due to meet in Strassbourg on 10 September 1952 as required by both the treaty and agreements among The Six, Foreign Secretary Eden revived a British proposal for submerging the ECSC in the much larger Council of Europe by attaching the secretariat of the latter to the Common Assembly of the former. To end the threat to his plans Monnet took the unusual step of organizing a meeting of the parliamentary secretaries of The Six, who agreed that the Common Assembly should meet at the chambers of the Council of Europe without, however, being bound to it administratively. Thus, the Common Assembly maintained a distinct identity as representative of the six treaty powers.[30]

Monnet's position was threatened less directly and deliberately by the enthusiastic parliamentarians of the European Assembly. At the inaugural meeting they quite spontaneously embarked upon highly ambitious plans of their own, proposing the creation of a future European Political Community (EPC). The EPC was supposed to bring together the existing ECSC and the planned European Defense Community (EDC), eventually developing into a federal government for Europe. Monnet was troubled by the European Assembly's assertion of independence and did not like the threat posed to his work by the proposed future federal government but could do little openly to discourage the EPC planning effort: It would have been embarrassing to raise public objections to the popularization of the European idea, not least of all because Adenauer strongly supported the Strassburg endeavor. Discussions concerning the European Political Community continued until the French rejected the EDC in August 1954. The EPC died unmourned and almost unnoticed but was not unimportant: Even before the ECSC had proved itself, many Europeans had begun to think seriously about planning the next stage of integration.[31]

The High Authority met thirty-seven times between August 1952 and January 1953, often in the intense, exhausting, almost wartime

---

[29] Confidential source, pp. 345–346.   [30] Monnet, *Memoirs*, pp. 380–381.
[31] Ibid, pp. 394, 430; Willis, *France, Germany, and the New Europe*, pp. 158–161.

atmosphere of crisis and emergency that Monnet liked to maintain. Yet it accomplished little. External relations consumed a huge amount of the HA's time, much of it going into the chimera of British association. Relations with OEEC, GATT, and the IAR were also prominent on the agendas of the meetings. Their only important result was agreement that IAR should continue to allocate German coal exports until the opening of the common market.[32]

Disagreements about procedures bedeviled High Authority operations. According to the treaty, the HA was to act by majority vote, but Monnet overlooked this stipulation: "[The] members seldom voted. Instead they talked out their problems, often late into the night, until at last they almost always agreed."[33] Neither "collective responsibility" nor the methods used to arrive at it were universally popular; by 2 October 1952 they had provoked a crisis serious enough for Monnet to have threatened to resign from the High Authority. Dressing the members down with the reminder that they constituted a deliberative body, not "a college for action," he asserted that the president needed a mandate to intervene whenever and wherever necessary and, unless allowed to exercise it, would have to carve the High Authority into separate bailiwicks, each under the control of an individual member. The crisis passed without Monnet making any concessions.[34]

Nor, though the matter came before the High Authority repeatedly in fall 1952, was the new president willing to compromise on matters of internal organization. Monnet did not want his hands tied by bureaucrats, least of all those working at cross-purposes to him; he intended to run the community either through his team or with experts brought in for special purposes, as he had done at the French Plan. "If one day there are more than 250 of us, we shall have failed," he said. Monnet hired as few officials as possible, insisted that they be retained on contract rather than given permanent status, and refused to commit himself to any overall organizational plan until his commando group was fully formed.[35]

The first appointments were made on 4 November, and by the end of December 1952 all the main posts were filled. The High Au-

[32] See JM AMH 7/1/1-7/1/37, Minutes of Sessions 1–37; JM AMH 7/1/1, "Procès-verbal de la séance de la Haute Autorité tenue de 17 août 1952."
[33] Confidential source, p. 351.
[34] JM AMH 7/1/12, "Compte-rendu de la quinzième séance de la Haute Autorité," 2 October 1952.
[35] Confidential source, p. 351; also, AN 81 AJ160, "Note E. Hirsch: Réflexions sur l'organisation de la Haute Autorité" (n.d.).

thority then consisted of 280 persons divided among five main branches (Production, Markets, Investment, Economics, and Labor) and five somewhat more specialized branches (Finance, Transportation, Control, Law, Statistics, and Administration).[36] The scheme was every bit as confusing as it appears to be; it is nearly impossible to determine who was competent for what. The Main Branch for Production was responsible for quotas, whereas the Main Branch for Markets dealt with sales and price issues, and the Main Branch for Investment dealt only with problems of readaption. How to deal with the one without at the same time touching on the other was unclear. Even though branches such as Statistics had well-defined functions and responsibilities, the remark often made about the camel applies equally to the organizational scheme of the High Authority: It was a horse designed by a committee.

More important than the offices themselves were the men in them. This was particularly true of the vaguely named Main Branch for Economics headed by Monnet's leading braintruster, Pierre Uri. Uri was responsible for harmonizing policy between Luxemburg and the member states. The men of WVESI feared Uri and were also uneasy about the control section responsible for cartels and mergers, whose director, the German-Jewish lawyer, Richard Hamburger, had spent four years hiding from the Nazis in a cramped Dutch attic.[37] Other than figures in Monnet's secretariat such as *chef de cabinet* Max Kohnstamm, another Dutch Jew, and François Duchêne and later Richard Mayne, both Englishmen, these were the only figures unequivocally committed to his leadership occupying top-level positions in the administrative machinery of the High Authority. Outside of this hard core, Monnet could count on a sympathetic support from nearly all representatives of labor as well as a great many from government. Otherwise, he was relatively isolated within the High Authority, and rather than work within its framework he operated outside of it.

Most of the key positions in the administration were held by non-Monnet men. Along with the German mining engineer Hermann Dehnen, the senior ARBED manager Tony Rollmann was the dominant influence in the Main Branch for Production. Wilhelm Salewski, the business manager for WVESI, was the director of its counterpart for investments. The Ruhr coal cartel's leading statistician, Rudolf Regul, played a strong second fiddle to Uri at the Main Branch for Economics. As for the less important branches, a French civil

[36] JM AMH 7/1/21, "Procès-verbal de la 23ème. séance de la Haute Autorité," 4 November 1952; WVESI, "Vermerk über die Vorstandsitzung am 13.1.1953."
[37] WVESI, "Vermerk über die Vorstandsitzung am 13.1.1953."

servant was in charge of finances, two senior railroad officials (one French, one German) were responsible for the transportation sector, an Italian politician headed the labor section, and an Italian nobleman directed the office staff. Two of the three influential legal counsels (Robert Krawliecki and Robert Much) were German as was the director of the important statistical section, the eminent SPD economist Rolf Wagenführ.

Statements made at the 13 January 1953 directors' conference of WVESI indicate that the Germans were actually quite pleased with the distribution of offices. They held sixty-three "leading positions," the French forty-one, the Belgians seventeen, the Italians nineteen, the Dutch thirteen, and the Luxemburgers, among whom office staff was recruited, sixty-four. The only change WVESI thought worth lobbying intensely for was the appointment of Ernst Wolf Mommsen, the director of the cartel for rolled steel products, as deputy director of the community's detrustification arm, the control branch headed by Richard Hamburger. This was pushing things a bit too far. Mommsen remained in the Ruhr.[38]

Monnet never admitted intending to eliminate the Advisory Committee (AC), but his actions speak clearly for the purpose behind them. He waited six months to convene it for the first time, and prior to that did everything possible to reduce its future role. On 24 September 1952 Monnet told the High Authority that its bureaucracy was getting too big to control. Though without explicitly stating that there was no need ever to convene the Advisory Committee, he proposed creating external committees to take over responsibilities foreseen for the AC in the treaty. They were to be formed for sales, investment, transportation, labor relations, common markets, and production. As implied in their name, the external committees were to serve as adjuncts to the High Authority, meet only at its request, and have no permanent staffs. Their powers were to be purely advisory.[39] Monnet's proposal occasioned considerable disagreement, with the German members of the High Authority recommending that the external bodies be subordinated to the still hypothetical Advisory Committee. He in turn objected that this was not permissible under the treaty.[40] The High Authority approved the creation of the proposed organizations only after the president

[38] Ibid.
[39] Ibid; KA/Henle EGKS HA Externe Ausschüsse, Wirtschaftsvereiningung, 1.9.52–31.10.52, "Vermerk über die Sitzung am 24.9.1952," and "Vermerk über die Besprechung in BWM," 30 August 1952.
[40] JM AMH 7/1/28, "Compte-rendu, 28ème. séance de la Haute Autorité," 9 December 1952.

had agreed that their members would also be allowed to sit on the
AC. The external committees eventually met and drafted numerous
studies, none of which had any influence because the new bodies
were no less business-dominated than those Monnet had hoped they
would replace.

On 15 January 1953 the Advisory Committee met for the first
time, the "last of the five important community organs to begin
operating."[41] It was a producers' club. The president, Hellmuth
Burckhardt, was a German-trained mining engineer, the general
director of the ARBED-controlled Eschweiler Bergwerke, a nephew
of the great Swiss historian Jacob, and a man of considerable personal charm. There could have been no better pick. The chief of the
French manufacturers' association, A. R. Métral, was elected vice-
president. To counterbalance the influence of this bitter enemy of
Monnet, a conservative Belgian unionist named André Renard was
chosen to fill the other vice-presidential slot.[42] The procedures for
delegate selection followed in the Federal Republic were typical of
those used by the other member nations. Representatives of heavy
industry were nominated by the producer associations for coal and
steel, those for dealers and consumers by the Bundesverband der
Deutschen Industrie, and those of labor by the unions. The government had agreed in advance to approve the choices made by these
bodies. The only problem in the selection process arose after German representation had been reduced from thirteen to eleven. This
cost Ruhr coal and steel interests one seat apiece.[43]

On 26 January 1953 Monnet greeted the delegates to the producers' parliament with a stiff reminder that they would have to prove
themselves before getting any real power to exercise – rise above
mere advocacy of special interests and make substantial progress in
getting the various national economies "out of the closet" and into
the "spacious European room." The opening of the common markets would, he promised, give them the chance to prove that they
could act in the spirit of the Schuman Plan idea.[44] It was too late for
this kind of rhetoric. Monnet had tried and failed to organize the

[41] WW, "Einsetzung des Beratenden Ausschusses," *Bulletin des Presse- und Informationsamtes der Bundesregierung*, Bonn, 27 January 1953.
[42] See Dichgans, *Montanunion*, passim.
[43] SP, "Sitzung des Engeren Vorstandes (WVESI)," 8 November 1952; WW "Die
Mitglieder des Beratenden Ausschusses der Europäischen Gemeinschaft für
Kohle und Stahl," *Bulletin des Presse- und Informationsamtes der Bundesregierung*,
Bonn, 28 February 1953.
[44] JM AMH 15/11/1, "Allocution de M. Jean Monnet à la Première Réunion du
Comité Consultatif," 26 January 1953.

High Authority as a commando unit imbued with his own ideals. It had taken six months to get the coal and steel community into operation, which was anything but smooth. Producers, supported by their governments, would have to handle the preparations for the work of opening the common markets. The most the High Authority could hope for would be to influence the process.

6.3 The common markets open

The creation of a single, communitywide area in which coal and steel products could be sold on a purely competitive basis, though a praiseworthy goal, was unrealistic, and there was not a ghost of a chance of meeting it in 1953: The risks of introducing free markets were prohibitive; the coal and steel commerce in Western Europe had been subject to regulation for decades, and each nation had its own complicated system of controls designed to meet national priorities and protect vested interests. Real costs were unknown, the effects of competition unpredictable. It made sense politically if not economically to begin the integration process by announcing the removal of tariffs and quotas. Tariffs were nonexistent for coal and unimportant in the European steel trade, and quotas obtained only in the case of solid fuel exports from the Ruhr. These barriers could be formally removed without seriously affecting the flow of commerce. The main causes of market distortions were artificial currency parities, the workings of officially mandated price controls, government subsidies, and differences in taxation, freight, and credit policies. To have ironed out such variables and created a single institutional environment for economic activity was the work of years if not decades. To have made even a start on harmonization was a substantial achievement.

Though officially committed to the unrealistic policy of creating a genuine common market, Monnet realized that the best one could hope for was to persuade producers to adopt "nondiscriminatory" pricing policies. In practice this meant introducing price listing and imposing price reduction. These were the two main subjects discussed during the preparations for the common markets in coal, ore and scrap (set to open on 10 February 1953), and steel (established on 1 May 1953). The scenario that Monnet had envisioned for opening the common market began with the collection of relevant sales and production data by producers, followed by the convening of expert commissions of industrialists chaired (and dominated) by representatives of the High Authority who would work this information into acceptable policy. The Advisory Committee was expected to endorse the conclusions of the expert commissions. Prices would then be officially reduced and publicly announced.

Not even these relatively modest objectives were met. Results varied by product. The event of 10 February had one important consequence for *scrap*, however: The High Authority found itself willy-nilly managing a cartel. *Ore* was relatively unimportant in either the politics or the economics of the community, and the opening of the common market for the commodity was largely of symbolic significance. Nothing changed in *coal* for the opposite reason: Fuel pricing was too sensitive an issue for the High Authority to handle. The opening of the common market in *steel* three months later did have one noteworthy effect, the partial decontrol of prices. Though in the Federal Republic deregulation had begun in August 1952, in France low ceilings remained in force until 31 May 1953. Decontrol nonetheless unexpectedly resulted in price increases instead of reductions. To mask its powerlessness, the High Authority developed face-saving policy rationales. For coal it established official price maximums comfortably above those contemplated by national authorities. In steel it "authorized" each district to sell according to its own price lists. Both decisions were presented to the public as evidence of a new policy.

Producers, or their governments, were in control at every stage of the price-setting and marketing process. Private discussions in connection with the openings had begun well before Monnet officially directed producers to consider the issue in January 1953. In November 1952 Salewski first officially warned the directors of WVESI that they were wrong to suppose that the event could be postponed until summer or winter of 1953. Monnet, he asserted, was irrevocably committed to eliminating all traces of price linkage in coal, and "in iron and steel we can expect to encounter ferocious resistance to all of our efforts to regulate our affairs."[45] Salewski intimated to his listeners that though the president of the High Authority might not really intend to "atomize" the internal market, as the Belgians were hysterically claiming, the man was indeed deadly serious: He and his associates were toiling from early morning until late into the night with incredible dedication and thoroughness. Even the elderly M. Daum was reported to have been working to exhaustion in order to keep pace with Monnet's eager young aides, and Salewski was highly

[45] SP, "Sitzung des Engeren Vorstandes (WVESI)," 8 November 1952.

impressed: "The level of intelligence is extraordinary . . . an elite has foregathered, . . . and the odds of their completing their big political plan are one-hundred percent."[46] WVESI's business manager nonetheless insisted that it was not yet too late to act, and by the end of November 1952 discussions were well under way within the industrial community. Success was not automatic. An early December proposal by General Director Ingen-Housz of Hoogovens for a European rolling program got nowhere even though it made economic sense: Steel was in short supply, economies could have been achieved in the use of available capacity, and the creation of savings through rationalization was supposed to be one of the main aims of the community.[47] Because such a program would have required close cooperation among producer associations, Monnet rejected it, as he again did in February when Salewski brought the matter to his attention for a second time.[48]

The steel producers' intraindustry discussions that took place in December 1952 focused on coal and scrap, for which the deadline was 10 February 1953. (That for steel, originally set for 10 April 1953, had been postponed until 1 May.) Ore, also with a deadline of 10 February, was only a minor industry concern because the Germans imported high-grade mineral from Sweden, and so long as coal costs did not increase they could continue to consume low-ferrous-content Salzgitter ore. French and Luxemburg minettes were competitive only locally. The Italians got the bulk of their ore from North Africa, from which they could be supplied at reasonable cost. A common market in ore thus represented little threat to established marketing patterns. The common market in coal, by comparison of immense potential significance, offered the High Authority little scope for action for several reasons: Outputs were subsidized everywhere, and any attempt to introduce market pricing would have been inflationary; the HA lacked the power to confront either the national cartels or the governments behind them; and the Ruhr was regulated by occupation controls that were to remain in effect until the opening of the common market. In the meantime, the International Authority for the Ruhr (IAR), the Deutsche Kohlen Bergbau Leitung (DKBL), and the Deutschen Kohlen Verkauf (DKV) continued to operate.

[46] Ibid.
[47] KA/Henle Europäische Gemeinschaften, EGKS, Hohe Behörde, Externe Ausschüsse WVESI, "Besprechung der 6 Eisenverbände in Rom am 28./29.11.1952."
[48] AMH 23/3/8, Hamburger to Monnet, 13 February 1953.

322   Coal, steel, and the rebirth of Europe

The High Authority had little direct influence in Ruhr coal policy. The DKV set domestic prices, and the West German government sanctioned them, a relationship between private and public authority in effect since the 1920s. The two decided on increases of 15 percent on 1 February 1953 sufficient to raise internal levels to those for foreign sales, thus eliminating dual pricing.[49] The Deutsche Kohlen Bergbau Leitung, a trusteeship agency organized by the British, fixed export prices for Ruhr fuel. These the United Kingdom had set high relative to domestic levels because it wanted to protect Germany's balance of payments. The International Authority for the Ruhr was responsible for the politically difficult job of allocating German coal exports. Monnet chose not to take over this function and confirmed IAR's responsibilities through the first quarter of 1953.[50]

The HA could also do little about the coal situation in France. Domestically mined combustible was under price control, with the government providing compensation for losses incurred in production. The Association Technique de l'Importation Charbonnière (ATIC) was the sole importer of coal. It exacted a tonnage fee for the consumption of foreign fuel that was then employed to compensate users of the domestic product. ATIC was a legal, semipublic corporation in which the nationalized mines were a major shareholder. Its elimination would have flooded domestic markets with foreign coal and forced French mines out of production.[51]

The changes introduced by the opening of the common market in coal were largely cosmetic. Though tariffs and quotas were removed, these existed only in Italy and came under a special exception listed in the treaty. The High Authority's price ceilings had no effect on prices themselves, and the highly publicized new policy of "zonal pricing" proclaimed as a corrective to old marketing abuses did not stop cartels and governments from setting prices but merely forbade a district from aligning its prices on those of competitors at distant basing points.[52] Zonal pricing would theoretically have encouraged competition and lowered costs if introduced when demand

[49] Abelshauser, *Der Ruhrkohlenbergbau*, p. 70–71.
[50] JM AMH 7/1/1, "Procès-verbal de la séance de la Haute Autorité tenue le 17 1952."
[51] Archives of the European Community, Brussels (EC), CEAB 7/41/1953, 59–73, "Organisation de la Distribution du Charbon en France," 14 October 1953; EC CEAB 7/41/1953, 42–48, "Note concernant l'association technique de l'importation charbonnière, (ATIC)," 14 September 1953.
[52] See Diebold, *Schuman Plan*, pp. 242–253; Lister, *Europe's Coal and Steel Community*, pp. 283–309; Meade, Liesner, and Wells, *Case Studies*, pp. 215–232.

was low. It was generally high, however. The actual effect of the ban on price alignment was to make it difficult for mines to sell coal produced at marginal cost to distant purchasers. Zonal pricing thus forced Aachen coal off of North German markets, made it impossible for the Saar to compete with British exports to Normandy, and had additional unanticipated consequences. The net effect of such disruptions, though far from catastrophic, was still sufficiently severe to cause the High Authority to back away from its new doctrine. Freight rate adjustments were soon allowed in special cases. Zonal pricing was eventually abandoned.[53]

Scrap was a troublesome issue, a notoriously volatile commodity, supplies of which varied enormously from place to place and demand for which was normally very inelastic. Siemens-Martin open-hearth furnaces, the main type to be found in Italy and the only type existing in the Netherlands, could not operate without scrap. This was not true of Thomas (Bessemer) converters, which predominated in France and Belgium and were the only type of unit operating in Luxemburg. These normally consumed pig iron, which was more coal-efficient than scrap. If, however, scrap prices were very low relative to either coal or pig, then the recycled material could be substituted for the latter. West Germany had 40 percent Thomas and 60 percent Siemens-Martin units and so was comparatively well positioned to vary consumption in response to price fluctuations. Moreover, Germany had a large scrap surplus and France a small one; together they roughly equaled the combined deficits of Italy and the Netherlands during the period 1952–1954.[54] Since the only important noncommunity supplier of the distinctly unglamorous raw material to Europe, the United States, was not yet competitive on Continental markets the Federal Republic held the upper hand in the community's politics of scrap.

Volitility plus vulnerability equaled government-authorized regulation of the scrap market in each of The Six. The Schrottvermittlungsvereinigung GmbH (SVG) was the exclusive buyer and seller in West Germany, as was CAMPSIDER in Italy, and the Union des Consommateurs de Ferailles in France. On 2 December 1952 the heads of the six community steel producers' associations met in Rome for the first time to discuss scrap-related issues arising from the planned opening of the common market. This was not a happy meeting because the Italians and the Dutch, the two main importers of the material, were angry because the Germans and

53 Diebold, *Schuman Plan*, pp. 251–252.
54 Ibid, pp. 287–288.

French arrived in Rome having just agreed upon a common policy in Paris; the importers demanded the formation of a committee to devise a communitywide policy. The Germans at first refused to join, but the French declared themselves willing if their partner was, and eventually the necessary arrangements were made.[55] The issue was price, which ranged from fifty-eight dollars a ton in Italy at the beginning of 1953 to an officially controlled level of twenty-two dollars a ton in the Netherlands, with France and West Germany in between, at twenty-eight dollars and thirty-seven dollars, respectively. For the French, whose mills used little scrap, exportation to Italy was highly lucrative. For the West Germans it was this and more: Low prices benefited Ruhr producers by both reducing the cost of steel production and, through stimulation of domestic consumption, raising export prices. The lifting of communitywide restrictions on the movement of used metal would have been a bonanza for the Bundesrepublik.[56]

At meetings during January 1953 the Germans and the French made major concessions to scrap importers, and on 10 February 1953 the steel producers had a package deal ready that the High Authority found difficult to reject and would eventually accept after the opening of the common market for the material on 15 March 1953. Its main features were cost sharing (*peréquation*) for importers of supplies from the United States, communitywide regulation of the used metal market through a new organization patterned on the German SVG, and the limitation of exports to third countries as well as the regulation of consumption and the imposition of maximum prices at slightly above the German level. The arrangements effectively shifted the costs of importing US scrap to the Germans.[57]

The Advisory Committee recommended the adoption of the industry plan on 26 January 1953. Overlooking flagrant violations of

[55] KA/Henle Europäische Gemeinschaften, EGKS, Hohe Behörde, Externe Ausschüsse WVESI, "Besprechung der 6 Eisenverbände in Rom am 28./ 29.11.1952."

[56] Diebold, *Schuman Plan*, p. 291.

[57] WVESI Sitzungsberichte "Vorstandssitzung am 10.2.1953 Bericht über die Sitzung des Ministerrats vom 2. Februar, der Vollversammlung des Beratenden Ausschusses vom 5. Februar, der dazwischenliegenden Unterkommissionssitzung des Beratenden Ausschusses und der weiteren Ministerratssitzung am 7.d.M"; BA B146/682, "Erwägungen der einzelnen nationalen Sachverständigen-Delegationen betr. die Errichtung des Gemeinsamen Marktes für Kohle, Eisenerz, und Schrott," 15 January 1952; WVESI, "Vorstandssitzung am 10.3.1953"; WVESI Sitzungsberichte, "Der Delegierten der Stahlindustrien der Montanunion (1953) in Düsseldorf"; EC CEAB 4/1953/156, 26–33, "Note sur les entretiens relatifs à la feraille," 10 March 1953.

the treaty, the High Authority permitted temporary allocation and sanctioned price equalization as well as joint foreign purchase of used metal. Though occasionally restraining national scrap cartels, it also later "sponsored a voluntary organization of merchants and producers to discuss common problems."[58] As steel demand cooled in late 1953, pressure on prices abated; it became possible for the High Authority to lower the theoretical price ceilings gradually and abolish them altogether in March 1954.

"The establishment of the common market for steel," the High Authority announced magisterially on the eve of its opening, "signifies the suppression of restrictions on exchange and the transference to the High Authority in [this] domain of powers exercised up to now by the governments of the six member-nations."[59] Customs, quotas, and currency barriers were all to be removed and, the communiqué added, external tariffs harmonized. In addition, prices would be freed: The HA would not set ceilings as in coal unless collusion by producers brought price increases; if this should happen, however, it promised to intervene decisively. The realities of the 1 May 1953 event were considerably less impressive than suggested in the High Authority's dramatic press release. The establishment of the common market was not of great significance economically; as in coal, the officially removed barriers had little influence on the steel trade. (The French quota for overall steel imports, which worked to the detriment of Belgium, was an exception.) External tariffs were harmonized but not lowered. Nor did a substantial transference of power from the national governments to Luxemburg take place; this was a convenient fiction. Finally, though the High Authority did free prices, it liberated them not from cartels but from governmental authorities in those nations where official price control had been in effect up to then, France and Belgium. The coal–steel pool exercised no influence over price formation; this power was in the hands of producers.

Discussions conducted by representatives of the steel associations of The Six began in January 1953 and were mainly technical. Basing points was one of the first topics considered. No industry representative dared rock the boat by reducing the number of them or changing their locations. Agreement was quickly reached at a 27 February meeting to adopt a German proposal for retaining one basing point per district, while simultaneously working toward the

[58] Diebold, *Schuman Plan*, p. 291 ff.
[59] EC CEAB 53/1953, 174–179, "Communiqué de la Haute Autorité sur l'etablissement du Marché Commun de l'Acier," 30 April 1953.

organization of a freight pool.[60] In such an arrangement a low-cost producer selling in a nearby district where prices were high would split the generous margin of profit. The harmonization of pricing differentials between different districts was worked out in subsequent meetings. In the one market where substantial competition seemed possible, South Germany, the French agreed without hesitation to align their prices on those of the Ruhr. The CFPS had tried too hard to raise prices, M. Ferry said, and to begin undercutting them in an important export market was unthinkable. The producers for whom he spoke preferred the security of pooling to a risky attempt to conquer new market shares.[61]

The Germans encountered little enthusiasm for the suggestion that the other delegations adopt wholesaling methods used by their steel industry, in particular the granting of rebates to large purchasers, as this practice often brought prices below even those charged for direct sale from the works and further raised unmanageable administrative problems for Belgium, which was overstocked with no less than 240 authorized sales agents. The Germans concluded privately that these objections were of a practical rather than a theoretical nature; the others, noted the WVESI report from the meeting, actually preferred German wholesalers to those of their own countries.[62]

The most complicated issue raised by the opening of the common market in steel was nomenclature for specialized manufactures. In basic products, definitions were standard, price lists published, and prices themselves stable. For custom jobs, however, producers had to establish mutually satisfactory criteria. Impressive technical progress was made on this laborious process.[63] Pricing was a controversial matter. Pierre Uri, representing Monnet, directed that the added charges (*Aufpreisen*) for custom manufactures be determined by ac-

[60] WVESI Sitzungsberichte, "Bericht über die Besprechung der Delegierten der Stahlindustrien der Montanunion am 27. Februar (1953) in Düsseldorf."

[61] KA/Henle Europäische Gemeinschaften, EGKS, Hohe Behörde, Externe Ausschüsse WVESI 1.2.53–31.3.53, E. W. Mommsen to Henle, 6 March 1953 (enclosure: "Besprechung mit Vertretern des Chambre Syndicale Paris," 28 February 1953).

[62] Ibid.

[63] EC CEAB 4/81/1952, 107–111, "Problème relatifs a l'etablissement du Marché Commun de l'Acier," 14 April 1953; WVESI Sitzungsberichte, "Vorstandssitzung am 14.4.1953"; KA/Henle Europäische Gemeinschaften, Ausschüsse WVESI 12.53–31.3.53, Mommsen to Henle, 6 March 1953 (enclosure: "Vermerk über die Besprechung mit Vertretern des Chambre Syndicale Paris," 28 February 1953).

tual production costs, which he wanted individual works to communicate to the High Authority. The producers, who would never willingly have provided such information to that *dirigiste* body, had a different policy in mind. According to Ernst Wolf Mommsen's notes from a 28 February 1953 meeting at the Chambre Syndicale in Paris, the French, and the others, were ready to adopt the German nomenclature for added charges as standards for the community.[64]

On 10 and 11 April 1953 the delegates of the six steel producers' associations presented their results to Uri, Rollmann, Dehnen, Vinck, Gueldner, and Mouget representing the High Authority. This, the last meeting prior to the opening of the common market, had an unsatisfactory outcome. Though the HA praised the new nomenclature for custom manufactures as a contribution to the transparency of steel pricing, it attacked the unified, German-based system for quoting added charges as being cartellike. One spokesman for the HA threatened to subvert it by encouraging underbidding. The steel delegates retorted that competition should be "channeled down" to basic products: If allowed to take place at the level of specialties it would result in confusion and prevent the High Authority from detecting discriminatory practices. The Germans also protested that unified lists of finished prices for exported products had long existed, maintaining that these had not interfered with competition and remained unchanged even when plummeting demand had caused decreases in base prices. The High Authority had to let the matter stand.[65]

Discussion of wholesaling lasted for another several hours before the big 10–11 April meeting degenerated into bitter recrimination. The issue divided producers from each other as well as from the High Authority. After complaining at length abut the unfairness of the rebating system in the Federal Republic, the exhausted delegates of the other steel associations finally "indicated that they had no objections to the Ruhr system as such so long as it had no negative consequences in their sales areas."[66] At the same time, they emphasized that "other dangers . . . had to be dealt with if they were to continue to endorse [the Germans'] right to maintain it." At this

[64] KA/Henle Europäische Gemeinschaften, Ausschüsse WVESI 12.53–31.3.53, Mommsen to Henle, 6 March 1953.
[65] KA/Henle Europäische Gemeinschaften, EGKS, Hohe Behörde, Externe Ausschüsse WVESI 1.4.53–31.5.53, "Ergebnisniederschrift über eine zweitägige Besprechung der Eisenerzeugerverbände bei der Hohen Behörde am 10. und 11. April 1953."
[66] Ibid.

point Uri exploded, "ripping into the entire [German] marketing system, using arguments . . . employed earlier by Allied negotiators, and coming out with a proposal that would have meant its destruction."[67] Refusing categorically to discuss the matter further, the two delegates from the Federal Republic "placed responsibility for the grave consequences that such a policy would have in the hands of the High Authority." The matter, they warned, would be brought to the attention of German wholesalers, German consumers, and the German Ministry of Economics. Uri capitulated to this ultimatum, claiming that he had only been expressing a personal point of view. Rollmann, Dehnen, and Vinck, "who had remained silent through the entire discussion," agreed that Uri's remarks in no way represented the High Authority's policy.[68]

These men were wrong: There would be no backing away from reform. On 20 April 1953 Monnet told the press that price collusion would not be tolerated. "If it appears," he said, "that prices which up to now have been different are becoming the same, this will be proof of a producer accord to evade competition," warning that "if prices are going to be fixed, this job will be done by the High Authority, not by the interests."[69] The following day Monnet accused producers of disloyalty at a meeting of the Council of Ministers, adding that they needed "an apprenticeship in liberty."[70] The president of the High Authority claimed that their backsliding had come as no surprise to him, citing the recent formation of the European export cartel as an example of their lack of atonement; yet rules were rules, he insisted, and he would enforce them no matter what the costs.

Jawboning had little effect. On 22 April 1953 Monnet summoned the Advisory Committee (AC) to express his extreme displeasure that prices were rising in preparation for the 1 May event and to reject the added price system devised by the producers that the AC had by then already endorsed. These, he indicated, were the work of cartels and in violation of the treaty, and as a result he was going to slap on a price ceiling.[71] Monnet's remarks touched off the first real savaging he would receive from the producers. The meeting opened

---

[67] Ibid.   [68] Ibid.

[69] JM AMH 23/4/24, "La Haute Autorité ne pourra accepter la formation d'ententes entre producteurs pour augmenter leurs prix," *L'information*, 21 April 1953.

[70] JM AMH 20/3/16, "Les industriels protestent contre les griefs que leur adresse M. Monnet au sujet des ententes de prix," 21 April 1953.

[71] JM AMH 15/13/1, "Comité Consultatif: Compte-rendu in extenso, séance du 22 April 1953."

at 11:46 in the morning and would last for fifteen hours. Ricard spoke first. Why, he asked, had Monnet intimated to the press that the pool was in danger? Did he want to wreck it? Was he unaware that price increases could be the work of the market rather than the product of collusion between producers? Was he ignorant of the fact that in the United States, where cartels were taboo, there was something called price leadership that amounted to roughly the same thing? Ricard was not about to be stopped: He even gratuitously raised the explosive issue of the export cartel that the producers had formed in the very weeks of preparation for the common market in steel, refusing to admit that it was wrong in either law or economics. The convention, he protested, merely forbade producers from selling below a range of seventy-two to seventy-five dollars per ton at a time when the market could command eighty-five dollars. Beyond this, the maintenance of high export prices brought foreign exchange into the community and made it possible to lower costs within The Six. What was the problem? Ricard apologized sarcastically for not having sent Monnet a copy of the convention: Maybe then he would appreciate it more! And, the steelmaster pitilessly ground away, Monnet might also begin to demonstrate an understanding of the problems that he faced as chief of the Comptoir Français des Produits Sidérurgiques in trying to unite eighty members behind any single agreement. How was he to impose sanctions on them for violations of the treaty if Monnet continued to undermine him by attacking his every action with accusations of engagement in illicit practices? Did the president imagine that he and Mr. Goergen of WVESI could arrange everything over a glass of beer? Monnet should be left in no doubt about what was coming, Ricard warned: Across-the-board "adjustments" were in the works because prices had been held down by the Direction des Prix to some 10 to 15 percent below German levels. The industry spokesman finally dismissed the High Authority's worries about price gouging as unfounded. CFPS order backlogs were down from 560,000 to 270,000 tons over the past twelve months. He, Ricard, hardly needed reminders from Monnet that France had to stay competitive.[72]

Representatives from the other steel industries amplified the CFPS chief's remarks. Bruns of WVESI stated unashamedly that the Germans were participating in conversations "being conducted in Luxemburg and elsewhere" aiming at the setup of joint export machinery and had so informed the government of the Federal Repub-

[72] Ibid.

lic. The Belgian van der Rest chimed in that the producers of his association also had every intention of raising prices, and for the same reason as the French, because of previous controls imposed by their government. Chome of Luxemburg echoed the remarks of the other industrialists, and the delegates for both French and German manufacturers, Métral and Lange, associated themselves with them as well.[73]

Barely able to defend himself against this barrage of accusation, and "speaking with open heart," Monnet admitted to "being troubled, very troubled," by the remarks of the producers. His ordeal was not yet over, however. Ricard expressed amazement that the president had still failed to develop any appreciation of the kinds of problems facing the steel industry, and he lashed out bitterly at Monnet's main defender, the Belgian union leader André Renard, for his "outrageously partisan approach." The meeting adjourned at 6:25 for dinner but soon resumed, eventually ending at 2:53 the following morning. Monnet would not convoke the Advisory Committee for another six months.[74]

In the first quarter of 1953 twelve new European steel syndicates formed for cooperation on third markets, two for the community, four for France, and one for Germany (see Table 6.1). Potentially, the most important of them was the International Steel Export Cartel set up in Brussels in March 1953. It set up price minimums for about 15 percent of total community production – actually a higher percentage than that of the old ISC of the 1930s. There was very little that Monnet could do about this. On 13 February 1953 the High Authority's all-but-powerless cartel cop, Richard Hamburger, threatened to resign unless given a permanent labor contract, unless the other branches of administration ceased withholding essential information from him and stopped making informal contacts with cartels, and unless Monnet himself desisted from "following a procedure in which you tell me 'there exists such and such a cartel, what do you propose in the way of immediate action?'"[75] Hamburger would remain in office, but his task was hopeless. The last thing Monnet had wanted to do was make Europe safe for cartels. The future of the Coal and Steel Community was nonetheless now in their hands.

[73] Ibid.   [74] Ibid.
[75] JM AMH 23/10/1, Hamburger to Monnet, 13 February 1953.

Table 6.1. *Associations and cartels founded after the creation of the Schuman Plan (purposes given in parentheses)*

Associations	Cartels
France Association of Steel Producers (study of treaty)	International Entente for the Steel Processing Industries (allocation of orders)
Association of Iron Ore Producers (study of treaty)	International Steel Export Cartel (price fixing)
French Office for the Study of Laminated Steel [Strassbourg] (joint producer representative)	Belgo-German Steel Cartel (price fixing, market allocation in ECSC)
Steel Syndicate of Centre-Midi (representation of regional interests)	German Steel Cartel for Rolled Products (price fixing)
Federal Republic of Germany Scrap Union West (joint purchase, sale, distribution)	Belgian Scrap Entente (general common policy promotion)
	"Gentlemen's Agreement on Wire Products" (price fixing)
European Coal and Steel Community Union of Industries of the Six Nations of the European Community (general advisory of overall interests)	Wire Cartel [Belgium, Luxemburg, France, Germany] (quota setting)
Joint Office of the Federation of Mechanical Industries (including delegation of the French casting industry)	Belgo-Luxemburg Wire Cartels (quota setting)
	Netherlands Wire Cartel (quota setting)
	Benelux Wire Cartel (quota setting)
	Dutch Steel Wholesalers' Cartel (rebating)
	Domestic Price Cartels [one per producing district] (price fixing)

Source: JM AMH 23/3/9[b] "Liste des associations formées après l'institution du Plan Schuman," 22 May 1953, and AMH 23/3/9[c] "Liste des cartels conclus après l'institution du Plan Schuman," 22 May 1953.

6.4   Action and inaction at the ECSC

The High Authority never surmounted the problems that appeared at the opening of the common markets. The organization suffered from lack of leadership at the top, a bloated bureaucracy, and factionalism. Inexhaustible in the production of investigations, analyses, statistical surveys, reports, and memos but paralyzed in action, the HA was a mere house of words. In fall 1953 conflict between President Monnet and industry broke out over price policy, and it could be only papered over. The president was also helpless when it came to cartels, his men being unable to break up "excessive concentrations of power" already in existence and unwilling to prevent the formation of new ones. Nor did a huge loan provided by the United States give him the leverage over producers he sought, and their anger gave way to ill-concealed contempt. On 2 December 1953 Pierre Ricard remarked sarcastically to an approving Advisory Committee that the ECSC could be scrapped the following day without making a difference: There had been no coordination of investment, no rationalizing of fabrication programs, no harmonization of labor conditions, and no cross-investment among the member nations of the community. This judgment was unnecessarily harsh. The ECSC stimulated progress in a number of areas ancillary to its central focus of activities. It made beginnings in the development of a consistent communitywide tax policy; brought about reductions in certain freight rate inequities; forced governments and industries to confront problems of readjustment and worker housing; and regulated the community scrap economy. These advances were the work of ad hoc study groups, teams of expert investigators, temporary boards, task forces, and intergovernmental commissions. From such improvised cooperation would arise a new approach to integrating Europe.

Monnet was no more willing to share power with the Advisory Committee after the opening of the common market than previously. He seldom convoked the body and whenever possible withheld information from it. As of January 1954 the High Authority had refused either to issue bylaws for the organization or otherwise to define the statutory relationship between itself and the "producer parliament." Monnet even denied, contrary to Article 60 of the treaty, being under any obligation to report to it. On 25 January 1954 the president of the High Authority warned the eighth session of the Advisory Committee "quite frankly and quite amicably that everyone has his place and it is not by mixing in that one solves problems." "Your formal role," he reminded the delegates, "is expressly that we present you with the general objectives of policy prior to publication,"[76] adding that if they wanted to become councilors to the High Authority it first would be necessary to liberate themselves from the advocacy of special interests.

No progress was made in the matter of the AC's bylaws (*règlement intérieur*) until after Monnet declined reappointment as president in November 1954. Unresolved were the composition and procedures of the body, the attendance of its members at plenums and expert commissions, and its authority, if any, to make nominations to such gatherings. At the fourteenth session on 4 November 1954 Ricard ridiculed Monnet's attempt to operate without producer cooperation as well as the idea that delegates could be independent. Surely, he said, "if tomorrow I am no longer President of the Chambre Syndicale . . . there will be no further reason for me to remain in this room."[77] Two years of meetings had done nothing to reconcile French steel to Monnet's leadership.

The first problem to confront the High Authority after the opening of the common market concerned steel pricing. The event inexplicably caught buyers unprepared, causing a near cessation of order placement and throwing markets temporarily into turmoil. Within three weeks it had become clear that there were few opportunities to shop around. As Gerhard Bruns happily reported to WVESI, price leadership had taken hold. "To the astonishment of many steel consumers, the prices of our neighbors to the west are rising up to our own at this very moment. The reasons for this [are] that steel prices [there] have been kept artificially low by governments up to now."[78] German levels, he added, had changed only moderately, with those for basic products declining while charges for custom finishing increased.

Considerable confusion surrounded the workings of price leadership, both then and now. According to Ernst Wolf Mommsen, it was unique in being exercised by "product group" rather than individual firm or national association, a result, presumably, of the arrangements made by producer associations in preparation for the com-

---

[76] JM AMH 15/13/4, "Comité Consultatif: 8ème session, compte-rendu analytique," 25 January 1954.

[77] JM AMH 15/13/10, "Comité Consultatif: 14ème session, compte-rendu analytique," 4 November 1954.

[78] EC CEAB 3/1953/116, 1–22, "Rapport fait par le President du groupement 'Walzstahl' de [WVESI] M. le Directeur Bruns," (n.d.).

mon market.[79] Writing in 1960 and struck by the fact that "for the first [and only] time official domestic prices in 1953–1954 were nearly equivalent to official export prices," the economist Louis Lister concluded that community producers had merely aligned the former on the latter.[80] This too is possible, even though the export cartel did not officially regulate prices but only established minimums. The High Authority, for its part, lacked the information to act upon its suspicions of collusion. Richard Hamburger reported on 23 June that he "did not know whether the published prices were [merely] harmonized or whether there exist far-reaching price conventions . . . underpinning them . . . which would be forbidden under the treaty."[81] In Italy, at least, the old evils of secret rebating and price gouging still thrived. Before as well as after the opening of the common market, according to a Monnet informant from the manufacturing industry, steel prices went up both openly and clandestinely, with profit margins spreading to unprecedented extremes. He claimed that tied-in purchasers were benefiting from huge rebates while unfortunate independent consumers were too intimidated to protest openly.[82] Broad price increases soon gave way to sharp reductions. On 24 November Günter Henle wrote Adenauer that because of the fall in demand for steel that set in after July 1953 "confusion prevails on the German market worse than any felt for a generation."[83] Henle doubted that the High Authority could solve the problem and recommended that the government step in with a large program of public works to keep the mills operating.

The unexpected recession caught the HA in a serious bind. Though suspecting that collusion had been involved in the harmonization of the published lists, Monnet's directorate could not tolerate sales below authorized levels lest it lose what remained of its price-setting authority. Insisting that producers respect official schedules, the HA thus found itself in the absurd position of de-

[79] KA/Henle Europäische Gemeinschaften, EGKS, Hohe Behörde, Externe Ausschüsse WVESI 1.6.53–31.12.53, Mommsen to Henle, 11 June 1953 (Anhang zur Kartellfrage).
[80] Lister, *Europe's Coal and Steel Community*, p. 224.
[81] KA/Henle Europäische Gemeinschaften, EGKS, Hohe Behörde, Externe Ausschüsse WVESI 1.6.53–31.12.53, "Note [Hamburger] an die Herren Mitglieder der Hohen Behörde bezüglich der Preisbildung und Vereinbarung der Preise nach Eröffnung des Gemeinsamen Marktes," 23 June 1953.
[82] JM AMH 29/3/36, Marigili to Monnet, "La vérité sur le marché sidérurgique italien" (n.d.).
[83] KA/Henle Europäische Gemeinschaften, EGKS, Gemeinsame Versammlung, Korrespondenz 9.52–1.54, Henle to Adenauer, 24 November 1953.

fending pricing practices of which it had previously disapproved and objecting to the market-conforming behavior to which it was ideologically committed. Enjoying the High Authority's consternation, the producers claimed that revising the lists would be administratively difficult and demanded the freedom to set their own prices.[84]

The debate over price policy raged from September 1953 until the beginning of the new year. The president of the Advisory Committee, *Bergassessor* Burckhardt found it difficult to take the High Authority's steel-pricing policy seriously. Burckhardt claimed that the HA had never really intended to let the market work, criticized its interventions as ineffective, and found that "market order" was all but absent and "discrimination" as rampant as before. He described the rigid coal-pricing system by comparison as at least enforceable and as such preferable to prevailing practices in steel. The Bergassessor pleaded for the restoration of pricing "elasticity," which in his view required the establishment of an officially sanctioned price cartel resembling the old steel syndicate, as "it had handled such problems with only a small staff at its disposal."[85]

In January 1954 the High Authority adopted the "Monnet margin," a compromise allowing producers to deviate up to 2.5 percent from the listed prices for sixty days. Over the year demand rose steadily, and prices, which had sagged at the lower limit for the first four months of the year, were by December pressing the upper one, comfortably above where they had been in the opening months of the common market. They then rose steadily as the boom continued through 1955. Though after Monnet's departure the Advisory Committee assumed official responsibility for making adjustments to the price lists, then as earlier the community had only the appearance of control; whether in line with market trends or in defiance of them, producers still decided what to charge.[86] No one knew precisely how much official prices were being respected at any given time. Ernst Wolf Mommsen informed Henle on 11 November 1953 that "Article 60 of the treaty is being circumvented to some extent by every producer. Numerous delivery contracts over the past few months deviate not only from a firm's own [private] price lists but

[84] See Diebold, *Schuman Plan*, pp. 258–259; KA/Henle Europäische Gemeinschaften, EGKS, Gemeinsame Versammlung, "Ausschuss für Fragen des Gemeinsamen Marktes, Kurzbericht der Sitzung," 13 November 1953.

[85] JM AMH 15/8/8, "Kritik der Preisvorschriften im Gemeinsamen Markt," by Burckhardt, 4 November 1953.

[86] Lister, *Europe's Coal and Steel Community*, pp. 226–227, 233–234, 237–238.

from those [that have been] published. The actual extent of rebating cannot be fathomed, but the estimate of Herr Etzel that this amounts to from ten to twenty percent is exaggerated."[87] Mommsen also thought a figure of 2 to 3 percent, the approximate level of the later Monnet margin, a gross underestimate.

Cartel policy was another area in which ECSC was ineffective. After April 1953 it aimed at reform rather than prohibition. In July 1953 cartel registration was required. Of the eighty syndicates reported by 1958, the ECSC had dissolved only three, all concerned with scrap and none of them important.[88] It is difficult to prove that by holding open a watching brief the HA either prevented the formation of cartels or significantly changed their operating methods. The creation of the Entente de Bruxelles, an international export cartel formed in March 1953 shortly before the opening of the common market for steel, was a distinct slap in the face to the HA, yet on 19 June 1953 the High Authority announced that it would not challenge the upstart organization. Turning the other cheek did no good: Responding in September to the weakening of export markets, the producers transformed their gentlemen's agreement into a permanent institution, introducing quotas, fines, and bonuses and otherwise "re-establishing the bases of an agreement analogous to the international steel convention of the interwar years." The United States protested this move as a violation of the treaty as well as of commitments made to GATT, which had waived the prohibition on the ESCS's establishment of a communitywide tariff on the understanding that the pool would promote fair and expansive trade policies.[89]

For its part, the High Authority got tied up in its own regulations and again ended up supporting the very producer position it should have opposed. Article 65 of the treaty forbade agreements restraining trade and competition on home markets. The industry claimed

[87] KA/Henle Europäische Gemeinschaften, EGKS, E. W. Mommsen 1.7.53–3.12.53, Mommsen to Henle, November 11, 1953 (Vermerk II).

[88] Diebold, Schuman Plan, p. 379.

[89] Lister, Europe's Coal and Steel Community, p. 200–201; JM AMH 23/4/la, "Exportation" (n.d.); EC CEAB 2/1953/184, 13–18, "Note pour M. Hamburger, Le Cartel à l'exportation des produits sidérurgiques" (with enclosures), 8 September 1953; KA/Henle Europäische Gemeinschaften, EGKS, E. W. Mommsen, "Akten der Hohen Behörde Internationale Exportvereinigung," 23 November 1953; JM AMH 23/4/56, "Annexe à la Note no. 6719/f pour Messieurs les Membres de la Haute Autorité, Le Cartel à l'exportation des produits sidérurgiques," 16 September 1953.

that the HA's competence did not extend to the export cartel because it did not raise prices within the community. This was nonsense, as Richard Hamburger protested in a lengthy memorandum to Monnet: A quarter of ECSC steel output was normally sold on third-country markets, much of it by heavily export-dependent firms. A sharp fall in foreign prices to below community levels would, he reasoned, force such producers to sell within the ECSC, a practice that under normal conditions would force prices down on the common market. But conditions on the common market were not in fact normal: Producers who reduced prices, Hamburger despaired, would find themselves in violation of the community's rules forbidding rebates and unauthorized discounts. He concluded sadly that the ECSC could not ban the export cartel because it was needed to enforce community pricing policy![90]

Under pressure from the United States and GATT, the High Authority tried in December 1953 to win support from both the Council of Ministers and the Advisory Committee for the adoption of export price ceilings. After the failure of this purely face-saving proposal the ECSC declared itself incompetent to act, allegedly because export price maintenance did not "influence competition on the common market." With negotiations for a big U.S. loan pending and the cartel showing signs of weakness, the Ententes and Concentration Division tried to reopen the issue in spring 1954. The Germans, an official named Mr. Blondeel had learned from informants, were aggressively price-cutting on foreign markets. He thought that testimony could easily be gathered of syndicate attempts to suppress the practice. Nothing happened. The cartel survived, riding the gentle upward drift in export prices that lasted until 1958.[91]

The attempt to break up GEORG (Gemeinschaftsorganisation Ruhrkohle) was like the rerun of a bad movie on a faulty projector. GEORG replaced the Deutschen Kohlen Verkauf (DKV) when the common market for coal opened on 10 February 1953 but differed from its predecessor in only one important particular: It was not a central marketing office but a coordinating mechanism for six legally distinct sales agencies. Like DKV, GEORG directed marketing, allocated orders, ran compensation schemes, and coordinated rail and

---

[90] JM AMH 23/4/11, Note (Hamburger), "La concurrence sur le marché commun est-elle limitée ou faussée par la fixation de prix pour les marchés tiers par l'entrée en matière d'exportation," 29 October 1953.

[91] JM AMH 23/4/16, "Note [Blondeel] pour M. Monnet," 8 March 1954.

barge movements. Since control had officially been lifted when the common market was opened, it also set prices. GEORG was a clear-cut violation of Article 65.[92]

The Ententes and Concentration Division began inspecting GEORG in summer 1953, but because of opposition from the legal and market sections the investigation soon got bogged down. The film of the decartelization drama then slowed up, as it were, to the point that the individual frames appeared almost like stills to the viewer, only to resume speed suddenly in May 1954 when, with scenes flashing by and characters twitching spastically, the president announced to the Common Assembly that the Ruhr coal syndicate was unacceptable and would have to be eliminated. The reels started spinning more and more erratically after Monnet named HA Vice-President Etzel chief negotiator. Etzel, who like all Germans objected in principle to tampering with the Ruhr's coal marketing machinery, recommended that "The High Authority . . . sketch a framework within which the mining enterprises could organize their sales without contravening the Treaty". This cryptic remark baffled nearly everyone. Images blurred as spindles whirled on and senses deadened. It took another five months for the Germans to present their well-known view that breaking up the coal syndicate was like the commission of a particularly loathsome sin and a further seven for the High Authority to devise a proposal consistent with Etzel's approach. It could hardly have been more banal: "to keep the six sales agencies that existed but to make them really independent of GEORG."[93] Someone thereupon appears to have reversed reels. In November 1955 Etzel proposed that three fleshed-out sales organizations would more effectively weaken the central marketing mechanism than the existing six skinny ones. A promise to apply uniform rules throughout the community secured the agreement of the German representatives at the Common Assembly to Etzel's plan, which went formally into effect on 1 April 1956. It duplicated rather than eliminating or reengineering the centralized marketing machinery and in practice amounted less to a reform scheme than a work-creation program (*Arbeitsbeschaffungs-*

---

[92] Diebold, *Schuman Plan*, pp. 380–393; JM AMH 23/13/1, "Récapitulation des activités des sept entités juridiques de la Ruhr contraires aux dispositions de l'Article 65. (annexe I/2) au document 1442/54f," 15 February 1954; EC CEAB 4/1954/338, "Note [Hamburger] pour la préparation du point de vue à adopter par la Haute Autorité lors des entretiens avec les délégués de la Ruhr," 26 October 1954.

[93] Both quotations from Diebold, *Schuman Plan*, pp. 282, 383.

massnahme). The welcome words "THE END" never flashed across the screen. The economic upswing that set in after the introduction of Etzel's plan sapped the remaining determination to get on with the business at hand: Everyone recognized that the cartel's demise would be followed by a jump in coal prices. Yet GEORG was still subject to the listless flogging of the reorganizers when the coal crisis of 1958 unexpectedly intervened. The emergency required a government takeover of the mines. The decade-long decartelization campaign was over without having ever been concluded.[94]

To lend credence to Etzel's promise of uniform treatment the High Authority felt obliged to challenge ATIC (Association Technique de l'Importation Charbonnière), the only authorized import agency for coal in France. In 1954 negotiations began with the French to bring ATIC into conformity with the treaty. Shortly after the GEORG settlement in 1956 the High Authority ordered the statutory elimination of the French agency's monopolistic features. This was a perfunctory demand. ATIC was a mechanism for equalizing costs to purchasers of domestic coal, which was priced higher than Ruhr imports. As such it was an essential component of a long-standing subsidy system that the French government could not have abandoned without putting its coal industry, modernized at great cost in the 1940s, in jeopardy. After challenging the HA in court, the French agreed to license independent coal importers, but since they were authorized to bill only through ATIC the effect of the measure was nil. Though in December 1957 the High Authority ordered further reorganization measures, the French government successfully tied the matter up in court. The effort to change France's coal-marketing methods met with no more success than in the case of the Ruhr.[95]

The High Authority acted even more circumspectly in the matter of concentration; its "policy" was to have a policy that needed no enforcement. Required under Article 66 of the treaty to approve all mergers in cases where the new firm might have an unfair competitive advantage, the HA waited a full year before establishing procedures for evaluating concentrations. It limited its subsequent interventions to important cases involving more than 1.2 million tons of steel or its equivalent. Only ten firms in the community pro-

94 Abelshauser, *Der Ruhrkohlenbergbau*, p. 96.
95 Diebold, *Schuman Plan*, pp. 396–398; EC CEAB 7/1953/41, 59–73, "Organisation de la distribution du charbon en France," 14 October 1953; CEAB 7/1953/41, 42–48, "Note concernant l'Association technique de l'importation charbonnière (ATIC)," 14 September 1953.

duced this amount. The HA further agreed to approve all mergers promoting either efficiency or competitiveness. As of 1958 it had not prohibited a single one of the more than a hundred such agreements submitted to it for review.[96]

The real meaning of deconcentration policy was made clear on 4 August 1953 in an address presented by HA Vice-President Albert Coppé to the annual meeting of Hüttenwerke Phoenix AG. In a fulsome, giddily enthusiastic article W. O. Reichelt reported in the normally sober *Die Zeit* that "Coppé, who spoke in fluent German, won the hearts of his audience faster than could ever have been hoped for. His speech revealed a profound appreciation of the importance of [the Ruhr's] tied-in economy, for entrepreneurial initiative, for the dynamic of progress, and for the overall economic structure of the union and its relationship to the rest of the world."[97] Coppé offered assurances that the High Authority wanted to promote economic development, not restrict it. "We are not hostile to mergers," he is reported to have said; "Where they promote productivity they will always be approved." To eliminate any remaining doubts the speaker specified that efficiency was to be understood in a very broad sense, encompassing not only machinery but capital investment. He added that the HA would further encourage transnational merges on an American scale. Coppé's words, the reporter continued, "require no [*sic*] commentary . . . they are the first time that anyone from outside our borders has discussed matters such as these in such a precise way . . . he has made an impressive contribution to reducing the understandable anxiety and skepticism that here in Germany surrounds the need to restore firms of a healthy size."[98] The reason for the author's frenzied tone "requires no commentary": The High Authority had given the green light to the reconcentration of Ruhr industry.

Monnet felt confident that even if all else failed the availability of American money would put him in a position to direct the development of heavy industry as he had formerly done at the French Plan, by making low-interest loans to influence investment decisions. He also hoped that an initial U.S. loan would be the forerunner of large additional flows of both public and private capital that Luxemburg could channel into the coal and steel interests of the community.[99] The idea of a big U.S. loan had been long in the works. In March 1951 Monnet's unofficial young American commando, Tommy

[96] Diebold, *Schuman Plan*, pp. 356–374.
[97] WW, "Erfreulicher Wandel im Montanbereich: Vizepräsident Coppe gewann an der Ruhr einen Vertrauenssieg für Luxemburg," *Die Zeit*, 5 August 1954.
[98] Ibid.   [99] Diebold, *Schuman Plan*, p. 320.

Tomlinson, drafted a memo stating that "from the early days of the Schuman Plan negotiations it has been understood by all concerned that it was the HA's financial powers – its ability to tax, to borrow, and to lend in support of the investment program – which constituted its major possibilities of contributing constructively to the development of the industries in its charge." "If," Tomlinson proceeded, "the United States Government could make available to the High Authority, from the beginning of its operations, a substantial sum of grant funds ... the prestige of the High Authority could be immediately established beyond any doubt in the minds of producers, the workers, and the governments of the six countries."[100]

In the same month, Monnet handed Ambassador Bruce a memorandum requesting "substantial" aid during the early months of ECSC operation. Though the State Department in strongly endorsing the request recommended making ERP funds available, the subject was touched upon only briefly during the visit of President Auriol and Foreign Minister Schuman to Washington on the twenty-sixth of the month. In 1952 the ECSC became eligible for aid under the Mutual Defense Assistance Act, which had replaced the Marshall Plan, but the actual financial requirements of the coal–steel pool could not be determined until it had begun to operate, and the loan issue thus did not arise again until after the November 1952 presidential election.[101]

The appointment of John Foster Dulles as Eisenhower's secretary of state gave Monnet ample reassurance that the money would soon be available. The two had been friends for over thirty years. Back in 1934 Dulles, a very rich man, had made Monnet, who was then broke, a large personal loan enabling him to form a financial consultancy in partnership with the banker George Murnane. Now, nearly twenty years later, a brief note sufficed to gain Monnet the secretary of state's influential backing for U.S. financial assistance to the community. In April 1954 President Eisenhower signed an agreement for a twenty-five-year loan of $100 million on very generous conditions, 3⅞ percent interest with repayment not scheduled to begin until 1958. The funds were earmarked for modernization (and were not needed for operating expenses, which were generously covered by receipts from the tax on turnover).[102]

[100] JM AMG 15/4/5, Tomlinson memorandum (n.d.); also JM AMG 15/4/2, 14/4/4, 15/4/6.
[101] See FRUS 1951/IV, "Editorial Note," p. 106.
[102] John Foster Dulles Papers (Mudd Library, Princeton University), Letters to Monnet of 2 February 1935, 21 March 1959.

Though the loan provision was hedged in by the customary legal and financial requirements for accountability, for all practical purposes it amounted to a personal slush fund. Insisting upon exercising sole responsibility for allocation, Monnet rejected the recommendation of the leading German banker, Hermann J. Abs, that the High Authority restrict its role to the issuance of general directives and leave credit decisions up to private bodies in the participating countries. Monnet initially set aside a quarter of the funds for worker housing, but this could not be spent because repayment was required in dollars and no mechanism was available for offsetting the foreign exchange risk. The entire sum was eventually made available at cost to enable plant modernizers to mobilize outside capital. The loan eventually covered about 27 percent of total outlays for the projects thus financed. This provided only 3.5–4.5 percent of total community investments in heavy industry between 1954 and 1957, too little to have given the HA the wherewithal to steer industrial growth.[103]

The ECSC never developed into an investment bank for coal and steel. Both public and private lenders preferred to loan directly rather than through the intermediary of the High Authority, and big industrial borrowers hesitated to accept funds from a creditor with a reputation for attaching political strings to deals. Moreover, by the mid-1950s acute dollar shortages were over, capital was no longer scarce, and industrial borrowers had little difficulty establishing creditworthiness. The ECSC was not needed as a financial intermediary.[104] The High Authority's activities in this sphere remained small-scale, with borrowings totaling only $215 million as of 1958. Yet it had loaned even less by that date, $173 million, some $9.2 million of which were from its own resources. The HA also ran up a huge surplus from its levy on turnover, which in July 1957 was reduced from 1 percent to 0.35 percent. Less than a quarter of the $207 million collected through this tax had actually been spent by the end of 1957. The ECSC had less difficulty raising money than in putting it to work constructively.[105]

Though the community had not reached a dead end, most of its main accomplishments occurred outside the framework of the treaty and the formal institutions created by it. Little progress was made on wage harmonization, even though this subject, a favorite of the

[103] Diebold, *Schuman Plan*, pp. 320–326; 330–334.
[104] Ibid., pp. 335–350; Lister, *Europe's Coal and Steel Community*, pp. 88–89.
[105] Diebold, *Schuman Plan*, pp. 317–318, 330, 331.

Belgian unions, continued to spawn study groups and intergovernmental commissions and otherwise command huge resources of
time, money, and paper. The free movement of labor promised in
Article 69 was also discussed to exhaustion; it took no less than four
years of negotiating to arrive at an agreement twice as long as the
treaty itself which, as William Diebold archly put it, only "[provided]
free movement within the Community for certain workers in certain
circumstances."[106]

Readaptation, an issue of great potential importance, provided
another instance of stumbling and staggering. Its first test was the
community's attempt, undertaken in early 1953 at the request of the
French government, to relocate 5,000 miners from the Cevennes to
Lorraine by 1956. In spite of prodigious planning and generous al-
lowances only 500 of them had accepted the offer by the end of
1955, when it was shelved. The High Authority had managed to
spend only a third of the $12 million earmarked for readaptation by
1958. Of the total, 90 percent went to Italy, which already enjoyed
special protection, and was assigned to projects having no evident
relationship to dislocation caused by technological or economic
change.[107] The financing of housing for workers, an activity not
foreseen in the treaty, was the ECSC's main accomplishment in the
field of social welfare. The HA had loaned $65 million by 1958
which made possible the construction of some 12,500 dwelling units
(with another 24,000 then under way), nearly two-thirds of them in
housing-scarce West Germany. ECSC funds also seeded additional
building projects, thereby drawing in large amounts of outside
capital.[108]

Scrap was another area in which community intervention was
constructive and unexpectedly important. Prices for the unglamorous but critical raw material were exceptionally volatile: Sharp increases stimulated by the upswing in steel demand felt in the fall of
1954 aggravated by a huge jump in transatlantic freight rates
threatened to spring the complicated equalization scheme worked
out under community auspices the previous year. The ECSC soon
found itself even more deeply involved than before in regulating the
economy of recycled steel. These interventions were not crowned
with glory and in no sense solved the scrap problem. They did, however, make it manageable; by 1958 the consumption of used metal

[106] Ibid., p. 437.   [107] Ibid., pp. 404–426.   [108] Ibid., p. 433.

was increasing no faster than steel output.[109] The ECSC provided leadership the scrap economy would otherwise have lacked. It propped up the community's cartel when price rises threatened the compensation scheme; placed a levy on consumption of the product as a premium for the use of pig iron in the Siemens-Martin process; and launched a long-term planning effort directed at weaning the European mills industry from dependence on the recyclable commodity.

The ECSC played an indispensable role in reducing transport rates by curbing several flagrant cases of discrimination when the common market in coal was opened. It eliminated a 25 percent differential between rates charged by the French railroads for shipping Lorraine ore to North France and the adjacent regions of Belgium as well as the 30 percent surcharge Luxemburg had to pay in order to ship to Strassbourg; required the Belgians to charge uniform rates for shipments of domestic and foreign origin; and made the Germans apply a single rate for domestic and foreign coke and coal, which brought a 20 percent reduction in costs to importers.

In 1955 the ECSC suppressed a particularly glaring form of protectionism, terminal charges (*rupture de charge*) imposed at the border to cover the (fictive) costs of breaking up and reorganizing trains. The purpose behind this practice was to impede rate tapering (long-haul reductions) advantageous to foreign producers. The abolition of this abuse reduced the cost of shipping Ruhr coke to France (Gelsenkirchen–Homécourt) and Luxemburg (Gelsenkirchen–Esch) by nearly a third and similarly lowered costs along two other heavily used routes, for iron ore from France to Belgium and coal from the Saar to South Germany. The ECSC made little additional progress on transport matters. The High Authority's attempt to harmonize freight rates met with insuperable economic and political objections. The national governments simply ignored its orders to eliminate subsidized rates to hardship areas.[110] The reductions imposed in 1953 and 1955 were nevertheless a pleasant, uncontroversial demonstration of how the ECSC was supposed to work, providing a good example of how rails, roads, and

[109] Ibid., pp. 292–293; EC CEAB 4/1956/156, 26–33, "Note sur les entretiens relatifs à la feraille," 10 March 1953; See JM AMH 7/1/46, 7/1/49, 7/1/59, "Procès-verbaux des séances de la Haute Autorité," 2 February 1953, 5 March 1953.

[110] Lister, *Europe's Coal and Steel Community*, pp. 362–363.   [111] Ibid, p. 368.

waterways could unite rather than divide. In the future transport rates would feature prominently on the European agenda.

Taxation would unexpectedly be the most important new field of ECSC activity. This inherently unpleasant subject was neither discussed in the Schuman Plan negotiations nor mentioned in the treaty. The exclusion was not deliberate: No one really appreciated the significance of tax issues until the outbreak of a dispute over them serious enough to have delayed the opening of the common market in steel to 31 May 1953. The episode alienated the German economics ministries permanently from the coal–steel pool but at the same time indirectly engendered a deeper and broader commitment to the integration process. The background to the tax dispute (*Steuerstreit*) was as follows. France observed the normal, GATT-sanctioned trade practice of subjecting imports to compensating levies equal to the excise taxes paid by domestic producers, while exempting exports from them. The Germans objected to this treatment as discriminatory: If, as the Treaty prescribed, all consumers were to be considered equal, nationality should not determine the amount of tax on products under the purview of the community; why, they reasoned, should French consumers be forced (in the case at hand) to pay a higher rate on steel than German? There was more at stake than even the serious matter of principle. Excises were set substantially higher in France than in either the Federal Republic or the rest of the community, where different revenue-collecting methods predominated; the compensatory taxes charged imports from community products sold in France was twice the rate French steel paid on exports to the Federal Republic. By raising Ruhr steel levels from 9 percent below French prices to 2 percent above them in both national markets, German products were put at a serious competitive disadvantage.[112] The HA's solution was to appoint a committee under Jan Tinbergen, a Nobel laureate in economics, composed of experts from each nation of the community. The appointment of this prestigious board was a politically astute move on Monnet's part; it decided in favor of France, as he had wished. The Germans were stuck with its conclusion – to leave things as they were.

In arriving at its findings the Tinbergen Committee's investigations raised a question that cut to the heart of the sectoral integra-

[112] Diebold, *Schuman Plan*, pp. 223–228; Papers of Hans von der Groeben (PG), Montanunion, Steuerstreit, "Stellungnahme zu den Thesen der Hauptabteilung Wirtschaft der Hohen Behörde in der Fassung vom 7. Dezember 1953."

tion theory that had provided the intellectual justification for the
ECSC. It reasoned that the alternative proposed by the Germans of
taxing at origin rather than destination would be trade-distorting
unless it could be extended to all products in all markets rather than
merely the coal and steel sold in the community. Under existing
conditions heavy industry outputs would have been taxed either
more heavily or more lightly than other production, thus stimulating or retarding foreign trade artificially, resulting in various anoma-
lies and absurdities, which the committee report gleefully cited.
Those unable to follow its complicated mathematics were easily persuaded that something was out of joint if, for instance, after the
German proposal were enacted the prices for French nails exported
to the Federal Republic fell below the price of the wire imported
from there and from which they had originally been fabricated.[113]
Outraged by the "calculation witchery" of the Tinbergen Committee, the Bundestag immediately granted the federal government
standby authority to raise the import-equalization tax on French
steel to 12 percent. It was never necessary to use it, however; preferring profit to new market share, French producers raised export
prices to German levels at the opening of the common market. On
3 June 1953 the Schuman Plan desk of the German Ministry of Economics began its own assessment of the Coal and Steel Community.
This would have significant consequences.[114]

The dozen reports submitted unanimously condemned both the
politics and the economics of the High Authority. One paper written by Section Chief Hans von der Groeben accused the High Authority of "exercising more independence than [authorized] by the
treaty" and criticized the lack of statutory authority enjoyed by the
Common Assembly and Council of Ministers, which he blamed the
HA for having treated "rudely" and as inconsequential. The HA,
the report proceeded, ruled from above without benefit of input
from below, and this too was contrary to the intent of the negotiators, who had expected that the regional groups would be a part of
the governing machinery. Current administrative methods were also
damned; worst of all, the "economic doctrinaire" Pierre Uri was the
number-two figure at the HA. The author's conclusion was that the
government had agreed to the treaty creating the ECSC under pres-

---

[113] Diebold, *Schuman Plan*, pp. 228–229.
[114] Papers of Hans von der Groeben (GP), Montanunion, Steuerstreit, "Erfahrungsbericht über die Rechtstellung, den Verwaltungsaufbau, und die Aufgaben der
Organe der Europäischen Gemeinschaft für Kohle und Stahl (III D1/7/481/
53)," 3 June 1953.

sure and not as a forerunner or model for more far-reaching integration. He strongly objected to extending its competence: "Any carry-over of the methods of the High Authority to other sectors of the European economy must be rejected from the outset as unworkable and damaging."[115]

The economic section of the same report had a similar tone. It inveighed against the rigidity of the treaty; ridiculed the campaign against discrimination as an unrealistic attempt of bureaucrats to impose uniformity upon a natural diversity of condition; and dismissed the effort to regulate competition as a waste of time, even noting that the cartel-obsessed Americans were now beginning to doubt the wisdom of their policies. The author was pleased that the attempt seemed to have been given up to harmonize wages, taxes, and so on because such efforts to impose total solutions in a community that was by definition partial could only lead to bad results. Here he pointed to the tax dispute as having "demonstrated the futility of a step-by-step coordination of . . . structures."[116] After a year of study, the joint industry–government "Special Commission for the Study of the Tax Question" also concluded "that a purely financial and economic study of the tax problem is impossible," emphasizing that a satisfactory settlement of it would require far more integration than anything yet attempted: It would have to be all or nothing.[117]

The *Steuerstreit* led not to a German rejection of integration but to a renewed determination to make it work. As of 27 January 1954 the policy of the Ministry of Economics toward the ECSC was one of *sauve qui peut*. Notes prepared for Erhard's impending meeting with HA Vice-President Etzel describe the economic articles of the treaty as outmoded and unworkable and the effort to enforce them more as more an annoyance than a threat; the real dangers facing it were the weaknesses on coal and steel markets. The tasks ahead were "keeping the High Authority out of danger" and "preparing for the next steps toward economic integration."[118] Meant by this was what

---

[115] GP Montanunion, Steuerstreit, "Erfahrungsbericht Hohe Behörde I, Organisatorischer Teil, Konstruktion und Organisation der Hohen Behörde," 8 June 1953.

[116] GP Montanunion, Steuerstreit, "Erfahrungsbericht Hohe Behörde II, Wirtschaftlicher Teil," 11 June 1953.

[117] GP Montanunion, Steuerstreit, WVESI to Bundesministerium der Finanzen, "Engerer Ausschuss zur Überprüfung der Steuerfrage," 16 June 1954.

[118] GP Montanunion, Steuerstreit, "Gedanken für eine Besprechung zwischen Herrn Minister Professor Erhard und Herrn Vizepräsident Etzel," 24 January 1954.

the paper called "a real common market." Its prerequisites were
"genuine" currency parities, a European law regulating competition, a European expansion policy with fiscal and monetary components, and a common external trade policy; without these there
would be no increases in productivity and labor specialization, no
rationalization, and no improvements in the geographical placement
of industry.

This ambitious program was not easy to get started. The still-
pending ratification of the treaty of the European Defense Community presented an immediate political obstacle to further progress,
and, the study emphasized, other problems loomed ahead, currency
parities prominent among them. And in what venue should discussions begin? The High Authority did not work properly and was in
need of thorough reorganization. The Council of Ministers was
weak. These things would have to change, in the judgment of the
paper's author: "The discussions concerning the European Political
Community have made it clear that the progress of the common
market – the merger of the six national economies – will require the
close and constant cooperation of the national governments and the
European executive organ."[119] It actually called for considerably
more than this. The advancement of the integration process required that the other members of The Six be ready to dwell in an
economic Europe dominated by the Federal Republic of Germany.

6.5   Bonn, boom, bang: the new Europe

During the years when Monnet was president of the High Authority
West German industry was reorganized, recapitalized, refurbished,
retooled, and expanded. The old trusts reappeared, and the welfare
of Europe was no less dependent upon them than before. Yet a
Ruhr revival never became the bogeyman of the 1950s: The industry complex had lost most of its power to terrify, and fear of it could
be conjured up only with intense effort: Between 1950 and 1955 the
Federal Republic became the power source of a new prosperity that
restored a sense of stability and well-being to European life and
changed the context of its politics; it provided the public with what
seemed like tangible and incontrovertible proof that integration actually worked. The advancement of the process involved more than
just rising standards of living: Economic interdependence increased;
the threat of nuclear annihilation made age-old disputes seem obso-

[119] Ibid.

lete; and the right lessons were learned from Europe's first real experiment with unification.

The attempt to create a European Defense Community (EDC) was the main diplomatic issue in Western Europe in the three years between the conclusion of the Schuman Plan in April 1951 and the ultimate rejection of the EDC treaty by the French in August 1954. Though this proposal for a multinational military force never got off the drawing boards, it was an episode with several important consequences for the history of integration. The EDC negotiations provided a way to nurse European federalism, which had been ailing since the Westminster Conference of 1949, back to life. Its supporters did not necessarily agree that Monnet's functional integration approach was either the best or the only therapy for the division of the Continent; their diagnosis called for political remedies. In September 1951 the European Union of Federalists, with the Belgian socialist Paul-Henri Spaak serving as *spiritus mentor*, had therefore proposed creating a parliamentary body, the European Political Community (EPC), to which both the ECSC and the EDC would be responsible. EPC was also supposed to serve as lead integrator in the future.[120]

It took many months to sort out the conflicting jurisdictional claims of the several supranational organizations eager to participate in drafting a constitution for the European Political Community. In September 1952 discussions began within the Common Assembly of the ECSC, which had been specially enlarged for the purpose of drawing up the document. Though on 10 March 1953 the body approved the draft unanimously, the governments viewed it more critically. The review process was still under way when the demise of the European Defense Community rendered further effort superfluous.[121] The EPC episode was not, however, an exercise in wheel spinning; it kept cadres in being, dialogue moving, and served as a learning experience. Spaak and other federalists associated with the EPC would have better luck during the "relaunch." A more immediate and important by-product of the European Defense Community was the contractual accords of 26 and 27 May 1952, which ended the occupation of Germany in everything but name. The accords were concluded only after many months of negotiation over force structures, defense contributions, and relationship to the Atlantic alliance. In return for a promise to rearm and join the defense community and NATO, the Federal Republic ac-

---

[120] Willis, *France, Germany, and the New Europe*, pp. 159–160.   [121] Ibid.

quired the main attributes of sovereign power. The contractual accords specified repeal of the occupation statute and abolition of the High Commission, the Allies retaining only rights in Berlin and the authority to complete decartelization.[122]

The EDC was also important because it overshadowed the ECSC in overall American integration policy. As priority shifted from economic cooperation to mutual defense, the United States forgot about the coal–steel pool with astonishing rapidity. In Secretary of State Acheson's case, this amounted to near amnesia. On 14 December 1952 Acheson, who by then was a lame duck, paid a final visit to Monnet before leaving office. In response to the Frenchman's pressing queries, the secretary of state admitted that "informed American opinion" had no idea about what was going on in Luxemburg and that he personally remembered only that the foreign ministers' meeting of August 1952 had been "inconclusive and discouraging"; he recalled hearing "nothing whatever" about ECSC operations and had the impression that discussions of political integration were in "cold storage."[123] Monnet wasted no time in disabusing Acheson of the notion that the Schuman Plan machinery had stalled. It was, he boasted, "operating successfully . . . and already having an impact upon business and economic affairs." Adding that the common market in coal was due to open soon, he reported with what was surely forced enthusiasm that "manufacturers were already adapting their business to it – French manufacturers getting ready to sell in Germany, Germans signing contracts in North Africa, and Belgians lining up to buy minette in Lorraine." Monnet finally made the fantastic claim that "the operations of the plan had already completely changed the method of thinking of both producers and consumers of coal and steel and of parliamentarians in the six countries."[124] The sales job worked. Addressing the NATO Council on 18 December 1952 for the last time before leaving office, Acheson described the movement toward unity as "not [altogether] new in the world . . . but new in the last five, six, or seven centuries" depicting the Schuman Plan, along with Bevin's proposal for a Western European union and the founding of NATO itself, as one of the three great steps toward the development of an Atlantic community. Acheson confessed as if overawed to not knowing "how vast will be the change in the thinking of Europe and of the countries outside

---

[122] Schwartz, "Occupation to Alliance," pp. 586–630.
[123] HST, Acheson Papers, Memorandums of conversations, Meeting between Acheson and Monnet, 14 December 1952.
[124] Ibid.; see also Memorandum of conversation, 15 December 1952.

of Europe as the Schuman Plan actually operates" but predicted that in the short run the ECSC along with the EDC would be "essential to the political association . . . now under discussion."[125]

Except for helping arrange the big loan to the community, Secretary of State Dulles also paid little attention to developments in Luxemburg, notwithstanding his friendship with Monnet and deep personal interest in European federalism.[126] This surprising disengagement was caused by the rumblings on the right that arose in response to the frustrations caused by the Korean War. Though defeated in his bid for presidential nomination in 1952, the isolationist Senator Taft remained a formidable figure within Republican politics. He represented the respectable opposition. "Tail-gunner Joe" McCarthy, the ill-educated, unscrupulous, often intoxicated senator from Wisconsin, caused paranoia to grip the nation. For two years the American public would be treated to the preposterous spectacle of televised witch hunts in which even the secretary of the army was accused of protecting Communists. Integration policy was also suspect as a "giveaway" to weak-kneed pinko peoples with morals, values, and habits reportedly very unlike those of the American Midwest. In this atmosphere of renewed fear and persecution, the most painless policy for a Republican administration was to talk tough and limit the scope of foreign policy to the defense of the free world against communism.[127] The Schuman Plan organization was of only minor concern to Dulles while in office. Its existence did nothing to deter him from the notorious December 1953 threat of an "agonizing reappraisal" of American support for European integration made in high dudgeon at the French refusal to ratify the EDC treaty. Dulles's general downgrading of the ECSC, and of the nonmilitary side of integration in general, nonetheless turned out to be a blessing in disguise. It gave the Europeans a chance to deal with the issue in their own way.

The EDC also had a third important consequence, touching off what Raymond Aron overheatedly described as "the greatest ideologico-political quarrel that France has known, probably since the Dreyfus affair . . . a quarrel whose most apparent issue was German

[125] HST, Acheson Papers, Memorandums of conversations, "Text of Secretary Acheson's Statement to the North Atlantic Council," 18 December 1952; see also untitled memorandum of conversation, 23 December 1952, with John J. McCloy.

[126] See JM AMG 20/3/1, "A Federal System for Germany: Mr. Dulles on Control from Without," *Times*, 20 January 1947; J. F. Dulles, *War or Peace?* (New York, 1950); MAE Y 1944–1949, 399, Telegram Bonnet, 1 February 1948.

[127] Kaplan, *NATO: Enduring Alliance*, p. 56.

rearmament, but whose deepest significance affected the very principle of French existence, the French state."[128] In more prosaic
terms, the EDC gave the politicians of France nearly two and a half
years in which to blow off anti-German steam before bowing to the
inevitability of a new Bundeswehr and a new German-centered European economy. As debate between the "partisans of Europe" and
the "partisans of France" reached a rhetorical crescendo on 29 August 1954 a deputy named Lebon condemned "Germany" for invading France not only in 1940, 1914, and 1870, but in 1815, 1814, and
1792 as well. His choleric colleague, Taillade, thereupon interjected
that the story really went back to Attila the Hun.[129]

The problem of integrating the Federal Republic militarily into
the Atlantic alliance was actually soon solved with ease and on terms
less favorable to France than those of the EDC. In September British Foreign Minister Eden proposed that in lieu of the defunct
military community the Federal Republic be allowed to sign the
Brussels Treaty of 1948, a mutual defense pact of the European
nations superseded by NATO. After the addition of a few purely
cosmetic clauses to the draft agreement, Premier Mendès-France
initialed it. Within six weeks the National Assembly would over-
whelmingly ratify a treaty that endorsed the militarization of the
Federal Republic subject only to the control of NATO and German
self-restraint.[130] In another six months the relaunching of Europe
would be under way.

The early 1950s were a time when the "re-" prefix was much in
use in the Ruhr. There was reinvestment, reassembling (*Remontage*),
and reconcentration (*Rekonzentration*) thanks to which, in part, by
the middle of the decade the Ruhr had regained its traditional
prominence in Europe. Another word with the same prefix, recartelization, was never used in official German (where the circum-
locutious *Wiederverflechtung*, or "knitting back together," was much
preferred). Yet it also occurred. Finally there was *Restauration*, a
doubled-edged term with an English cognate; if to conservatives it
had a pleasant ring, to partisans of radical change it sounded more
like a death knell.

Remontage was completed by 1955. The repairing of plant was a
very productive way to employ capital. Land, labor, management,
stocks, and tools were often all on hand, and to resume production
often required only the replacement of a missing piece of critical
machinery. Raising the necessary investment, by comparison, was

---

[128] Quoted in Willis, *France, Germany, and the New Europe*, p. 138.
[129] Ibid, p. 183.   [130] Ibid, pp. 187–188.

very difficult in 1950. Self-financing was impossible, private capital
markets were still thin, and government compensation schemes difficult to administer. Though the Investitionshilfegesetz (IHG) was a
welcome start toward solving the problem, additional funding soon
proved necessary. The Land Nordrhein-Westfalen raised a 228-
million-Deutschmark *Remontagekredit* in 1952 and 1953, and the federal government provided another 1.2 billion Deutschmarks. A
special producer-controlled Emergency Association of Reparations-
damaged Industry (Notgemeinschaft für Reparationsgeschädigte Industrie) administered the federal funds.[131]

The August Thyssen Hütte in Hamborn provides an example of
how the reconstruction process took place at one firm. The first
blast furnace at this key installation was lit on 7 May 1951. Operations resumed without benefit of bank financing, and though the
firm got a portion of the 297 million Deutschmarks made available
through the IHG, its largest borrowing was 217 million Reichsmarks
from the Bund. On 11 July 1955 Chancellor Adenauer dedicated
the Ruhr's new flagship, its first wide-strip continuous rolling mill.
Construction of the leviathan had been financed wholly by the
heavy equipment manufacturer DEMAG, which earlier had been
linked fraternally to ATH in Vereinigte Stahlwerke. In 1956 ATH
became the first German firm to secure a loan from the U.S. Export–Import Bank ($10 million for the purchase of American ma-
chinery), but this was for expansion. Financially as well as physically,
1955 ended the period of reconstruction at ATH: For the first time
since the war it was able to pay a dividend, a generous 8 percent. By
1959 it had repaid the last installment of its loan from the Bund.[132]

Reconcentration went hand in hand with *Remontage*. One had to
cross difficult legal terrain in order to get the trusts reorganized,
but by the end of 1953 those passing through had gained the confidence that it could be negotiated without peril. On 16 November
1953 the influential trade journal *Industriekurier* editorialized that
the federal government had the right, and should attempt, to block
all further Allied decartelization through the High Authority.[133]
Nine days later HICOM tore up the paper-bag cartel (Gemeinschaft
Papiersackindustrie Wiesbaden).[134] This ineffective show of force in-

[131] WW, "Wiederaufbaufragen der demontierten Werke," *Der Volkswirt*, 1 November 1952, pp. 44–45.
[132] Sohl, *Notizen*, p. 145.
[133] JM AMH 23/5/6, "The Cartels Return" by Terence Prittie, *Manchester Guardian*, 5 December 1953.
[134] Ibid.

timidated no one. On 22 December 1953 the financial monthly *Volkswirt* prophesied that "Law 27 will soon be in the file cabinet . . . The next few years will . . . demonstrate the imperative necessity of integrating the Federal Republic into the West [because] the Allies can no more afford deficient industrial organization in the Ruhr than Germany."[135]

Most of the reshuffling took place in 1955. For three of the medium-sized trusts (Mannesmann, Gutehoffnungshütte, and Hoesch) the process involved relatively unproblematic tendered stock swaps through which the old headquarters companies,[136] with the assistance of the big investment banks, reacquired former affiliates. The fourth medium-sized trust, Klöckner, presented a special problem. A Dutch holding company set up by the firm's founder in the 1920s controlled about half the firm's outstanding shares, which after the war had been confiscated as enemy property. The government of the Netherlands refused to discuss the matter until the early 1950s, according to Managing Director Henle's memoirs. Agreement was eventually reached to set up a holding company for the German assets that was partly controlled by the Dutch government through the leading private Amsterdam bank, Hope and Company. The arrangement, according to Henle, "continues to be beneficial in every respect for the descendants of [Peter Klöckner], and the obligation to leave these assets in Holland is one with which we gladly comply." Henle described the Dutch transaction as "the most important one for the Klöckner group since the end of the war" but mysteriously refused to more than "hint at its full significance."[137]

The breakup of Vereinigte Stahlwerke (VS) resulted in the formation of three new companies. To prevent control of the properties from falling into the same hands as before, the Allied had ordered that large shareholders be required to accept compensation in only one successor company. The effect of this was to give Fritz Thyssen's widow sole control of Rheinische Röhrenwerke and his daughter that of Rheinstahl-Union. By 1957 two astute lawyers, the Adenauer adviser Kurt Birrenbach and the less visible Robert Ellscheid, had succeeded through exchanges of shares in gaining the Thyssens control over the first- and third-largest Ruhr trusts, ATH and

[135] WW, "Die Rückverflechtung in der Montanindustrie," *Der Volkswirt*, 22 December 1953.

[136] WW, "Deconcentration and Reconcentration of the West German Iron and Steel Industry," *Continental Iron* (special ed.), 18 August 1959.

[137] Henle, *Three Spheres*, p. 140.

Rheinstahl. The third new company to emerge from Vereinigte Stahlwerke, Dortmund-Hörder-Hütte, was in the hands of another former shareholder in the giant trust, Hoogovens of the Netherlands.[138]

Krupp was a special case. The "lex Krupp," dating from the days of the empire, had declared the "armorer of the Reich" indivisible and inalienable for all time. The trust's owner, Alfried Krupp von Bohlen und Halbach, had been found guilty of war crimes at Nürnberg, remained in prison until 1951, and was released only on the condition that he hold no responsibilities of any kind in industry. Yet Krupp too would survive. The ambitious detrustification plan of 1953 was never carried out, and by selling off its coal mines the firm soon raised enough capital to take over a traditional competitor, Bochumer Verein. Shares in this, another former component of VS, had fallen into the hand of the Swedish industrialist Axel Wenner-Gren, a trusted Krupp ally, and were acquired without difficulty. Krupp would emerge from the years of reorganizing with less coal but more steel than before the war.[139]

Except at Krupp, the reconcentration process was over by 1957. Günter Sieber described the results at the time as follows: "Today, as before World War II, there are again eight *Konzerne* . . . dominating the west German steel industry. These new firms control 79 percent of [the Ruhr's] pig iron production, 75 percent of its steel production, 60 percent of its rolled steel production, and 33 percent of its coal production."[140] Though the figure for steel production, 75 percent, was down from the 95 percent that obtained before the war, Sieber still described the market as being oligopolistic: Though Vereinigte Stahlwerke had disappeared, the two largest firms, Thyssen and Krupp, still controlled a quarter and a seventh of steel outputs, respectively, with the shares of the four middle-sized trusts (Gutehoffnungshütte, Klöckner, Mannesmann, and Hoesch) remaining the same as before the war. Though Friedrich Flick reconstituted his empire in other sectors of industry and Otto Wolff, which had never been a major force in steel production, left the business altogether in favor of merchanting, Sieber concluded in early 1958 that "within five years the iron and steel industry has completed a process that has led back to prewar conditions. The Allied detrustification program is no more than a reminiscence."[141]

[138] Sohl, *Notizen*, pp. 151–152.
[139] WW, Günter Sieber, "Die Rekonzentration der eisenschaffenden Industrie in Westdeutschland," *Stahl und Eisen* 1:1958, pp. 46–55.
[140] Ibid., p. 55.   [141] Ibid.

An increase in the investment rate paralleled the progress of re-concentration. Between 1950 and 1955 heavy industry investment grew three-and-a-half-fold, the fastest rate in any sector of West German industry. In the judgment of the Bundesrepublik's leading institute for economic research, this offset both the lack of investment between 1945 and 1950 and the *Demontagen*, thus "normalizing" the structure of industry. Coal would prove a laggard in Germany as in Europe generally, but steel outputs increased dramatically, at the rate of 15 percent annually between 1948 and 1953 – half again as fast as the overall expansion of national income during these years. By the end of 1954 the value of industrial output was twice that of 1936, the last normal prewar year.[142] Industrial exports to developing nations fueled much of the growth. In prices adjusted to 1950 levels they increased from 3.5 billion Deutschmarks in 1949 to 18.6 billion in 1954. Their significance is evident in the record-breaking development of the optical, machine tool, and chemical industries during these years as well as in the new prominence of turnkey projects in the growth strategies of the industrial leadership.[143]

The industrial restoration also had a social dimension. Co-determination proved to be a source of stability that some critics such as Dr. Heinrich Deist regarded as amounting to a betrayal of unionism.[144] Others praised the conservative mood of labor relations. A discreet three-month study conducted under the auspices of ECSC concluded that the new policy had revived the old-fashioned paternalism of the German firm. The outlook of the labor force, the study reported, was very positive; the men were willing to work overtime in order to buy consumer gadgetry for their wives and seemed to believe that striking was unnecessary because the grievance machinery worked well. The unions stood foursquare behind the system, the study concluded, with Walter Freytag of the metalworkers having personally suppressed the one important wildcat strike of recent years. In the judgment of the report, co-determination had improved productivity, facilitated reassignment and restructuring of workplaces, and by improving the on-job atmo-

[142] WW, Deutsches Institut für Wirtschaftsforschung, *Wochenbericht*, 12 April 1952.

[143] WW, Wirtschaftswissenschaftliches Institut der Gewerkschaften, Cologne, "Der Schwerpunkt in die Nachkriegsexpansion des Industrieexports," by Ernst-Georg Lange, March 1955, pp. 64–74.

[144] Heinrich Deist, *Die Neuordnung in der Montanwirtschaft und die Mitbestimmung in den Holding-Gesellschaften*, (Dortmund, 29 July 1954), pp. 1–24.

sphere also apparently helped keep wages in line. It deserved credit, in sum, for having coopted the unions into assuming responsibility for the welfare of the firm, created "an astonishing cooperation toward common goals," and given Ruhr steel a competitive edge over its ECSC counterparts.[145]

The economic miracle unfolding in West Germany increased the Federal Republic's importance within the ECSC substantially. Between 1949 and 1953 German industrial production increased at an annual rate of 15 percent as opposed to 9.2 percent for the community as a whole, leveling off at just over 10 percent, or about a fifth above the community average until 1957. By 1953 West German national income was over twice that of France and reckoned to be equal to 48.6 percent of the ECSC total. Steel outputs increased by 55 percent between 1952 and 1957, as opposed an overall 42.7 percent for The Six, with only those of Italy growing faster.[146] The upsurge in demand made the Federal Republic particularly important as an import market for ECSC-produced steel. Impressive though the gains in German output were, they fell short of meeting the needs of the domestic manufacturing industry. Of the total net change of 1.6 million tons in the communitywide trade balance between 1952 and 1955, no less than 1.2 million was due to an increase in German steel imports. France was the main beneficiary: French steel exports to Germany quintupled, an amount seven times as great as the amount of ferrous product imported into France from the Federal Republic.[147] West Germany was, in other words, the only really important new ECSC steel market to open between 1950 and 1955.

The ECSC coal trade showed markedly less interpenetration then steel; intracommunity traffic did not increase significantly after 1952 and varied little from year to year. In fuel as well as in metal the welfare of the coal–steel pool depended on the Federal Republic. The Six consistently ran deficits in the coal balance because controls kept prices artificially low; if allowed to rise they would have resulted in the substitution of more efficient energy sources. As it was, Europe found itself having to import increasing amounts of American coal which, though cheap to mine, normally cost more in Europe than local combustible because of the prevailing high trans-

[145] EC CEAB 3/1952/115, 21–31, Remarques sur le fonctionnement de la cogestion dans la sidérurgie de la Ruhr," November 1952.
[146] See WW, Bruno Gleitze, "Verstärkte Industrialisierung Westdeutschlands," *Wirtschaftswissenschaftliche Mitteilungen*, March 1955, pp. 49–63.
[147] Diebold, *Schuman Plan*, p. 580.

atlantic freight rates. By agreeing to continue exporting solid fuel at a time when it had no surplus to spare, the Federal Republic bore most of the community's costs of consuming the expensive U.S. product.[148] Except for 1954, when they jumped to 9.7 million tons, West German coal and coke sales to the other nations of the community ranged between 7.2 million and 8.3 million tons between 1952 and 1957. At the same time, imports from countries outside the union doubled from 6.4 million tons to 12.9 million tons.[149]

If the French had one stake in West Germany in the form of a new steel export market and another in the form of subsidized coal imports, in 1954 they acquired a third in the form of the biggest and best mine in the Ruhr, Harpener Bergbau. The output of this former Flick property approximately equaled the total coke importing requirements of the Lorraine steel industry and, or so claimed the purchasers, would end its historical dependence on the Ruhr. Negotiations on the French side were handled by René Mayer on behalf of a consortium of French mills, the Société Sidérurgique de Participation et d'Approvisionement en Charbon directed by L. Charvet.[150] The deal took a year to arrange, but the happy French estimated that they had acquired an important asset at one-quarter of construction cost.

On 22 October 1954 the Bundesverband der Deutschen Industrie and the Conseil National du Patronat Franŗis concluded an agreement meant to commemorate the opening of a new era in Franco-German business relations. It called for "closest possible cooperation," for "the enlargement of commercial accords into long-term agreements aimed at expansion," and, because "the anarchic commercial practices on international markets have enfeebled [our] national economies," for "coordination of policy among the NATO nations."[151] The agreement envisaged regular consultation in order to coordinate fabrication programs in manufacturing, the electrical equipment and other industries, as well as a common program of armaments manufacture. The text emphasized that "it seems particularly advisable to promote the construction of common aeronauti-

[148] Ibid., pp. 575, 582.   [149] Ibid., p. 587.

[150] EC CEAB 2/1954/322, Sidechar to President of the High Authority (3052/54f), 3 May 1954; AN 457 AP31, Telegram (Franŗis-Poncet), 29 June 1953; and "Note du Ministère de l'Industrie et de l'Energie, Achat de la Mine Harpener . . . ," 26 June 1953.

[151] JM AMH 22/5/1b, "Procès-verbal concernant un accord convenu entre de Bundesverband der Deutschen Industrie et le Conseil National du Patronat Français," 22 October 1954.

cal factories." Rationalization and specialization accords were foreseen and, additionally, "the joint exploitation of patents."[152] The French and German signatories also committed themselves to mutual aid in the struggle against nationalization. The text was quite specific about "modalities of collaboration." ECSC-like *dirigisme* was to be fought whenever it turned up; though no new institutional machinery was thought to be necessary to establish the new cooperation, work was to be intensified at "the committee of Franco-German industrialists." With this in view, the chiefs of the two producer associations were to meet more frequently than before in order to develop common policies with which to lobby their national governments and facilitate further joint actions.

This neo-Düsseldorf Agreement proved to be more an anachronistic curiosity than a guide to the future. The main supposition upon which it was based – that the European Defense Community would stimulate a rearmament boom exploitable in common – proved to be false for more than one reason: The EDC failed; the kind of centralized European military procurement agency that would have been needed to initiate and administer common manufacturing and purchasing programs never came into existence; and the notion that, as in the late 1930s, rearmament would be the prime mover of economic progress was wrong. Contrary to what the chiefs of the BDI and the CNPF expected, consumer spending would soon play this role. Like their contemporaries, they hardly realized that a new era in European capitalism was dawning in which there would be little need for machinery of business diplomacy that had been designed for use in the depression decade.

By the mid-1950s old fears in Western Europe were giving way to new ones. The steel glut dreaded in 1949 had failed to materialize by 1955. Recovery-, export-, and consumer-fired demand in West Germany took up the slack in Europe. Except in Italy, there was little expansion of plant within The Six; the emphasis was rather on cost reduction, expansion into finishing, and diversification toward the consumer. A race to add capacity as after World War I did not occur. Coal developments were complicated and temporarily less reassuring. Shortages caused by government price policy were persistent, but allocation worked in a politically though not economically satisfactory manner. Sales of non-European coal increased on the markets of The Six as did the relative consumption of other fuel sources. A sharp drop in transatlantic freight rates after the Suez

[152] Ibid.

crisis of 1956 brought U.S. coal flooding into European markets and soon brutally revealed that, like Belgium, the Ruhr was not, and could not become, competitive.[153] The once feared coal weapon was now fit only for a heraldic museum. The resource over which the French and Germans had struggled for fifty years had been reduced to strategic insignificance.

Few indeed were those living on either side of the Rhine in the late 1940s who realized that changing technology had consigned Franco-German heavy industry struggles to irrelevance in war. Yet the American victories over Germany and Japan had depended far less on coal and steel than on a host of other materials and energy sources. Strategic perceptions do not often change instantaneously and until 1953 did not affect the fundamentals of European diplomacy. There were too few atomic bombs after 1945 to have transformed the nature of warfare overnight; military planning in both East and West assumed that the next war in Europe, though considerably more destructive than the last one, would be fought essentially like it, with ground forces as the decisive element of battle.

In the early 1950s the serial manufacture of weaponry, miniaturization, and the perfection of the H-bomb together raised the ratio of firepower to cost exponentially. With "more bang for the buck" as its byword, President Eisenhower adopted the "new-look strategy" in January 1954 calling for the deployment in Europe of nuclear weapons, large and small, on a huge scale. Its diplomatic accompaniment was the "massive retaliation" that Secretary of State Dulles championed as a deterrent to Soviet aggression. The fear that in the next war Europe would become a nuclear battleground for the United States and the USSR relegated the Ruhr Problem to the old curio chest.

Though Monnet's motives for resigning the presidency of the ECSC on 9 November 1954 have never been fully explained, the official reasons given may well be largely true: He was disappointed and even disgusted after the failure of the European Defense Community – an issue that had preoccupied him during his entire period in office – and wanted to be in a position to give new leadership to the European unification movement.[154] This explanation overlooks his dissatisfaction with the ECSC, however, which as then constituted was worthless as a vehicle for integration. Years later he hinted at this reason: The High Authority, he admitted, "had only

[153] Abelshauser, *Der Ruhrkohlenbergbau*, p. 90.
[154] Pascal Fontaine, *Le Comité d'Action pour les Etats-Unis d'Europe de Jean Monnet* (Lausanne, 1974), pp. 21–22.

limited functions which it had to fulfill and could not go beyond."[155] In electing not to seek another term he wanted to break with the ECSC and, as he phrased it, "build Europe from the outside."

He got to work at once, in December 1954, organizing the European Front, an unfortunate name with neo-Fascist connotations soon changed to the Action Committee for the United States of Europe. This label was also misleading because the new organization had no specific program. The Action Committee consisted of a permanent staff of former commandos from the community. To it subscribed members representing pro-European political parties and labor unions, including a strong element but from the SPD and DGB. Business associations were pointedly excluded. The Action Committee served as a lobby and think tank for the integration movement, initially championing a single cause, Euratom.[156]

It appeared to be a good one. In December 1953 President Eisenhower had told Prime Minister Churchill and Premier Joseph Laniel that the United States was ready to share nuclear technologies with Great Britain and France. In September of the following year the U.S. Senate passed the necessary enabling legislation. The atom symbolized the opening of a limitless technological horizon, and nuclear energy had the power of an *idée force*. Inspired by the politicoscientific vision of Louis Armand, Monnet wanted the new technology to be as central to strategies of peace maintenance as it had become to those of war making. Intending to make it the focus of policy in energy, industrial development, and research, he hoped that it would provide an even more stunning example than the Schuman Plan of how men could turn the instruments of destruction into implements of civilization. Monnet assumed that both France, which was having trouble bearing the capital costs of nuclear development, and Germany, which had forsworn the construction and possession of atomic weaponry, would be eager to enter a new pooling arrangement.[157]

Yet his was not the blueprint that guided the next stage of European integration: Monnet would be politely but firmly shunted aside by tacit agreement of the national governments. Though he had initially planned with Paul-Henri Spaak's support, to relaunch Europe by means of a public campaign, a change of government in early March 1955 that brought Schuman, Antoine Pinay, Pierre Pflimlin, Pierre-Henri Teitgen, and other committed Europeans

[155] Monnet, *Memoirs*, p. 411.
[156] Fontaine, *Le Comité d'Action*, passim; confidential source, pp. 412–413.
[157] Fontaine, *Le Comité d'Action*, pp. 26–27.

into the French cabinet persuaded him to push for consideration of the integration issue by a conference of the foreign ministers. This was a fatal mistake. Adenauer insisted upon postponing the gathering, and on 6 April 1955 a German Foreign Office spokesman let it be known that the Federal Republic would not agree to enter an atomic pool unless negotiations were broadened to include a general common market for all goods. This much larger proposal overwhelmed Euratom, only brief mention of which would eventually appear in the communiqué summarizing the results of the Messina meeting at which Europe was formally relaunched.[158]

Apparently fearing that the work of integration was about to proceed without him, Monnet than made an embarrassing faux pas, at the end of May rescinding his resignation from the High Authority. It was too late for this: René Mayer had already been appointed as his successor. The gray eminence of Europe was finally out, replaced by a man who though a personal friend had very strong ties to business interests and a distinctly different notion of how the coal–steel pool should operate. Under President Mayer conflict would be all but absent from the High Authority, which soon turned into a businessman's club whose activities provided pleasant relief from the daily grind of work at the office.[159] The ECSC would eventually be hooked up to the more efficient and powerful machinery designed at Messina.

The tone of the Messina Resolution was matter-of-fact and even monotonous: "It is necessary to work for the establishment of a united Europe by the development of common institutions, the progressive fusion of national economies, the creation of a common market and the progressive harmonization of . . . social policies." It called for the development of communitywide energy, transport, and nuclear policy, the setup of joint financing mechanisms, and cooperation in research. Its main thrust was, however, toward the organization of "a European market free of all customs duties and all quantitative restrictions," which was to be achieved in stages. In pursuance of liberalization, tariffs against third countries were to be removed, "financial, economic and social" policies harmonized, monetary policies coordinated, and the free movement of manpower introduced.[160]

Of equal importance to the resolution were the procedures used at Messina – those actually of the old Interim Committee – which

[158] Confidential source, pp. 416–417.
[159] Willis, France, Germany, and the New Europe, p. 229.
[160] Quoted in confidential source, pp. 426–427.

would be adopted not only in negotiating the Treaty of Rome establishing the European Economic Community but as its operating methods. The governments acted jointly, as a council, in pursuit of common aims but for specific purposes and through the mechanism of individual treaties. Supranationism like that of the ECSC was deliberately avoided; the unconditional transfer of sovereign powers was not required as up-front payment for admission to Europe. The Messina Resolution of 3 June 1955 was based upon the so-called Benelux memorandum drafted by Dutch Foreign Minister Johan Willem Beyen but echoed positions presented in numerous papers written under the auspices of the Schuman Plan desk of the West German Ministry of Economics, whose chief, Hans von der Groeben, represented his country at Messina. This underlying similarity of approach made it possible for The Six to reach agreement within the space of two days about how to proceed with the building of Europe.

The West Germans had arrived at their position by way of a scholarly assault on partial integration that began as an upshot of the tax dispute. With pedantic thoroughness it demonstrated to their own complete satisfaction that the sectoral approach represented by the ECSC created unacceptable economic distortions and concluded that the only feasible alternative was to establish a genuine common market creating a single environment for economic activity. In spite of disagreement between the Economics Ministry and the Foreign Office as to whether this should be done through treaty or by means of permanent political institutions, consensus existed within West German officialdom about how best to proceed along the path of integration.[161] It was the route taken at Messina. One cannot state on the basis of known fact that in the other ministries of The Six critiques similar to those in West Germany led to a similar result (although the outcome of the conference strongly suggests that they did) or necessarily conclude that a common vision of Europe's future arose from within the community as a result of a shared ECSC experience; agreement might well have been limited to procedures and techniques. In terms of law and politics, "Europe's first experiment in supranationalism" was nonetheless a latter-day equivalent to the American Articles of Confederation – a forerunner to a far stronger and more permanent union.

[161] GP Montanunion, Papers of June 1953, by Estne, Kaps, Czermak, Holthaus, Ehring, von Hoffmann, Treitschke, Rinck, Gaedke, von der Groeben, all entitled "Erfahrungsbericht betreffend Rechtstellung, Verwaltungsaufbau, und Aufgaben der Organe der Europäischen Gemeinschaft für Kohle und Stahl."

Conclusion

The Schuman Plan failed to provide an adequate framework for supranational government, did little to harmonize economic conditions, and by no means set in motion an inexorable process of unification, but it ended the competitive bids for heavy industry domination that had wrecked every previous large-scale attempt to reorganize the Continent since 1918, led to *Westintegration* and Franco-German partnership, and resulted in the creation of a new political entity, Europe. The coal–steel settlement is like a spherical wooden puzzle held together by interlocked parts. When taken apart and laid out piece by piece on a table, reassembly seems impossible; yet after patient, frustrating, mostly blind manipulation the thing will eventually, unexpectedly, and inexplicably snap into place, forming a tight round object almost as hard to pull asunder as it once had been to put together. The parts of this puzzle, through perplexing in shape, can be seen and thus depicted, and the assembly procedure can be partially reconstructed from tactile memory. The precise way in which the various pieces fit into one another, however, and the interconnections that give the puzzle inner strength can only be visualized. The solution will always remain partly a mystery.

The Americans represent one piece in the puzzle. The French were not strong enough to unify Europe by their own lights, the Germans tried twice to do so and failed, and the British had little interest in the subject. Only the United States could have taken the lead in restoring the commercial and financial system of the Continent to normal operation. A redefinition of national purpose was required before it could do so. A people that believed itself to be unique because separated geographically and historically from the vices of Europe, and the world, now faced having to dwell within the larger universe from which it had sought to remain separate.

The change did not take place overnight but began early in the century and is still under way. The critical turning point nonetheless occurred during the Truman years. The United States had withdrawn from Europe after 1918 and once again after 1929. Its single major interwar interventions was to underwrite the Dawes Plan in 1924, which though limited and private in character made an essential contribution to the prosperity of the next four years and provided an intimation of what might have been accomplished through sustained and official international financial cooperation. World War II thrust the United States into a prominence it had not sought and for which it was unprepared. At a time when much of Europe and Asia was being torn apart America underwent an economic revolution that secured its leadership for a generation and vastly increased its global responsibility. Even so, Roosevelt's Grand Design did not contemplate a fundamental shift in foreign policy. It was a formula not for intervention but for supervision of a postwar world organized on the basis of national self-determination, free trade, and international government. The United States, together with the Soviet Union and the other wartime Allies, was to act as guarantor of an international order based on law. The threat of conflict being thereby eliminated – it was hoped – the United States could disengage from its recent involvements and return to life as it once had been.

The Grand Design had little value as a reconstruction strategy, and the wartime domestic consensus behind it crumbled with the death of Roosevelt and the onset of peace. No master hand or idea guided America's postwar foreign policy; when not merely an emergency response, it was the product of interventions by representatives of factions split off from what had been the New Deal coalition. After 1945 America was at war with itself in Germany, Europe and the world at large. U.S. military government in the defeated Reich was chaotic. From one side practical men pushed for recovery without reform, on the other radicals demanded precisely the opposite. Such conflicts had to be resolved and coherence introduced into U.S. policy before conditions in Germany and in Europe could be normalized.

There was nothing elegant about the political stabilization process in the United States. It began with the expulsion of the Left and was accompanied by a thundering anti-Communist rhetoric that would eventually cause brain disorders in the body politic. Yet crude methods were needed to consolidate bipartisan support for the decisive departure in foreign policy known as the Marshall Plan. It was a break with isolationism, noninterventionism, and globalism, a program for a unified Europe sponsored by the United States. Beyond this its purposes were vague. Washington's initiatives did not produce integration. The terrain (tariffs and currencies) was ill chosen as was the *masse de manoeuvre*, the British. The Marshall Plan organization for the European states, the CEEC, was immobilized by a de facto refusal of the United States to free it from its leading strings as well as a lack of European initiative. Real progress in fitting the unification puzzle together began only with the Schuman Plan.

German industrial corporatism represents another key piece of the puzzle. An oddly chiseled thing of hooks and notches calculated to perplex and frustrate, its relationship to the other parts is hard to figure out. The tradition rested on cartels, whose elimination the United States demanded and whose disappearance the Germans promised. Upon the syndicates of the Ruhr had been built housings of private associations and public offices enabling vast sectors of industry to function as a single unit as well as to act, for better or worse, as instruments of national policy. Without officially admitting the fact, the Allies relied upon the syndicates to allocate scarce goods, earn foreign exchange, and otherwise direct recovery, and in truth the industrial equivalent of the kidney machine would have been required to restore Germany without them. Yet, while not daring to excise the critical organ, the doctor never acknowledged diagnostic error. Expert opinion is therefore of little use in explaining the patient's survival. One must similarly grope toward an understanding of how the piece that is the German industrial tradition fit into the puzzle representing the overall settlement.

Contrary to contemporary American belief and prevailing historical opinion, the net impact of Ruhr corporatism has been positive both internationally and domestically, diverging rather than paralleling or converging with the rise of political extremism. In foreign affairs, the heavy industry fanaticism of 1914 was an extreme from which retreat occurred thereafter; instead of trying to dispossess foreign counterparts, German coal and steel producers developed successive and broadening patterns of cooperation with them. The formation of the International Steel Cartel in 1926 was the first great advance along this path and the renewed agreement of the 1930s a second one. During the war, mutual aid between the heavy industries of Germany and those in the conquered nations of Western Europe was the rule rather than the exception. Defeat and Allied occupation brought an official proscription of international coal and steel cartels, but their eventual return in some form was unanimously desired by managers of mill and mine throughout Western Europe.

Economic modernization has been both cause and consequence of the rise of industrial corporatism within Germany proper. Its milestones are the rationalization movement of the 1920s, the national manpower policy of the 1930s, the mobilization before and during the war, and the generalized interpenetration of business–business, state–industry, and labor–management relationships that give credence to the smug, saucy, and sarcastic expression "Germany Inc." The importance of these developments was not immediately apparent; only the economic performance of the Federal Republic over the past thirty years has provided truly compelling evidence for the success of the German model. The postwar generation can therefore hardly be damned if in trying to extirpate the economic roots of national socialism it threatened to pull out the good along with the bad. At the same time, contemporaries failed to appreciate fully that ultimate responsibility for the conduct of business in the affairs of state rests with the politicians. The same judgment must be applied to the situation both after as well as before 1945. Though Hitler surely deserves more blame than usually given for the economic crimes of the Third Reich, Adenauer should be granted more credit than he normally gets for having reformed and strengthened industry. He ordered the enactment of co-determination in coal and steel and made the concessions that saved the Schuman Plan. His shrewdness and wisdom were essential in mobilizing Germany's industrial assets for Europe.

Jean Monnet was another integral part of the settlement. He is represented by a cruciform piece, with a long arm being a relationship to France and a short one that to the United States. Monnet turned modernization from an intellectual cult into the French national pastime. More prophet than prototechnocrat, his plan to reequip and renovate the nation's factories ran wild, and in industry, civil service, and politics his enemies far outnumbered his friends. Though hardly the first of his countrymen to be obsessed with the need to overcome industrial decadence, he deserves exceptional credit for having preached tirelessly that national sovereignty was meaningless without industrial competitiveness. His ultimate triumph was due less to the success of the institutions he created than to determination, vision, and American support. The latter, provided chiefly in the form of money, made him an independent power in a country that otherwise would not have given him a chance.

Without extraordinary backing from the United States the Schuman Plan would have aborted, and without Jean Monnet it would not have been given: No foreigner has ever occupied a position of trust comparable to Jean Monnet's in U.S. foreign affairs. At a time when American involvement with the rest of the world was increasing with explosive force but when the nation's experience was largely limited to contact with places where English was spoken, Monnet was extraordinarily well placed to speak on behalf of Europe. An international civil servant and financier with strong ties to Wall Street as well as great expertise in the management of war economies, Monnet spent most of World War II in Washington where he exercised immense unofficial influence in key areas of decision making. His close friendship with the policy makers of that city stemmed from something more than common service in the Allied cause; he shared with the new internationalists on the Potomac a vision of Europe's future. Monnet believed profoundly that the revolution under way in the American war economy could take hold in the demoralized and disintegrating civilization from which he had been forced to part. He wanted a New Deal for French, and European, industry and planned to launch and land it with the help the handful of like-minded men who wielded decisive power in the post war world. The founding of the Commissariat Général du Plan in France marked the beginning of Monnet's modernization campaign, and the proposal of 9 May 1950 to form a coal–steel pool extended it to Europe.

Monnet was nonetheless the agent of a primarily French rather than an American policy. France's attempts to deal with the Ruhr Problem within the framework of an overall European economic settlement date from 1918, and similar proposals were revived after 1945. The French accepted the inevitability and necessity of economic coexistence with Germany but were preoccupied with turning it to advantage. The Schuman Plan proposal was consistent with this approach. Monnet wanted to make a bargain with the Federal Republic before the opportunity to do so on favorable terms disappeared. Control was not to be replaced by cooperation overnight but to merge gradually with it. The European Coal and Steel Community was thus also a logical extension of traditional French *Ruhrpolitik*. The Schuman Plan initiative precipitated fundamental changes in *U.S.* policy, however. The Americans had to endorse integration without the British as participants, something previously unthinkable, as well as compromise traditional principles in order to remain faithful to the overriding commitment to letting Europeans work out the process by themselves. Though the French planning commissioner's personal views were reassuring to Washington, U.S. policy makers doubted that he could organize a European heavy industry pool without restoring the interwar cartels. Yet the United States backed Monnet's conduct of the Schuman Plan negotiations, gave full endorsement to his proposal to create a European Defense Community, revived the Ruhr decartelization campaign at his behest, and even delayed German rearmament when at war in Korea in order to promote his integration attempt. The Americans agreed to let Monnet run, and he set the course for Europe.

No one knew on 9 May 1950 how the various parts of the coal–steel settlement would fit together, and an answer was not soon forthcoming. The problem remained unsolved when negotiations formally concluded on 18 April of the following year, and progress continued to be fumbling even after the formal inauguration of the European Coal and Steel Community in August 1952. The puzzle was eventually put together in ways difficult to predict. The French *Document de Travail* prepared by Monnet set the agenda for the coal–steel talks. Amounting to a draft constitution for the community, it made provision for a powerful central directorate, the High Authority (HA), through which a regulatory code strongly influenced by U.S. antitrust ideas could be imposed on industry. Though operating in conjunction with both legislative and judicial bodies, the HA was meant to be run as Monnet imagined the wartime boards of American industry had been. The reform-minded economic provisions of the French draft forbade cartelization, concentration, price and market rigging, called for openness in business methods, and championed an enlightened policy toward labor. Monnet wanted to do more than merely protect French economic security from a resurgent Ruhr: He wanted to create a new, progressive, and truly European sector of industry that would serve as a first great step toward closer economic integration.

From the outset the French were on the defensive in the Paris coal–steel negotiations, not least of all because, being preoccupied with Germany, they had failed to consider adequately the particular concerns of the remaining Six, each of whom eventually exacted exceptional treatment as quid pro quo for participating in the talks. The main common interest of the negotiating parties soon proved to be in stimulating the economic revival of West Germany. Military defeat notwithstanding, the Ruhr had become more critical to the welfare of Europe than ever before. German industry had expanded and modernized during the war whereas that of France and the Low Countries had languished under occupation and the already serious ailments of Britain's producers had grown still more severe. The rest of Europe had no choice but to look to the Ruhr for coal, markets, and leadership even after the war. Requiring deft diplomacy rather than major structural adjustments, the reestablishment of old ties between Western European producers and the economics officials occurred in the course of the coal–steel deliberations. The delegates left the French proposal formally intact but denuded it of substance; behind the scenes producers tightened up domestic cartels and wove international syndicates back together. With the Korean war accelerating German recovery, only American agreement to crank up the long-idle decartelization campaign in the Ruhr and support the French proposal for a European army prevented Monnet's withdrawal from the talks. In the interim period between the initialing of the treaty and the commencement of ECSC operations in Luxemburg, the Germans expanded and tightened up the corporate structures of their economy, which soon spurted to a high rate of sustained growth; industrialists who had previously acted behind the scenes openly proclaimed an intention to eliminate Monnet and restore the prewar partnership; and the civil servants and diplomats who had negotiated the treaty met somewhat surreptitiously to assure that they rather than Monnet and his team would run the heavy industry community.

In Luxemburg the president of the High Authority did battle against increasingly heavy odds. The new Eisenhower administration championed a doctrine of mutual security in place of the economic cooperation that had primacy under Truman, and in Europe the forces of restoration grew in strength. The first six months at the ECSC were the scene of uninterrupted power struggles between President Monnet and traditionalists representing the interests of national industry associations. When the common markets opened the following spring producers determined what would be sold where, at which prices, to whom, and under which conditions. The ingenious regulations devised by Monnet to govern business activity were little more than artful camouflage, the reform program a dead letter. The old gang was back.

When Monnet resigned as president of the High Authority in November 1954, it had only begun to be clear that the coal–steel settlement, and the integration process, would hold together rather than split apart. How the puzzle was ultimately solved and why it has proved to be so strong are questions calling for intelligent speculation rather than rigorous structural analysis. Certain small parts seem to have had an unsuspected importance, and others have fit together in wholly improbable ways. Whereas pieces sturdy enough to have borne much weight apparently did not, less impressive ones must have had unexpected strength. The solidity of the overall structure surely also rested on the curved outer surfaces of the puzzle rather than its interlocking inner members.

One source of strength was the often remarked and all too easily trivialized deep longing for unity felt to some extent by all Europeans. It cannot be weighed, measured, otherwise quantified, or shown to have had demonstrable effect in advancing integration from one point to the next but worked like a powerful undercurrent slowly shifting and moving events. The post–World War I reparation schemes of Clémentel and Seydoux aimed at restoring a sense of wholeness to Western Europe, as did the Mayrisch committee, the world economic conferences of the late 1920s, and the business diplomacy of the late 1930s. Speer tried with indifferent success to exploit the same sentiment during the war, and it figured prominently in the platforms of all political parties after 1945. European unity was the concern of numerous formal and informal postwar gatherings, congresses, conferences, study groups, and spiritual uplift sessions; it created a constituency which though lacking a coherent program was receptive to all initiatives serving the common cause and comprised a reserve force that could be called upon to substitute personnel and ideas for exhausted or failing actives. European federalists, insisting upon the institutionalization of political union, gave new life to the integration movement at a time when the ECSC and the EDC were sinking.

Another source of strength was the willingness of West German industry to make short-term concessions in the interest of long-term gains. The Ruhr leaders disagreed bitterly with Monnet but lived with their misgivings rather than wreck the negotiations, apparently having much earlier learned to think strategetically as well as tactically. In order to keep cooperative relationships intact, the German delegation to the International Steel Cartel allocated foreign markets domestic producers could not supply to other Western European producers. The wartime counterpart of German heavy industry to collaboration was reinsurance, in the absence of which, to cite but one example, Klöckner would have become Dutch after 1945. Without the sometimes grudging but important sacrifices of the smokestack barons, the ECSC would also have been gridlocked: the Germans subsidized the Belgian mines directly, bore the expense of importing American coal, sold fuel at below market prices to the rest of The Six, exported scrap to the community instead of hoarding it, and uncomplainingly shared domestic markets with foreign sellers. They made partnership profitable.

A third source of strength was the existence of a learning process that changed the very nature of European statecraft. One should not claim that either Monnet or any other figures involved in creating the coal–steel pool invented a new kind of diplomacy: Attempts to restore the broken unity of Europe though economic cooperation date from 1918 and include both the business initiatives of the interwar period and the Marshall Plan. Fresh ideas were nonetheless needed for the ultimate result. The Schuman Plan announcement moved international cooperation from the domain of diplomats, bureaucrats, and businessmen into the public realm, turning a cherished ambition of policy-making elites into the great hope of the masses. The proposal of the French foreign minister was more than a giant public relations coup, however. Containing the gist of a new approach to the relationship between government and industry, the New Deal notion that business could become democratically responsible fundamentally changed the integration context. No longer would discussions of political federalism have primacy; in their place came debate over implementing economic policy. It continued even after the proposals incorporated in the *Document de Travail*, written into the treaty and institutionalized in the community itself, had proved to be largely unworkable. The results of this inquiry were impressive. Significant theoretical progress was made toward eliminating such obstacles to integration as inequities in rail, road, and waterway rates, taxes, social legislation, and other nontariff barriers to trade. Issues that five years earlier had not aroused even scholarly interest were by 1955 at the forefront of the European agenda. Functional integration had somehow been transformed from frustrated ambition to practical methodology.

Circumstance – the curved outer edges of the puzzle – also contributed to the strength of the heavy industry settlement. By midcentury old hopes and fears were retreating before new ones. Europe was awash in a consumerism that acted as solvent to conflict between class and nation and resulted in power shifts between states and within economies. The mines and mills were no longer critical determinants of national power. The automobile had replaced the blast furnace as symbol of economic might. Oil was rapidly pushing coal off the market. The strategic importance of heavy industry was also in rapid decline: Not tanks and ships but atomic bombs were now the decisive element of battle, and Moscow and Washington, not Paris and Berlin, were the places where the decisions would be made to use them. The Schuman Plan succeeded in part because it gave the nations of Europe an opportunity to surmount historic antagonisms that were fast becoming irrelevant.

Bibliography

Abelshauser, W. "Ein Briefwechsel zwischen John J. McCloy und Konrad Adenauer während der Korea-Kirse." *Vierteljahrschefte für Zeitgeschichte* 30/4:1982, pp. 715–751.

Abelshauser, W. "The First Post-Liberal Nation: Stages in the Development of Modern Corporatism in Germany." *European History Quarterly* 14:1984, p. 285–318.

Abelshauser, W. "Freiheitlicher Sozialismus oder soziale Marktwirtschaft? Die Gutachtertagung über Grundfragen der Wirtschaftsplanung und Wirtschaftslenkung am 21. und 22. Juni 1946, *Vierteljahrshefte für Zeitgeschichte* 24/19:1976, pp. 415–449.

Abelshauser, W. "Korea, die Ruhr, und Erhards Marktwirtschaft: Die Energiekrise von 1950/51." *Rheinische Vierteljahrsblätter* 45:1981.

Abelshauser, W. "Problem des Wiederaufbaus der westdeutschen Wirtschaft, 1945–1953." In H. A. Winkler (ed.), *Politische Weichenstellungen im Nachkriegdeutschland, 1945–1953*. Göttingen, 1979.

Abelshauser, W. *Der Ruhrkohlenbergbau seit 1945: Wiederaufbau, Krise, Anpassung*. Munich, 1984.

Abelshauser, W. "Staat, Wirtschaft, Wachstum: Das deutsche 'Wirtschaftswunder' im Zeichen der Diktatur." Unpublished ms., September 1987.

Abelshauser, W. "Wiederaufbau vor dem Marshall-Plan: Westeuropas Wachstumschancen und die Wirtschaftsordnungspolitik in der zweiten Hälfte der vierziger Jahre." *Vierteljahrshefte für Zeitgeschichte* 29/4:1981, pp. 545–578.

Abelshauser, W. *Wirtschaft in Westdeutschland, 1945–1948: Rekonstruktion und Wachstumsbedingungen in der amerikanischen und britischen Zone*. Stuttgart, 1975.

Acheson, D. G. *Present at the Creation*. New York, 1969.

Acheson, D. G. *Sketches from Life of Men I Have Known*. New York, 1961.

Adamsen, H. R. *Investitionshilfe für die Ruhr: Wiederaufbau, Verbände, und Soziale Marktwirtschaft 1948–1952*. Wuppertal, 1981.

Adamthwaite, A. *France and the Coming of the Second World War*. London, 1977.

Adenauer, K. *Erinnerungen*. Stuttgart, 1965–1968.

Ambrose, S. E. *Rise to Globalism: American Foreign Policy since 1938.* 3rd ed. New York, 1984.

Ambrosius, G. "Marktwirtschaft oder Planwirtschaft? Planwirtschaftliche Ansätze der bizonalen deutschen Selbstverwaltung, 1946–1949." *Vierteljahrschrift für Sozial- und Wirtschaftsgeschichte* 66/1:1979, pp. 74–110.

Arkes, H. *Bureaucracy, the Marshall Plan, and the National Interest.* Princeton NJ, 1972.

Baade, F. "Eisen und Stahl in der langfristigen Europaplanung." *Stahl und Eisen 69:* 1949, pp. 836–838.

Baade, F. "Das Ruhrrevier in der Weltwirtschaft." Brochure. Essen, 1949.

Backer, J. H. *The Decision to Divide Germany: American Foreign Policy in Transition.* Durham NC, 1978.

Backer, J. H. *Priming the German Economy: American Occupational Policies, 1945–1948.* Durham NC, 1971.

Balabkins, N. *Germany under Direct Controls: Economic Aspects of Industrial Disarmament, 1945–1948.* New Brunswick NJ, 1964.

Balfour, M. *Four-Power Control in Germany and Austria, 1945–1946.* Survey of International Affairs, 1939–1946, ed. Arnold J. Toynbee. London, 1956.

Ball, G. *The Past Has Another Pattern: Memoirs.* New York, 1982.

Bariéty, J. "Der Tardieu-Plan zur Sanierung des Donauraums (Februar–Mai 1932)." In J. Becker and K. Hildebrand (eds.), *Internationale Beziehungen in der Weltwirtschaftskrise, 1929–1933.* Munich, 1980.

Bariéty, J. "Das Zustandekommen der Internationalen Rohstahlgemeinschaft (1926) als Alternative zum misslungenen 'Schwerindustriellen Projekt' des Versailler Vertrages." In H. Mommsen D. Petzina and B. Weisbrod (eds.), *Industrielles System und politische Entwicklung in der Weimarer Republik,* pp. 552–568. Düsseldorf, 1974.

Barnett, C. *The Pride and the Fall: The Dream and Illusion of Britain as a Great Nation.* New York, 1986.

Baudhuin, F. *Histoire économique de la Belgique, 1945–56.* Brussels, 1958.

Baum, W. C. *The French Economy and the State.* Princeton NJ, 1958.

Beloff, M. *The United States and the Unity of Europe.* Washington DC, 1963.

Benz, W. *Von der Besatzungsherrschaft zur Bundesrepublik: Stationen einer Staatsgründung, 1946–1949.* Frankfurt/Main, 1984.

Benz, W. *Die Gründung der Bundesrepublik: Von der Bizone zum souveränen Staat.* Munich, 1984.

Berding, H. (ed.) *Wirtschaftliche und politische Integration in Europa im 19. und 20.* Göttingen, 1984.

Berghahn, V. R. *The Americanization of West German Industry, 1945–1973.* Cambridge, 1980.

Bidault, G. *Resistance: The Political Biography of Georges Bidault.* London, 1965.

Blum, J. M. (ed.). *From the Morgenthau Diaries: Years of War, 1941–1945.* Boston, 1967.

Boelcke, W. A. *Die Kosten von Hitlers Krieg: Kriegsfinanzierung und finanzielles Kriegserbe in Deutschland, 1933–1948.* Paderborn, 1985.

Bok, D. C. *The First Three Years of the Schuman Plan.* Princeton Studies in International Finance, 8. Princeton NJ, 1955.

Boltho, A. *The European Economy: Growth and Crisis.* Oxford, 1982.

Bonn, M. J. *Whither Europe – Union or Partnership?* New York, 1952.

Bonnell, A. T. *German Control over International Relations.* Urbana IL, 1940.

Bossuet, G. "Le poids de l'aide américaine sur la politique économique et financière de la France en 1948." *Relations Internationales* 37:1984, pp. 17–36.

Bracher, K. D., Sauer, W., and Schulz, G. *Die nationalsozialistische Machtergreifung.* 2nd ed. Cologne, 1962.

Brady, R. A. *The Rationalization Movement in German Industry: A Study in the Evolution of Economic Planning.* Berkeley CA, 1933.

Brady, R. A. *The Spirit and Structure of German Fascism.* London, 1937.

Braunthal, G. *The Federation of German Industry in Politics.* Ithaca NY, 1965.

Buchanen, N. S., and Friedrich, L. *Rebuilding the World Economy: America's Role in Foreign Trade and Investment.* New York, 1947.

Bührer, W. "Der Marshallplan und die deutsche Rückkehr auf die diplomatische Bühne, 1948–1949." *Viertaljahrshefte für Zeitgeschichte* 3/19:1988, pp. 529–556.

Bullen, R. "The British Government and the Schuman Plan, May 1950–March 1951." In Schwabe (ed.), *Die Anfänge des Schuman-Plans, 1950/51 [The Beginnings of the Schuman Plan].* Baden-Baden, 1988.

Bullock, A. *Ernest Bevin: Foreign Secretary, 1945–1951.* London, 1983.

Burn, D. *The Steel Industry, 1939–1959: A Study in Competition and Planning.* Cambridge, 1961.

Byrnes, J. F. *Speaking Frankly.* New York, 1947.

Cairncross, A. *The Price of War: British Policy on German Reparations, 1941–1949.* Oxford, 1986.

Cairncross, A. *Years of Recovery: British Economic Policy, 1945–1951.* London, 1985.

Calleo, D. *The German Problem Reconsidered.* New York, 1978.

Calleo, D. *The Imperious Economy.* Cambridge MA, 1982.

Catton, B. *The Warlords of Washington.* New York, 1948.

Centre de Recherches européenes. *Emile Mayrisch: Précurseur de la construction de l'Europe.* Lausanne, 1967.

Clarke, R. *Anglo-American Economic Collaboration in War and Peace, 1942–1949,* ed. A. Cairncross. Oxford, 1982.

Clémentel, E. *La France et la politique économique interalliée.* Paris, 1931.

Collins, R. M. *The Business Response to Keynes, 1929–1964.* New York, 1981.

Court, W. H. B. *Coal.* London, 1945.

Davis, J. S. *The World between the Wars, 1919–1939: An Economist's View.* Baltimore, 1975.

de Cuttoli-Uhel, C. "La politique allemande de la France (1945–1948): Symbole de son impuissance?" Unpublished ms., April 1954.

Deutsches Institut für Wirtschaftsforschung. *Die deutsche Wirtschaft zwei Jahre nach dem Zusammenbruch.* Berlin, 1947.

Dichgans, H. *Montanunion: Menschen und Institutionen.* Düsseldorf, 1980.

Diebold, W. "Imponderables of the Schuman Plan." *Foreign Affairs* 29/ 1:1950, pp. 114–129.

Diebold, W. *The Schuman Plan: A Study in the Economic Cooperation, 1950– 1959.* New York, 1959.

Diebold, W. *Trade and Payments in Western Europe: A Study in Economic Cooperation, 1947–1951.* New York, 1952.

Ehrmann, H. W. "The French Trade Associations and the Ratification of the Schuman Plan." *World Politics* 6/1:1953, pp. 453–481.

Ehrmann, H. W. *Organized Business in France.* Princeton NJ, 1957.

Eichholtz, D. *Geschichte der deutschen Kriegwirtschaft.* Vol. 2. Berlin (DDR), 1971, 1985.

Erhard, L. (ed.). *Deutschlands Rückkehr zum Weltmarkt.* Düsseldorf, 1954.

Fedder, E. H. *NATO: The Dynamics of Alliance in the Postwar World.* New York, 1973.

Feis, H. *The Diplomacy of the Dollar: First Era, 1919–1932.* Baltimore, 1950.

Feis, H. *From Trust to Terror: The Onset of the Cold War, 1945–1950.* New York, 1970.

Feldman, G. *Iron and Steel in the German Inflation.* Princeton NJ, 1977.

Feldman, G. D., and Nocken, U. "Trade Associations and Economic Power: Interest Group Development in the German Iron and Steel and Machine Building Industries, 1900–1933." *Business History Review* 49/ 4:1975, pp. 413–445.

Ferrell, R. H. *George C. Marshall.* New York, 1966.

Ferrell, R. H. *Harry S. Truman and the Modern American Presidency.* Boston, 1983.

Fischer, F. *Germany's Aims in the First World War.* New York, 1967. [*Griff nach der Weltmacht.* Düsseldorf, 1961.]

Fontaine, F. *Plus loin avec Jean Monnet.* Lausanne, 1983.

Fontaine, P. *Le Comité d'Action pour les Etats-Unis d'Europe de Jean Monnet.* Lausanne, 1974.

Först, W. "Die Politik der Demontage." In W. Först (ed.), *Entscheidungen im Westen,* p. 138. Cologne, 1979.

Först, W. (ed.) *Zwischen Ruhrkontrolle und Mitbestimmung.* Cologne, 1982.

Friedensburg, F. *Politik und Wirtschaft. Aufsätze und Vorträge.* Berlin, 1961.

Gaddis, J. L. *Strategies of Containment: A Critical Appraisal of Postwar American National Security Policy.* New York, 1982.

Gaddis, J. L. *The United States and the Origins of the Cold War, 1941–1947.* New York, 1972.

Gatzke, H. W. *Germany and the United States: "A Special Relationship"?* Cambridge, 1980.

Gatzke, H. W. *Germany's Drive to the West.* Baltimore, 1950.

Geer, J. S. *Der Markt der geschlossenen Nachfrage: Eine morphologische Studie über die Eisenkontingentierung in Deutschland, 1937–1945.* Nürnberger Abhandlungen zu den Wirtschafts- und Sozialwissenschaften, 14. Berlin, 1961.

Gerbet, P. "La Genèse du Plan Schuman." *Revue Française de Science Politique*

6:1956, pp. 525–553. Republished as *La Genèse du Plan Schuman: Des origines à la declaration du 9 Mai 1950*. Lausanne, 1962.

Gillingham, J. *Belgian Business in the Nazi New Order*. Ghent, 1977.

Gillingham, J. "Coal and Steel Diplomacy in Interwar Europe." In A. Wurm (ed.), *Internationale Kartelle und Aussenpolitik [International Cartels and Foreign Policy]*. Stuttgart, 1989.

Gillingham, J. "The Deproletarianization of German Society: Vocational Training in the Third Reich." *Journal of Social History* 9/3:1986, pp. 423–432.

Gillingham, J. "Die Europäisierung des Ruhrgebietes: Von Hitler bis zum Schuman-Plan." In K. Düwell and W. Kollmann, (eds.), *Rheinland-Westfalen im Industriezeitalter*. Vol. 3: *Vom Ende der Weimarer Republik bis zum Land Nordrhein-Westfalen*, pp. 179–189. Wuppertal, 1984.

Gillingham, J. "Die französische Ruhrpolitik und die Ursprunge des Schuman-Plans." *Viertljahrschefte für Zeitgeschichte*, 35:1987, pp. 1–24.

Gillingham, J. *Industry and Politics in the Third Reich: Ruhr Coal, Hitler, and Europe*. New York, 1985.

Gillingham, J. "Zur Vorgeschichte der Montan-Union: Westeuropas Kohle und Stahl in Depression und Krieg." *Vierteljahrshefte für Zeitgeschichte* 34:1986, pp. 381–405.

Gimbel, J. *The American Occupation of Germany: Politics and the Military, 1945–1949*. Stanford CA, 1968.

Gimbel, J. *The Origins of the Marshall Plan*. Stanford CA, 1968.

Girault, R., and Frank, R. *Turbulente Europe et nouveaux mondes, 1914–1941*. Histoire des relations internationales contemporaines, 2. Paris, 1983.

Graubard, S. R. (ed.). *A New Europe?* Boston, 1964.

Griffiths, R. T., and Lynch, F. M. B. "The Fritalux-Finebel Negotiations, 1949–1950." European University Institute Working Paper 84/117, November, 1984.

Grossmann, E. *Methods of Economic Rapprochement*. Geneva, 1927.

Haas, E. B. *The Uniting of Europe: Political, Social and Economic Forces, 1950–1957*. Stanford CA, 1958; 2nd ed., 1968.

Hahn, C. H. *Der Schuman Plan: Eine Untersuchung in besonderen Hinblick auf die deutschfranzösische Stahlindustrie*. Munich, 1953.

Hall, H. D. *North American Supply*. London, 1955.

Halle, L. J. *The Cold War as History*. New York, 1967.

Hallgarten, G. W. F., and Radkau, J. *Deutsche Industrie und Politik*. Frankfurt/Main, 1974.

Hamby, A. L. *Beyond the New Deal: Harry S. Truman and American Liberalism*. New York, 1973.

Hancock, W. K., and Gowing, M. M. *British War Economy*. London, 1949.

Haussmann, F. *Der Neuaufbau der deutschen Kohlenwirtschaft im internationalen Rahmen*. Munich, 1950.

Haussmann, F. *Der Schuman Plan im Europäischen Zwielicht: Ein Beitrag zu den Grundproblemen und zur Weiterentwicklung des Schuman-Planes*. Munich, 1952.

Hayward, J. E. S. *Steel.* In Raymond Vernon (ed.), *Big Business and the State: Changing Relations in Western Europe.* Cambridge MA, 1974.

Heller, F. H. (ed.). *Economics and the Truman Administration.* Lawrence KS, 1981.

Henle, G. "Der Schumanplan vor seiner Verwirklichung." In G. Henle (ed.), *Zusammenarbeit mit dem Rheinisch-Westfälischen Institut für Wirtschaftsforschung Essen.* Duisburg, 1951.

Henle, G. *Three Spheres.* Chicago, 1970.

Herbst, L. "Ludwig Erhard und die Nachkriegsplanungen am Ende des Zweiten Weltkrieges." *Vierteljahrschfte feur Zeitgeschichte,* 25/3:1977, pp. 305–340.

Herbst, L. *Option für den Westen: Vom Marshallplan bis zum deutsch-französischen Vertrag.* Munich, 1989.

Herbst, L. *Der Totale Krieg und die Ordnung der Wirtschaft: Die Kriegswirtschaft im Spannungsfeld von Politik, Ideologie, und Propaganda, 1939–1945.* Stuttgart, 1982.

Herbst, L. "Die zeitgenössische Integrationstheorie und die Anfänge der europäischen Einigung, 1947–1950." *Vierteljahrshefte für Zeitgeschichte* 34:1986, pp. 161–205.

Hermann, W. "Der Wiederaufbau der Selbstverwaltung der deutschen Wirtschaft nach 1945." *Zeitschrift für Unternehmensgeschichte* 23/2:1978, pp. 81–97.

Hexner, E. *International Cartels.* London, 1946.

Hexner, E. *The International Steel Cartel.* Chapel Hill NC, 1943.

Hirsch, A. "Cartels et ententes." in Albert Sauvy (ed.), *Histoire économique de la France entre les deux guerres,* vol. 4, pp. 49–78, Paris, 1975.

Hirschfeld, G. *Fremdherrschaft und Kollaboration: Die Niederlande unter deutscher Besatzung, 1940–1945.* Stuttgart, 1984.

Hoffman, P. G. *Peace Can Be Won.* New York, 1951.

Hoffman, S., et. al. *In Search of France.* Cambridge MA, 1963.

Hoffmann, S., and Maier, C. S. (eds.). *The Marshall Plan: A Retrospective.* Boulder CO, 1984.

Hogan, M. J. *Informal Entente: The Private Structure of Cooperation in Anglo-American Economic Diplomacy, 1918–1928.* Columbia MO, 1977.

Hogan, M. J. *The Marshall Plan: America, Britain, and the Reconstruction of Western Europe, 1947–1952.* Cambridge, 1987.

Hüttenberger, P. "Wirtschaftsordnung und Interessenpolitik in der Kartellgesetzgebung der Bundesrepublik, 1949–1957." *Vierteljahrshefte für Zeitgeschichte* 24:1976, pp. 287–307.

Jacobson, J. "Is There a New International History of the 1920's?" *American Historical Review* 88:1983, pp. 617–645.

Jacobson, J. *Locarno Diplomacy: Germany and the West, 1925–1929.* Princeton NJ, 1972.

Jerchow, F. *Deutschland in der Weltwirtschaft, 1945–1947: Alliierte Deutschland- und Reparationspolitik und die Anfänge der westdeutschen Aussenwirtschaft.* Düsseldorf, 1978.

Jones, J. M. *The Fifteen Weeks (Feb. 21–June 5, 1947): An Inside Account of the Genesis of the Marshall Plan.* New York, 1955.

Jürgensen, H. *Die Westeuropäische Montanindustrie und ihr Gemeinsamer Markt.* Göttingen, 1955.

Kaplan, L. S. *A Community of Interests: NATO and the Military Assistance Program, 1948–1951.* Washington DC: Government Printing Office (for the Office of the Secretary of Defense, Historical Office), 1980.

Kaplan, L. S. *The United States and NATO: The Formative Years.* Lexington KY, 1984.

Kaplan, L. S. *The United States and NATO: The Enduring Alliance.* Boston, 1988.

Kastl, L. (ed.). *Kartelle in der Wirklichkeit: Festschrift für Max Metzner.* Cologne, 1963.

Kehrl, H. *Kriegsmanager im Dritten Reich.* Düsseldorf, 1973.

Kehrl, H. *Zur Wirklichkeit des Dritten Reich.* Cologne, 1975.

Kennan, G. *Memoirs, 1925–1950.* Boston, 1967.

Keynes, J. M. *The Economic Consequences of the Peace.* London, 1919.

Kiersch, G. "Die früheren internationalen Stahlkartelle und der Schumanplan: Eine Gegenüberstellung der Grundlagen, Ziele, und Organisationsformen." *Sonderveröffentlichung des Rheinisch-Westfälischen Instituts für Wirtschaftsforschung Essen* No. 2 September, 1951.

Kiersch, G. *Internationale Eisen- und Stahlkartelle.* Essen, 1954.

Kindleberger, C. *Europe and the Dollar.* Cambridge, MA, 1966.

Kindleberger, C. "European Economic Integration. In *Money, Trade, and Economic Growth: Essays in Honour of John Henry Williams.* New York, 1951.

Kindleberger, C. *Marshall Plan Days.* Berkeley CA, 1973.

Klein, B. H. *Germany's Economic Preparations for War.* Cambridge MA, 1959.

Knapp, M. (ed.). *Von der Bizonengründung zur ökonomisch-politischen Westintegration: Studien zum Verhöltnis zwischen Aussenpolitik und Aussenwirtschaftsbeziehungen, der Entstehungsphase der Bundesrepublik Deutschland (1947–1952).* Frankfurt/Main, 1984.

Kuisel, R. F. *Capitalism and the State in Modern France: Renovation and Economic Management in the Twentieth Century.* Cambridge, 1983.

Kuls, R. "The Schuman Plan and the Ruhr: Attempting a de facto Peace Settlement." Unpublished ms., 1986.

Küsters, H. J. *Die Gründung der Europäischen Wirtschaftsgemeinschaft.* Baden-Baden, 1982.

Lademacher, H. "Die britische Sozialisierungspolitik im Rhein-Ruhr-Raum, 1945–1948." In C. Scharf and H. J. Schröder, (eds.), *Die Deutschlandpolitik Grossbritanniens und die Britische Zone, 1945–1949.* Wiesbaden, 1979.

Lammers, C. "The Cartel Question at the World Economic Conference." *American Academy of Political and Social Sciences* 129–134:1927, p. 147.

Laufenberger, H., and Pflimlin. P. *La nouvelle structure économique du Reich: Groupes, cartels, et politique de prix.* Paris, 1938.

Lerner, D., and Aron, R. *France Defeats EDC.* New York, 1957.

Link, A. S. *American Epoch: A History of the United States since the 1890's*. Vol. 3: 1938–1966. 3rd ed. New York, 1967.

Link, W. *Deutsche und amerikanische Gewerkschaften und Geschäftsleute, 1945–1947*. Düsseldorf, 1978.

Lipgens, W. *Die Anfänge der europäischen Einigungspolitik, 1945–1950. Part 1: 1945–1947*. Stuttgart, 1977.

Lipgens, W. *Europa-Föderationspläne des Widerstandes, 1940–1945*. Munich, 1968.

Lipgens, W. *A History of European Integration*. Vol. 1: *1945–1947: The Formation of the European Unity Movement*. Oxford, 1982.

Lister, L. *Europe's Coal and Steel Community: An Experiment in Economic Union*. New York, 1960.

Lochner, L. P. *Die Mächtigen und der Tyrann. Die deutsche Industrie von Hitler bis Adenauer*. Darmstadt, 1955.

Lorwin, V. R. *The French Labor Movement*. Cambridge MA, 1954.

Loth, W. *Sozialismus und Internationalismus: Die französischen Sozialisten und die Nachkriegsordnung Europas, 1940–1950*. Stuttgart, 1977.

Loth, W. *Die Teilung der Welt, 1941–1955*. Munich, 1980.

Lubell, H. "The French Investment Program: A Defense of the Monnet Plan." Special Report of Mutual Security Agency. Paris, 1952.

Lynch, F. M. B. "Resolving the Paradox of the Monnet Plan: National and International Planning in French Reconstruction." *Economic History Review*, 2nd set 2:1984, pp. 229–243.

MacGregor, D. H. *International Cartels*. Geneva, 1927.

Maier, C. "The Politics of Productivity: Foundations of American International Economic Policy after World War II." In P. Katzenstein (ed.), *Between Power and Plenty: The Foreign Economics Policies of Advanced Industrial States*. Madison WI, 1978.

Maier, C. *Recasting Bourgeois Europe: Stabilization in France, Germany, and Italy in the Decade after World War I*. Princeton NJ, 1975.

Maier, C. *In Search of Stability: Explorations in Historical Political Economy*. Cambridge, 1987.

Maier, C. "The Two Postwar Eras and the Conditions for Stability." *American Historical Review* 86/2:1981.

Manning, A. F. "Die Niederlande und Europa von 1945 bis zum Beginn der fünfziger Jahre." *Vierteljahreshefte für Zeitgeschichte* 29/1:1981, pp. 1–20.

Margairaz, M. "Autour des Accords Blum-Byrnes: Jean Monnet entre le consensus national et le consensus atlantique." *Histoire, Économie, Société* 1/2:1982, pp. 440–470.

Marjolin, R. *Europe and the United States in the World Economy*. Durham NC, 1953.

Marshall, B. *The Origins of Post-War German Politics*. London, 1988.

Martin, J. S. *All Honorable Men*. Boston, 1950.

Mason, E. S. *Controlling World Trade: Cartels and Commodity Agreements*. New York, 1946.

Mason, H. L. *The European Coal and Steel Community: Experiment in Supranationalism.* The Hague, 1955.

Mayne, R. *Postwar: The Dawn of Today's Europe.* London, 1983.

Mayne, R. *The Recovery of Europe.* London, 1970.

McArthur, J. H., and Scott, B. R. *Industrial Planning in France.* Boston, 1969.

McDougall, W. "Political Economy versus National Sovereignty: French Structures for German Economic Integration after Versailles." *Journal of Modern History* 51:1979, pp. 4–24.

McLellan, D. S. *Dean Acheson: The State Department Years.* New York, 1976.

McNeil, W. L. *American Money and the Weimar Republic: Economics and Politics on the Eve of the Great Depression.* New York, 1986.

Meade, J. E., Liesner, H. H., and Wells, S. J. *Case Studies in European Economic Union: The Mechanics of Integration.* Oxford, 1962.

Mee, C. F. *Meeting at Potsdam.* New York, 1975.

Mee, C. F. *The Marshall Plan: The Launching of the Pax Americana.* New York, 1984.

Mélandri, P. *Les Etats-Unis face à l'unification de l'Europe, 1945–1954.* Paris, 1980.

Mendershausen, H. "First Tests of the Schuman Plan." *Review of Economics and Statistics* 35/1:1953, pp. 269–288.

Mendershausen, H. "Prices, Money, and the Distribution of Goods in Postwar Germany." *American Economic Review* 39:1949, pp. 647–672.

Merkl, P. H. *The Origin of the West German Republic.* New York, 1963.

Milward, A. S. *The German Economy at War.* London, 1965.

Milward A. S. *The New Order and the French Economy.* Oxford, 1970.

Milward, A. S. *The Reconstruction of Western Europe, 1945–1951.* London, 1984.

Milward, A. S. *War, Economy, and Society, 1939–1945.* Berkeley CA, 1977.

Mommsen, H., Petzina, D., and Weisbrod, B. (eds.). *Industrielles System und politische Entwicklung in der Weimarer Republik.* Düsseldorf, 1974.

Monnet, J. *Memoirs.* London, 1978.

Morgan, R. *The United States and West Germany, 1945–1973.* New York, 1974.

Morgan, R. *West European Politics since 1945: The Shaping of the European Community.* London, 1972.

Murphy, R. *Diplomat among Warriors.* New York, 1964.

Neebe, R. *Grossindustrie, Staat, und NSDAP, 1930–1933: Paul Silverberg und der Reichsverband der Deutschen Industrie in der Krise der Weimarer Republik.* Göttingen, 1981.

Nelson, D. M. *Arsenal of Democracy: The Story of American War Production.* New York, 1946.

Overy, R. J. "Heavy Industry and the State in Nazi Germany: The Reichswerke Crisis." *European History Quarterly* 15:1985, pp. 313–337.

Parker, W. N. "The Schuman Plan: A Preliminary Prediction." *International Organization* 5/1:1952, pp. 381–395.

Parker, W. N. and Pounds, N. J. G. *Coal and Steel in Western Europe: The Influence of Resources and Techniques on Production.* London, 1957.

382   Bibliography

Pelling, H. L. *Britain and the Marshall Plan*. New York, 1988.

Petzina, D. *Autarkiepolitik im Dritten Reiche*. Stuttgart, 1968.

Petzina, D. "The Origin of the European Coal and Steel Community: Economic Forces and Political Interests." *Zeitschrift für die gesamte Staatswissenschaft* 137:1981, pp. 450–468.

Philip. A. *Der Schuman Plan: Ein entscheidender Schritt auf dem Wege zum Vereinigten Europa*. Ghent, 1951.

Piettre, A. *L'Economie Allemande Contemporaine (Allemagne Occidental), 1945–1952*. Paris, 1952.

Poensgen, E. "Hitler und die Ruhrindustriellen. Unpublished ms., 1946.

Pogue, Forrest C. *George C. Marshall: Statesman, 1945–1959*. New York, 1987.

Poidevin, R. "Der Faktor Europa in der Deutschland- und Ruhrpolitik Robert Schumans (Sommer 1948 bis Früjahr 1949)." *Vierteljahrshefte für Zeitgeschichte* 33:1985, pp. 406–419.

Poidevin, R. "Frankreich und die Ruhrfrage, 1945–1948." *Historische Zeitschrift* 228:1979, pp. 317–334.

Poidevin, R. *Robert Schuman: Homme d'état, 1886–1963*. Paris, 1986.

Pollard, R. A. *Economic Security and the Origins of the Cold War, 1945–1950*. New York, 1985.

Pollard, S. *The Development of the British Economy, 1914–1967*. London, 1969.

Pollard, S. *The Integration of the European Economy since 1815*. London, 1981.

Postan, M. M. *British War Production*. London, 1952.

Postan, M. M. *An Economic History of Western Europe, 1945–1964*. London, 1967.

Potthoff, H. *Vom Besatzungsstaat zur europäischen Gemeinschaft: Ruhrbehörde, Montanunion, Europäische Wirtschaftsgemeinschaft, Euratom*. Hannover, 1964.

Pounds, N. J. G. *The Ruhr: A Study in Historical and Economic Geography*. New York, 1965.

Price, H. B. *The Marshall Plan and Its Meaning*. Ithaca NY, 1955.

Pritzkoleit, K. *Männer, Mächte, Monopole. Hinter den Türen der westdeutschen Wirtschaft*. Düsseldorf, 1953.

Problem des Schuman-Plans: Eine Diskussion zwischen Walter Hallstein, Andreas Predoehl, und Fritz Baade. Kiel, 1951.

Pünder, T. *Das bizonale Interregnum: Die Geschichte des Vereinigten-Wirtschaftsgebiets, 1946–1949*. Rastatt, 1966.

Rabier, J. R. "Plan Monnet et Plan Marshall." *Esprit* 1:1948, pp. 575–597.

Ranieri, R. "The Italian Steel Industry and the Schuman Plan Negotiations." in K. Schwabe (ed.), *Die Anfänge des Schuman-Plans, 1950/51 [The Beginnings of the Schuman Plan]*. Baden-Baden, 1988.

Reichert, J. W. *Nationale und Internationale Kartelle*. Berlin, 1936.

Reuter, P. *La Communauté européenne du charbon et de l'acier*. Paris, 1953.

Riedel, M. *Eisen und Kohle für das Dritte Reiche: Paul Pleigers Stellung in der NS-Wirtschaft*. Göttingen, 1973.

Rohland, W. *Bewegte Zeiten: Erinnerungen eines Eisenhüttenmannes.* Stuttgart, 1978.

Rostow, W. W. *The Division of Europe after World War II: 1946.* Austin, 1981.

Rousiers, P. de. *Cartels, Trusts, and Their Development.* Geneva, 1927.

Rowe, J. W. F. *Markets and Men.* Cambridge, 1936.

Royal Institute of International Affairs. *Survey of International Affairs, 1939–1946: Four-Power Control in Germany and Austria, 1945–1946.* London, 1946.

Royal Institute of International Affairs. *Survey of International Affairs, 1947–1948.* London, 1952.

Royal Institute of International Affairs. *Survey of International Affairs, 1952.* London, 1955.

Royal Institute of International Affairs. *Survey of International Affairs, 1953.* London, 1956.

Ruge, B. "The United States and the Creation of the ECSC, 1950–1952." Master's thesis, University of North Carolina, 1987.

Sauvy, A. *La vie économique des Français de 1939 à 1945.* Paris, 1978.

Scharf, C. and Schröder H. -J. (eds.). *Die Deutschlandpolitik Frankreichs und die französische Zone, 1945–1949.* Wiesbaden, 1983.

Scharf, C., and Schröder, H. -J. (eds.). *Die Deutschlandpolitik Grossbritanniens und die Britische Zone, 1945–1949,* Wiesbaden, 1979.

Scharf, C., and Schöder, H. -J. (eds.). *Politische und Okonomische Stabilisierung Westdeutschlands, 1945–1949.* Wiesbaden, 1977.

Schmidt, G. (ed.). "Konstellationen internationaler Politik, 1924–1932: Politische und wirtschaftliche Faktoren." In *Den Beziehungen zwischen Westeuropa und den Vereinigten Staaten.* Bochum, 1983.

Schmidt, H. G. "Reorganization of the West German Coal and Steel Industries under the Allied High Commission for Germany, 1949–1952." Preliminary draft. Historical Division, HCCOM, September 1952.

Schmitt, H. A. *The Path to European Union: From the Marshall Plan to the Common Market.* Baton Rouge, 1962.

Schuker, S. "The Two Postwar Eras and the Conditions for Stability." *American Historical Review* 86/2:1981.

Schuman, R. "French Policy towards Germany since the War." Stevenson Memorial Lecture No. 4. London, 1954.

Schwabe, K. (ed.). *Die Anfänge des Schuman-Plans, 1950/51 The Beginnings of the Schuman Plan.* Baden-Baden, 1988.

Schwartz, T. A. "From Occupation to Alliance: John J. McCloy and the Allied High Commission in the Federal Republic of Germany." Ph.D. diss., Harvard University, 1985.

Schwarz, H. -P. "Adenaur und Europa." *Vierteljahrschfte für Zeitgeschichte* 27:1979, pp. 471–523.

Schwarz, H. -P. *Die Ara Adenauer: Gründerjahre der Republik, 1949–1957.* Stuttgart, 1981.

Schwarz, H. -P. "Die europäische Integration als Aufgabe der Zeitgeschichtsforschung." *Vierteljahrshefte für Zietgeschichte* 31:1983, pp. 555–572.

Schwarz, H. -P. *Vom Reich zur Bundesrepublik: Deutschland im Widerstreit der aussenpolitischen Konzeptionen in den Jahren der Besatzungsherrschaft, 1945– 1949.* Neuwied, 1966; 2nd ed., Stuttgart, 1980.

Schweitzer, A. *Big Business in the Third Reich.* Bloomington IN, 1964.

Seebold, G. -H. *Ein Stahlkonzern im Dritten Reich: Der Bochumer Verein, 1927– 1945.* Wuppertal, 1981.

Sethur, F. "The Schuman Plan and Ruhr Coal." *Political Science Quarterly* 67/4:1952, pp. 550–570.

Seubert, R. *Berufserziehung und Nationalsozialismus: Das Berufspädagogische Erbe und seine Betreuer.* Weinheim, 1977.

Shonfield, A. *Modern Capitalism: The Changing Balance of Public and Private Power.* London, 1965.

Silverman, D. P. *Reconstructing Europe after the Great War.* Cambridge MA, 1982.

Sohl, H. -G. *Notizen.* Bochum-Wattenscheid, 1985.

Speer, A. *Inside the Third Reich.* New York, 1970.

Spinelli, A. *The Eurocrats: Conflict and Crisis in the European Community.* Baltimore, 1966.

Steininger, R. "Reform und Realität: Ruhrfrage und Sozialisierung in der angloamerikanischen Deutschlandpolitik." *Vierteljahrshefte für Zeitgeschichte* 27:1979, pp. 169–240.

Steininger, R. *Die Ruhrfrage und die Entstehung des Landes Nordrhein-Westfalen: Britische, französische, und amerikanische Akten.* Düsseldorf, 1988.

Steininger, R. "Ruhrfrage und Sozialisierung in der anglo-amerikanischen Deutschlandpolitik, 1947/48." *Vierteljahrshefte für Zeitgeschichte* 7/2:1979, pp. 167–240.

Stimson, H. L. and Bundy, M. *On Active Service in War and Peace.* New York, 1947.

Stocking, G. W., and Watkins, M. W. *Cartels in Action: Case Studies in International Business Diplomacy.* New York, 1947.

Stocking, G. W., and Watkins, M. W. *Cartels or Competition? The Economics of International Controls by Business and Government.* New York, 1948.

Stocking, G. W., and Watkins, M. W. *Monopoly and Free Enterprise.* New York, 1951.

Stopler, G. *Die deutsche Wirklichkeit.* Hamburg, 1949.

Strange, S. *Sterling and British Policy: A Political Study of an International Currency in Decline.* New York, 1971.

Thum, H. *Mitbestimmung in der Montanindustrie: Der Mythos vom Sieg der Gewerkschaften.* Stuttgart, 1982.

Tinbergen, J. *International Economic Integration.* Amsterdam, 1954.

Trachtenberg. M. "Reparation at the Paris Peace Conference." *Journal of Modern History* 51:1983, pp. 24–55.

Trachtenberg, M. *Reparation in World Politics: France and European Economic Diplomacy, 1916–1923.* New York, 1980.

Treue, W. *Die Demontagepolitik der Westmächte nach dem Zweiten Weltkrieg.* Göttingen, 1967.

Triffin, R. *Europe and the Money Muddle: From Bilateralism to Near-Convertibility, 1947–1956.* New Haven, 1957.

Troeger, H. *Interregnum: Tagebuch des Generalsekretärs des Länderrats der Bizone, 1947–1949,* ed. W. Benz and K. Goschler. Munich, 1985.

Van der Beugel, E. H. *From Marshall Aid to Atlantic Partnership: European Integration as a Concern of American Foreign Policy.* Amsterdam, 1966.

Van der Wee, Herman. *Prosperity and Upheaval: The World Economy, 1945–1980.* Berkeley CA, 1984.

Vatter, H. G. *The US Economy in World War II.* New York, 1985.

Vaubel, L. *Zusammenbruch und Wiederaufbau: Tagebuch aus der Wirtschaft, 1945–1949,* ed. W. Benz. Munich, 1984.

Vogelsang, H. "Die Deutsche Kohlenwirtschaftsorganisation in ihrer historischen Entwicklung." Ph.D. diss., Munich, 1957.

Volkman, H. -E. "Autarkie, Grossraumwirtschaft, und Aggression: Zur ökonomischen Motivation der Bestzung Luxemburgs, Belgiens, und der Niederlande." *Militägeschichtliche Mitteilungen* 19:1976, pp. 51–76.

Von der Groeben, H. *Aufbaujahre der europäischen Gemeinschaft: Das Ringen um den Gemeinsamen Markt und die politische Union (1958–1966).* Baden-Baden, 1982.

Wagenführ, R. *Die deutsche Industrie im Kriege, 1939–1945.* 2nd ed. Berlin, 1963.

Wall, I. M. *L'Influence américaine sur la politique française, 1945–1954.* Paris, 1985.

Wallich, H. C. *Mainsprings of the German Revival.* New Haven, 1955.

Warner, I. "Allied-German Negotiations on the Deconcentration of the West German Steel Industry." In I. D. Turner (ed.), *Reconstruction in Post-War Germany; British Occupation Policy and the Western Zones,* pp. 155–183. Leamington Spa, 1989.

Watt, D. C. *Britain Looks to Germany.* London, 1965.

Watt, D. C. *Succeeding John Bull: America in Britain's Place, 1900–1975.* Cambridge, 1984.

Weisbrod, B. *Schwerindustrie in der Weimarer Republik.* Wuppertal, 1978.

Wendt, B. -J. *Economic Appeasement: Handel und Finanz in der britischen Deutschland-Politik, 1933–1939.* Düsseldorf, 1971.

Westdeutschlands Weg zur Bundesrepublik, 1945–1949. Beiträge von Mitarbeitern des Instituts für Zeitgeschichte. Munich, 1976.

Wexler, I. *The Marshall Plan Revisited: The European Recovery Program in Economic Perspective.* Westport CT, 1983.

Wheeler-Bennett, J., and Nicholls, A. *The Semblance of Peace: The Political Settlement after the Second World War.* New York, 1972.

White, T. H. *Fire in the Ashes: Europe in Mid-Century.* New York, 1953.

White, T. H. *In Search of History: A Personal Adventure.* New York, 1978.

Wiggershaus, N., and Foerster, R. (eds.). *Die westliche Sicherheitsgemeinschaft 1948–1950: Gemeinsame Probleme und gegensätzliche Nationalinteressen in der Gründungsphase der Nordatlantischen Allianz.* Boppard, 1988.

Willis, F. R. (ed.). *European Integration.* New York, 1975.

Willis, F. R. *The French in Germany, 1945–1949.* Stanford CA, 1962.
Willis, F. R. *France, Germany, and the New Europe, 1945–1967.* Stanford CA, 1968.
Wilson, J. H. *American Business and Foreign Policy, 1920–1933.* Lexington KY, 1971.
Winkler, H. A. (ed.). *Politische Weichenstellungen im Nachkriegsdeutschland, 1945–1953.* Göttingen, 1979.
Wolsing, T. *Untersuchungen zur Berufsbildung im Dritten Reich.* (Schriftenreihe zur Geschichte und Politischen Bildung, 24). Kastellaun, 1977.
Wurm, C. A. "Handelsdiplomatie in der Weltwirtschaftskrise: International Kartelle, Stahl, und Baumwolltextilien in der Aussenpolitik Grossbritanniens, 1924–1939." In Wurm (ed.), *Internationale Kartelle und Aussenpolitik.* Stuttgart, 1989.
Yergin, D. *Shattered Peace: The Origins of the Cold War and the National Security State.* Boston, 1978.
Young, J. W. *Britain, France, and the Unity of Europe, 1945–1951.* Leicester, 1984.
Youngson, A. J. *The British Economy, 1920–1957.* London, 1960.
Zawadzki, K. W. F. "The Economics of the Schuman Plan." *Oxford Economic Papers,* new ser. 5:1953, pp. 157–189.

Index

44; post-World War II policy in Europe, 97–100, 104, 112, 176; post-World War II trade policy, 123, 145; reaction to the Schuman Plan, 149, 235, 364; recession, 230; strategy in World War II, 45, 46, 86; U.K. economic control, 125; United States High Commission in Germany (HICOG), *see* McCloy, John J.; Victory Program, 51, 89; war economy, 47, 50, 86, 87, 93, 95
United States–United Kingdom coal talks (1947), 128
Uri, Pierre, 139, 244, 316, 327, 328, 346

van den Berg, 311
Van der Rest, Pierre, 292, 311, 330
Verbundwirtschaft, 197, 216
Vereinigte Stahlwerke (VS), 21–23, 41, 58, 59, 62, 63, 68, 192, 197, 201, 202, 215, 220, 230, 272, 301, 306, 353, 354, 355
Vereinigte Stahlwerke Plan (of 1949), 220
Vernon, Raymond, 254, 255
Versailles Peace settlement, 1–7, 15, 17, 26
Verwaltungsamt Stahl und Eisen (VSE), 193, 195, 226
Vichy France, *see* France: World War II
Victory Program, 51
Vignuzzi, 311
Villiers, Georges, 224, 289, 292, 309
Vinck, 38, 327
Viviani, René, 4
Vögler, Albert, 11, 63
vom Rath, 40
Vorhees, Tracy, 165

Wagenführ, Dr. Rolf, 243, 316
Wallace, Henry A., 103, 105, 112, 113, 125
Walter, Paul, 55
Walzstahlgruppe, 307

War aims, German, 9, 67–8, 366
War criminals, Germany, 112
War Production Board (WPB), 89
Washington Conference (April 1949), 210, 212, 213
Watkins, Myron, W., 19, 20
Wehrer, Albert, 313
Wehrmacht, 42
Wendt, Berndt-Jürgen, 31
Wenner-Gren, Axel, 355
Wenzel, Hermann, 215, 216, 219
Western Union Speech (January 1948), 135
Westintegration, 238, 279, 300, 363
Westminster Economic Conference of the European Movement, 222, 223, 230, 349
White, Theodore, 63, 292
Wiesbaden Accords, 8, 12, 13, 221
Willner, Sidney, 260, 306
Winant, John, 122, 123
Winkhaus, Hermann, 64
Wirtschaftswunder, *see* Federal Republic of Germany
Wirtschaftvereinigung der Eisen-und Stahlindustrien (WVESI), 193, 195, 198–204, 233, 226, 233, 243, 280, 285, 294, 295, 310, 315, 317, 320, 326, 329
Wolf Mission, 211–213
Wolff, Otto, 193, 200, 201, 355
World Bank, 142, 261
World Economic Conference, *see* League of Nations
World War I, 1, 2, 4, 8, 11, 17, 32, 50, 57, 81, 106, 111, 124, 152
World War II, 45–96, 104, 234, 365, 369; management, 61

Yalta Conference, 106

Zangen, Wilhelm, 192
Zentrale Planung, 58
Zollverein, 23